Underwater Electroacoustic Transducers

A Handbook for Users and Designers

D. Stansfield

PP PENINSULA PUBLISHING

Underwater Electroacoustic Transducers

Copyright ©1991 D. Stansfield
ISBN 0932146724
Library of Congress Control Number 2003101552

Reprint of the original edition published by:
Peninsula Publishing
Los Altos Hills, California 94022 USA
Telephone: 650-948-2511
Fax: 650-948-5004
Email: sales@peninsulapublishing.com
Web site: www.peninsulapublishing.com

Original edition published in Great Britain in1991 by:
Bath University Press
Claverton Down
Bath BA2 7AY

Institute of Acoustics
PO Box 320
St. Albans AL1 1PZ

Printed in the United States of America

FOREWORD

As can be testified by anyone with more than a cursory interest in the specification or design of underwater acoustic transducers, there is a singular lack of texts in the open literature which address the topic at a useful depth. It was this gap in the underwater literature which prompted me, both as an individual scientist and as the then Chairman of the Underwater Group of the Institute of Acoustics, to "persuade" Dennis Stansfield to set out his vast experience in the field for the benefit of others.

In supporting this publication the Institute of Acoustics and the Bath University Press have ensured that the underwater acoustics community have a comprehensive work which might otherwise have remained unavailable.

N.G. Pace
School of Physics
University of Bath, UK.

CONTENTS

viii

PREFACE

Although underwater transducers perform a vital function in any sonar set or underwater acoustic experiment, they have received relatively little attention in the scientific literature compared with other elements of the system. Yet they form the essential link between the electrical power generator and the acoustic output, or between the received acoustic pressures and the signal processing. This book is aimed at filling the gap by describing the principles and practical aspects involved in the design of underwater electroacoustic transducers. It is hoped that this will improve the user's understanding of the significance of the various characteristics which need to be defined in specifying his transducer requirement. For the transducer designer, the book is intended to serve as a handbook by describing the design methods which can be adopted for the more demanding applications. The successful conversion of the theoretical design to a reliable transducer requires considerable attention to the details of the assembly, and these are discussed throughout the book.

The role of the transducer as the link between electrical and acoustic parts of the overall system is a crucial feature in optimising its performance. Development of a transducer thus involves a mixture of electrical and acoustic considerations, together with the mechanical engineering aspects which are needed to convert it to a practical design. This rather unusual combination of skills is perhaps the reason why the number of transducer designers is fairly small, and transducer design is sometimes regarded as more of an art than a science. The main objective of this book is to widen the appreciation of the basic principles of transducer design, with the hope that this will lead users to formulate their demands on the basis of greater knowledge of what is possible, and hence encourage further improvements in design methods.

This book is based primarily on experience gained in the transducer and array design section at the Admiralty Underwater Weapons Establishment (now the Admiralty Research Establishment) at Portland. Many of the applications required wide band, high power transducers within the range 2–20kHz, which could often be satisfied by piston–type transducers using a piezoelectric ceramic driving stack. It is this type of design, operating within this frequency band, which is taken as the main focus of this book. This has the advantage that the analysis of the transducer elements may be simplified by treating them as lumped masses and springs, and also that elements of this type offer a wide range of opportunities to optimise the design. They therefore serve as convenient examples of the basic design principles. These basic principles are not restricted to such designs, however, and many of them are also applicable to other types of transducer and to a wider range of frequencies. The design of broad band hydrophones requires a quite different approach, but the treatment adopted is also based largely on piston–type elements, which can again be used to illustrate the fundamental principles.

As might be expected, the development of these design methods and expertise does not arise from the work of one individual, but is the result of contributions by many. Foremost amongst these have been my colleagues over the years in the transducer and array design section at AUWE, all of whom helped to develop these methods and put them into practice, and to all of whom I owe an enormous debt of gratitude. Two of them deserve especial mention. Firstly, my initial introduction to the use of equivalent circuits, and the application of filter methods to the analysis of the matching between transducer and driving amplifier, was due to Dr J H Mole, whose application of systematic and scientific methods to transducer design also provided an outstanding foundation and example. Secondly, Mr R J Gale has been responsible, not only for carrying on and continuing to develop these methods, but also for providing me with invaluable advice during the writing of this book. Over the years, I have also benefitted from discussions with many other co-workers in various establishments and countries, and their many and varied contributions are gratefully acknowledged.

This book would not have been started without the backing of the Underwater Acoustics Group of the Institute of Acoustics, and particularly Dr. N.G. Pace, to whom I am indebted for his continuing interest and encouragement. My thanks are due also to all the people who have helped in its production in a variety of ways. And finally, its successful completion could not have been achieved without the contribution of my wife, whose tolerance and unfailing support have been an essential element through all the stages of the undertaking.

PREFACE TO THE REPRINT EDITION

More than a decade has passed since the first edition of this book was written. In that time much work has been carried out on the development of underwater acoustic transducers and their active materials, mostly aimed at applications for hydrophones and hydrophone arrays. Unfortunately, it has not been feasible to survey these developments for this reprint edition. However, there appear to be no major changes in the fundamentals of transducer design and I believe that this book will continue to be welcomed by both designers and users of underwater transducers.

No significant changes to the original edition of this book have been made in this reprint but several typographical errors have been corrected.

D. Stansfield
January 2003

ACKNOWLEDGEMENTS

Grateful acknowledgement is made for permission to reproduce illustrations and diagrams as follows:-

Figs 2.1 (a) and 2.4	Ferranti-Thomson Sonar Systems UK Ltd
Fig 2.1 (b)	Ministry of Agriculture, Fisheries and Food
Fig 2.1 (c) and cover	Ministry of Defence
Fig 2.5	DI vs beamwidth; from Albers, *Underwater Acoustics Handbook - II*, Pennsylvania State Press, 1965, Fig 21,6. p 292
Fig 4.3	Admittance loop; from *Proc IRE*, March 1957, 353-358, IRE Standards on Piezoelectric Crystals – The Piezoelectric Vibrator, Fig 3
Fig 6.9	Beamwidth; from NDRC Div 6, Summary Tech Report Vol 13, Design and Construction of Magnetostrictive Transducers, 1946, p 128
Fig 6.13	Acoustic field in front of a radiator; from *J Acoust Soc Am*, in Zemanek,J., *JASA*, 49, 181-191, (1971)
Fig 8.6	Cray Sonics Ltd

Chapter 1
INTRODUCTION

The good transmission of acoustic energy through water has led to a wide range of applications, of which the many systems for sonar detection and underwater communications are typical. In all of these applications a vital role is played by the electroacoustic transducers, which convert electrical energy into acoustic energy or vice versa. The aim of this book is to serve as a handbook for designers of underwater electroacoustic transducers, and to present to the user information on their characteristics and limitations which should help in specifying realistic requirements.

Transducers in the general sense are devices for converting energy from one form to another. By concentrating on electroacoustic transducers we are concerned with the transfer between electrical and acoustical energy. This could include a wide range of devices, including for example loudspeakers or microphones, or surface wave devices, as well as those transducers used in water. It is these latter which are our concern in this book, although the principles described could often be applied to the other forms of transducers. We shall thus concentrate on electroacoustic transducers for use in water, and especially on designs for what might loosely be described as the audio-frequency band, from about 2 to 20kHz. By focussing on this frequency band, some simplification of the design relationships is possible, thus allowing us to pay attention especially to the basic design principles.

Because the function of transducers is solely to convert energy from one form to another, they have no inherent use in themselves, and it is a recurrent theme throughout this book that they should be regarded fundamentally as components in larger systems. It is the performance desired of a transducer in its particular system which should determine its required characteristics, these being specified by the user from an understanding of the interaction between the transducer and the other related parts of the system. The task of the transducer designer is then to realise these requirements in the most effective way, within whatever constraints are appropriate. The whole process is an unusual combination of electrical system design with mechanical engineering and physics, which sometimes leads to its being regarded as an arcane science − or a "black art". It is the author's hope that this book will go some way to correcting this impression by clarifying the basic principles of the design procedure, whilst not of course decrying the importance of practical experience to the designer.

If a transducer is used solely as a receiver, converting the acoustic pressure variations at its sensitive surface to electrical output signals, it is usually known as a **hydrophone**. A transducer which is used primarily as an acoustic source, by converting electrical to acoustic energy, is often referred to as a **projector**. It is usually desirable, and often essential, for a projector to have a high efficiency of conversion between electric and acoustic energy, and this is achieved by making the resonance frequency of the projector for any particular application near to the desired operating frequencies. Most electroacoustic transducers are reciprocal − ie they can function as either transmitters or receivers − and it is therefore often convenient and economical to use the projector in an active sonar system also as a receiver. However, if a receiving hydrophone is to be used over a wide frequency band, − for example in a passive sonar − it is usually preferable to avoid the marked variations of response which occur around its resonance. Since the power handled by a hydrophone is very small, its efficiency is much less important than for a projector, and it is not necessary for it to be operated near resonance. It is thus common for broad−band hydrophones to be designed to have their resonances well away from the operating frequency band. This allows the possibility of any particular hydrophone design being used for a variety of applications, whilst projectors tend to be more specifically tailored to their individual requirements. In this sense, and because of their higher power handling requirements, projectors pose the greatest design needs, and they will therefore receive the most attention in this book. However, the particular problems which arise in the design of hydrophones are discussed in Chapter 10.

Underwater electroacoustic transducers have a long history, dating from the early discoveries of the piezoelectric effect in various crystals such as quartz in about 1880 (Ref [1.1]). For a description of the progressive development of piezoelectric materials and underwater transducers the reader is referred to the references and list of suggested reading at the end of this Chapter. After the early work with piezoelectric crystals, magnetostrictive materials came to the fore, to be replaced in turn by the introduction of piezoelectric ceramics. These have been much the most commonly used active materials for transducers within the last twenty-five years, although there are still particular applications which are best realised by other means, and in recent years various new materials have received study.

This book will therefore deal mainly with transducers using piezoelectric ceramics as the active element. The important features of such ceramics are described in Chapter 3; for the moment it is worth noting the very wide dynamic range over which a piezoelectric transducer can be operated with an essentially linear response. A well designed transducer can, for example, measure pressures of the order of 20 MPa in a pulse from an explosive charge, and also be capable of detecting signals in ambient noise of around 10^{-2}Pa - a dynamic range of some 10 orders of magnitude, or 200dB. The dynamic range of amplifiers is generally less than that of the transducer, so that the full 200dB is not usually practical for a complete system, but the wide dynamic range of the transducer itself is matched by very few other sensors.

The first part of this book is concerned mainly with the design of projectors. Chapters 2,4, and 5 discuss the features which are primarily external to the transducer itself and relate to the performance of the transducer within its associated system; these are the chapters of most significance for those concerned with the use of transducers rather than their actual design. The characteristics of active materials, and the acoustic aspects, are described in Chapters 3 and 6. The design of the transducer itself is dealt with in more detail in Chapters 7 and 8, and its testing in Chapter 9. The special features of hydrophones are discussed in Chapter 10, and the final chapters give a brief description of some applications lying outside the main scope of the book, and some of the developments arising from newer materials.

1-1   BASIC CONCEPTS

We start with brief explanations of some of the basic concepts and mathematical relationships which will be used throughout the text.

Simple Harmonic Motion

Variables such as displacements, pressures, etc are often treated as sinusoidal functions of time. Thus, the displacement (x) of a particle may be written

$$x = x_0 \sin \omega t \qquad (1.1)$$

where x_0 is the amplitude of the displacement, and ω is the **angular frequency** (sometimes called the **pulsatance**), which is related to the frequency (f) by the relation $\omega = 2\pi f$. The instantaneous velocity of the particle is given by

$$u = dx/dt = \omega x_0 \cos \omega t \qquad (1.2)$$

The acceleration ($\dot{u}$) of the particle is obtained by differentiating again, ie

$$\dot{u} = -\omega^2 x_0 \sin \omega t \qquad (1.3)$$

The amplitude, or peak value (u_0), of the velocity is thus ω times the amplitude of the displacement, and the amplitude of the acceleration ($\dot{u}_0$) is ω^2 times that of the displacement, ie:-

$$u_0 = \omega x_0 \qquad (1.4a)$$

and

$$\dot{u}_0 = \omega^2 x_0 = \omega u_0 \qquad (1.4b)$$

These simple relationships will quite often be useful in the subsequent discussions. The phase of the velocity leads that of the displacement by 90°, whilst the acceleration is 180° out of phase with the displacement.

The sinusoidal variation with time of the variables may

alternatively be described in the exponential form, in which the displacement x is the real part of $x_0 \exp(j\omega t)$ Equation (1.1) may then be written in the form $x = x_0 \exp(j\omega t)$ where $j = \sqrt{(-1)}$. It is then easy to differentiate this equation, and hence obtain the expressions for the velocity and acceleration, ie:-

$$u = j\omega x_0 \exp(j\omega t) = j\omega x$$

$$u_0 = j\omega x_0 \tag{1.5a}$$

$$\dot{u}_0 = -\omega^2 x_0 = j\omega u_0 \tag{1.5b}$$

These correspond to equations (1.4a,b) above, but with the j-factors representing the phase differences between the variables. Similar relationships apply of course to any other sets of quantities describing sinusoidal variations; they will be applied for example to electrical quantities in subsequent sections.

Impedance and Admittance.

In an electrical circuit, the alternating current is characterised by its amplitude and its phase relative to some reference. These two quantities may be denoted by a phasor **i** which is a vector in the complex plane representing both amplitude and phase. The sinusoidal variation of the current with time may be represented by a continuous rotation of the vector about the origin, and this generally disappears from the equations because it is common to all the variables. However, the complex phasor **i** also contains a factor representing the phase of the rotating vector. Although there is no significance in the absolute phase of the variables, the phase *differences* between them are very important. It is therefore usual to take one of the variables as the phase reference, and include the phases of the other variables relative to this reference within their appropriate phasors. Similarly, the voltage may be represented by a phasor **V**, and so on for any other similar variables which vary sinusoidally with time. If a steady voltage is applied to a resistor, the value of the voltage across the resistor divided by the current through it defines its resistance (by Ohm's law). A similar relationship holds for alternating currents, except that the ratio of the complex **V** and **i** phasors is defined as the impedance (Z) of the circuit, and is itself in general a complex vector. Thus, $Z = \mathbf{V}/\mathbf{i}$. The impedance Z may be broken down into its real and imaginary components, its resistance (R) and reactance (X), so that $Z = R+jX$. The inverse of the impedance is the admittance (Y), ie $Y = 1/Z$. This may also be decomposed into its real and imaginary parts, its conductance (G) and its susceptance (B), ie $Y = G+jB$. (In some treatments, a convention is adopted in which $Y = G-jB$, but in this book the positive version $Y = G+jB$ will be used.)

A circuit may be dealt with either in terms of its impedance or its admittance, and we shall find it useful to be able to convert between one set of components (eg R,X) and the other (ie G,B). To do this, we make use of the principle that both approaches should lead to equivalent expressions. Thus,

$$Z = R + jX$$

$$= 1/Y$$

$$= \frac{1}{G + jB}$$

$$= \frac{G - jB}{G^2 + B^2}$$

Equating the two forms of real and imaginary parts, we must have

$$R = \frac{G}{G^2 + B^2} \quad \text{and} \quad X = \frac{-B}{G^2 + B^2} \qquad (1.6a)$$

Or, since by Pythagoras theorem $Y^2 = G^2 + B^2$,

$$R = G/Y^2 \text{ and } X = -B/Y^2 \qquad (1.6b)$$

Conversely, it is easy to show that

$$Y = 1/Z = \frac{R - jX}{R^2 + X^2}$$

and hence that

$$G = \frac{R}{R^2 + X^2} \quad \text{and} \quad B = \frac{-X}{R^2 + X^2} \qquad (1.7a)$$

Or, using in this case $Z^2 = R^2 + X^2$,

$$G = R/Z^2 \quad \text{and} \quad B = -X/Z^2 \qquad (1.7b)$$

The impedance form is most appropriate for a combination of components in series, for which the total impedance is obtained by summing the individual impedances. The current is the same through each component, so this form tends to be used when it is the input current which is known. When components are connected in parallel, however, the resultant impedance is obtained by calculating the sum of the admittances of the individual components; this gives the total admittance, which may if necessary be inverted to give the impedance. Because the components are connected in parallel, they have the same voltage applied to all, and this admittance form is thus more appropriate when the applied voltage is known. In analysing a circuit which contains both series and parallel combinations of components it is usual to work through the circuit converting from one form to the other as necessary.

If the input impedance of a circuit is much larger than that of the power source which is driving it, the voltage applied to the circuit by the source is approximately independent of the actual value of the circuit impedance. It is then described as a constant voltage source, and power into the load is given by V^2G. In this case, it is natural to adopt the admittance form for analysing performance. Conversely, for loads of low impedance – ie low compared with typical electrical sources – it is natural to adopt the series impedance form. We shall see that piezoelectric transducers are generally fairly high in impedance, and it is therefore more common to consider them in terms of their admittances, with a constant voltage drive assumed.

Similar concepts can be applied to mechanical systems. If a sinusoidal force is applied to a mechanical system, it may be represented by a phasor (**F**), as may also the resulting velocity of the displacement (**u**). Then, by analogy with the electrical case, the (complex) ratio of the force to the velocity may be defined as the mechanical impedance (Z_M) of the system; ie $Z_M = $ **F**/**u**. When dealing with problems in acoustic radiation, it is often convenient to express the force as the acoustic pressure multiplied by the area of the radiating surface. It is then possible to define a parameter called the **specific acoustic impedance** as the ratio of the acoustic pressure (instead of the force) to the particle velocity. This parameter is commonly used for problems involving, for example, radiation of a plane wave through the medium, since it is then independent of area, having a value (ρc) which is a characteristic of the medium itself. When the radiating area is more clearly defined, as for the face of a transducer, it is generally more appropriate to use the mechanical impedance.

Impedance and Admittance Diagrams.

The behaviour of a circuit as the frequency is varied may be illustrated in a variety of ways. For example, the conductance and susceptance may each be plotted as a function of frequency, or the resistance and reactance may be plotted as a function of frequency. A presentation which is often more informative, however, is an impedance (or admittance) curve. For an impedance curve, values of reactance are plotted against the corresponding values of resistance as the frequency is varied; for an admittance curve, susceptance values are plotted against conductance. As an example, consider how the impedance of a series L,C,R combination would vary with frequency. The impedance of an inductor L is given by $j\omega L$, where ω represents the angular frequency (ie $\omega = 2\pi f$, where f is the frequency of the current through L). The impedance of a capacitor C is $1/j\omega C = -j/\omega C$ The impedance of the inductor is thus a reactance of magnitude ωL, and that of the capacitor a reactance of $-1/\omega C$; the total reactance of the LC combination is thus the sum of the two, ie $(\omega L - 1/\omega C)$ At very low frequencies this reactance has a very large negative value. As the frequency is increased, the reactance increases until it becomes zero when $\omega L = 1/\omega C$ (ie at the resonance frequency), and at high frequencies the reactance becomes large positive. The resistive component of the impedance of the LCR combination is due solely to the resistor R, and has the value R independent of the frequency. The impedance curve for this LCR circuit is therefore a straight line in the complex R,jX plane as shown in *Fig 1.1a*, in which the impedance at any frequency is represented by a vector Z from the origin to a point on the vertical line so that Z has components $R + j(\omega L - 1/\omega C)$.

The corresponding shape of the admittance curve is less obvious. It can be derived by noting that Z may be decomposed into its modulus and argument, and written as $Z = |Z|\exp(j\theta)$. Then the admittance $Y = 1/Z = (1/|Z|\exp(j\theta)) = (1/|Z|)\exp(-j\theta)$. Mathematical texts (eg [1.2]) show that the conversion of a variable Z to its inverse has the effect of converting a straight line to a circle:– ie a straight line in the impedance plane maps

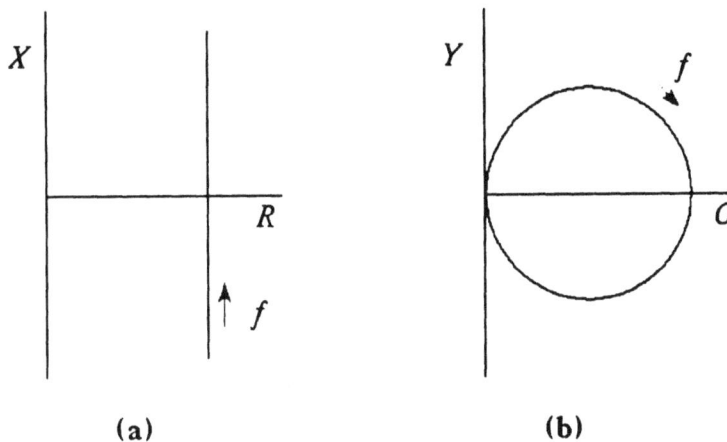

Fig 1.1 Impedance and admittance curves for series resonant circuit: (a) impedance curve; (b) admittance curve.

onto a circle in the admittance plane. At very high or very low frequencies, the admittance must be very small (since the impedance is large), so the admittance curve must start and finish at the origin. At the resonance frequency, the admittance must be real, with a value of $1/R$. It is thus clear, and can readily be confirmed mathematically, that the admittance curve for a simple series resonant circuit is a circle as illustrated in *Fig 1.1b*. Because of the change of sign of the phase factor in converting from impedance to admittance, the susceptance is positive for frequencies below resonance, and negative above resonance. The frequency parameter thus moves clockwise around the circle.

By similar arguments it is easy to show that the admittance curve for a parallel *LCR* combination is a straight line in the admittance (G, jB) plane, but a circle in the impedance plane. For more complicated circuits it is possible, though sometimes tedious, to derive the admittance or impedance curves by extending the treatment above. Such an analysis of the behaviour of the curves can be very helpful in gaining a qualitative understanding of the electrical characteristics of circuits, but for calculations of absolute impedance values it is usually preferable to derive the curves by the more rigorous mathematical circuit analysis techniques.

1-2   ACOUSTIC QUANTITIES

If a plane progressive acoustic wave of pressure amplitude p_0 is transmitted through a medium of density ρ and speed of sound c, then the **acoustic intensity**, or power flow per unit cross-sectional area, is given by $I = p_0^2/2\rho c$. (See eg [1.3]) For example, the value of ρc for water is approximately $1.5 \times 10^6 kg.m^{-2}s^{-1}$, and the peak pressure in an acoustic wave in water corresponding to a power flow of $10^3 W/m^2$ would be approximately $5.5 \times 10^4 Pa$.

The wide dynamic range of the quantities involved in acoustic problems makes it convenient to express them in logarithmic form. The basis for the system is to relate acoustic *intensities* to a reference intensity I_{ref}, an intensity I then being said to have a level of $10\log(I/I_{ref})$ decibels (dB) referred to the intensity I_{ref}. Since intensity is proportional to the square of the acoustic pressure, the corresponding expression for the acoustic pressure level is $20\log(p/p_{ref})$ dB. It is customary in underwater acoustics to take the value of p_{ref} as $1\mu Pa$, and acoustic pressures in water are therefore usually expressed in dB re $1\mu Pa$. (A different reference pressure is often used in air.) A peak pressure of $5.5 \times 10^4 Pa$ would thus be equivalent to a level of $20\log(5.5 \times 10^{10}) = 214.8$ dB re $1\mu Pa$.

The ratio of acoustic pressure to particle velocity (either peak or rms values, provided they are the same for each) for a plane wave is the specific acoustic impedance of the medium, equal to ρc. The peak or rms values of the particle displacements may then be derived by using equation (1.4), ie

$$x_0 = v_0/\omega = p_0/\omega\rho c. \qquad (1.8)$$

Examples of the resulting values are shown in *Fig 1.2* for both air and water. Although the specific acoustic impedance of the medium is independent of frequency, the displacement corresponding to a given value of sound pressure is inversely proportional to frequency, and therefore increases as the frequency is reduced. An acoustic pressure of 10^4 Pa would represent a high intensity wave in water; this would correspond, for example, to a particle displacement of 1.06×10^{-4} mm at 10kHz, and 1.06×10^{-2} mm at 100Hz. Since the acoustic impedance of air is less than that of water by a factor of about 3.6×10^3, an acoustic wave of the same pressure in air would require displacements of about 0.38mm at 10kHz and 38 mm at 100Hz, – a very large movement. These figures illustrate the major difference between transducers to work in air and those

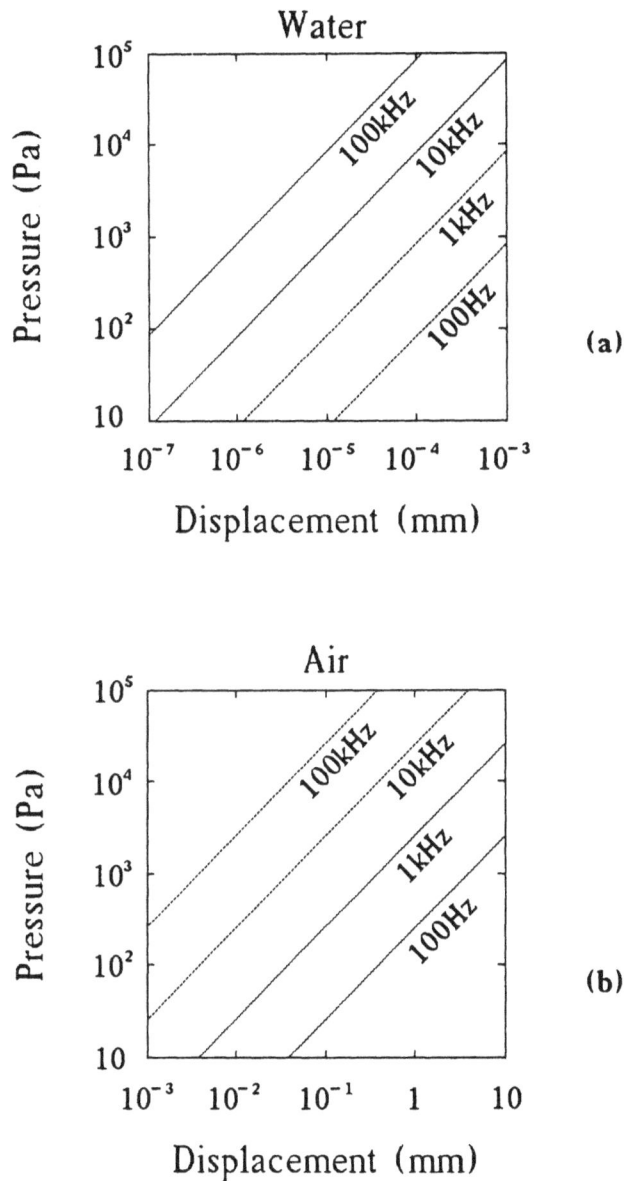

Fig 1.2 Relationship between particle velocity and acoustic pressure for plane wave: (a) in water; (b) in air.

for use in water, in that very much larger amplitudes of vibration are needed in air than in water for equivalent acoustic pressures. As a result of this demand for larger amplitudes, the mechanical stiffness of a transducer for use in air must generally be much lower than for an underwater design, as is clearly evident in a loudspeaker. At the other end of the scale, pressures as low as 10^{-2}Pa may be measured in the sea. The corresponding particle displacement is 1.06×10^{-10}mm at 10kHz, and 1.06×10^{-8}mm at 100Hz, – amplitudes rather smaller than atomic dimensions!

If a transducer radiates or receives sound equally in all directions, it is described as being omni-directional. In many applications, a transducer is required to concentrate the acoustic power by forming a beam pointing in a particular direction. The direction in which the acoustic response has its maximum value is called the **acoustic axis**, and this is usually taken as the reference in defining the acoustic characteristics of the transducer.

The **hydrophone sensitivity** (*M*) of a transducer is the ratio of the electrical output voltage, with the output open-circuited, to the applied sound pressure, usually expressed in decibel form, ie:-

$$M \text{ (dB re 1V/μPa)} = 20 \log_{10}(V/p)$$

The measurement frequency must be specified. Two forms of hydrophone sensitivity are used. The more common form is the "**free-field sensitivity**", for which the voltage is referred to the acoustic pressure in a plane wave which would have existed at the hydrophone position in the absence of the hydrophone; ie any alteration in the sound field due to the presence of the hydrophone itself is included within the sensitivity measurement. By convention, the hydrophone sensitivity for a directional transducer is taken as the output due to a plane wave incident along the acoustic axis. It is sometimes more convenient to relate the output voltage to the actual acoustic pressure which exists at the face of the hydrophone when it is in the field. This is referred to as the "**pressure sensitivity**" of the hydrophone, and is normally used only for small hydrophones, where the pressure is uniform over the sensitive surface. The two sensitivities are effectively equal if the hydrophone is acoustically hard and small compared with a wavelength at the measurement frequency. (See also Section 9.5 and Ref [1.4].)

The **projector sensitivity** (S_V) is the ratio of the sound pressure p produced at a nominal distance of 1 metre on the acoustic axis to the voltage V applied to the transducer, referred to the standard pressure of $1 \mu Pa$ and expressed in decibel form, ie:-

$$S_V \text{ (dB re } 1\mu Pa/V \text{ at } 1 \text{ m)} = 20 \log_{10}(p/V)$$

The acoustic field near to the source often exhibits rapid variations with position, which may make it difficult to ensure a reliable measurement of the pressure at 1m. It is therefore common for the acoustic pressure to be measured at a greater distance, where the field has settled down into its "far-field" behaviour (See section 2.3). This measured value is then corrected to its reference value at the nominal range of 1m by assuming spherical spreading (ie by adding a correction of $20\log r$).

S_V relates the acoustic pressure to the applied voltage, and it is sometimes known as the voltage projector sensitivity. There is an alternative definition of the projector sensitivity in terms of the input current (i) instead of the voltage, ie:-

$$S_I \text{ (dB re } 1\mu Pa/Amp \text{ at } 1 \text{ metre)} = 20 \log_{10}(p/i)$$

This current projector sensitivity is more commonly used for "current-driven" low impedance devices such as magnetostrictive transducers than for piezoelectric transducers, although it could in principle be applied to either type. In referring to the projector sensitivity of piezoelectric transducers, we shall assume it to be the voltage sensitivity unless stated otherwise. It will sometimes be convenient to refer to the linear forms of hydrophone or projector sensitivities, and in those cases they will be denoted by $\hat{M}$, $\hat{S}_V$, $\hat{S}_I$.

1-3   SENSITIVITY-FREQUENCY RELATIONSHIPS

Suppose that an alternating force ($F = F_0 \exp(j\omega t)$) acts on the sensitive face of an electroacoustic transducer. To a first approximation this may be represented by a lumped system as indicated in *Fig 1.3*, in which m represents the effective vibrating mass, K the stiffness of the spring to which it is attached, and r the resistive damping coefficient. The equation of motion of the system, derived by applying Newton's laws, is

$$m du/dt + ru + K\int u\,dt = F_0 \exp(j\omega t) \qquad (1.9)$$

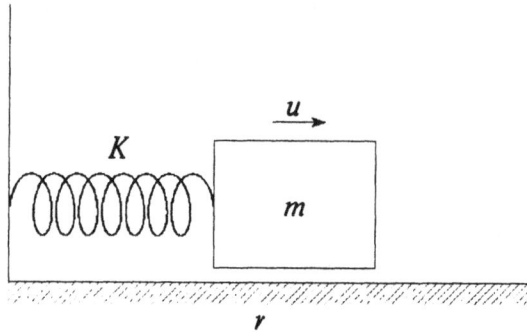

Fig 1.3 Mass-spring system representing mechanical resonator.

where u is the alternating velocity of the mass m, t represents the time variable, and ω is the angular frequency of the applied force.

Assuming the mass also to oscillate sinusoidally with a frequency ω, we may use the relationships (1.5) to convert equation (1.9) to the form

$$m.j\omega u \; + \; ru \; + \; K.(u/j\omega) \; = \; F \qquad (1.10)$$

Thus,

$$F/u \; = \; jm\omega \; + \; r \; + \; K/j\omega$$

$$= \; r \; + \; j\frac{K}{\omega}\left(\omega^2\frac{m}{K} - 1\right)$$

ie,

$$u/F \; = \; \left[r \; + \; j(K/\omega)\{\omega^2(m/K) \; - \; 1\}\right]^{-1} \qquad (1.11)$$

For a given F, u is a maximum when $\omega^2 m/K = 1$, ie when $\omega^2 = K/m$, and this value of ω is the (angular) resonance frequency (ω_s). It is useful to consider the response in three frequency regions:-

A)   Well below the resonance frequency, the third term of equation (1.10) predominates, and $u \simeq j\omega F/K$. This is the "stiffness-controlled" region, in which $u \propto \omega F$.

B) Near to the resonance frequency, there is a band within which the second term of (1.10) predominates. In this region, $u \simeq F/r$, ie $u \propto F$.

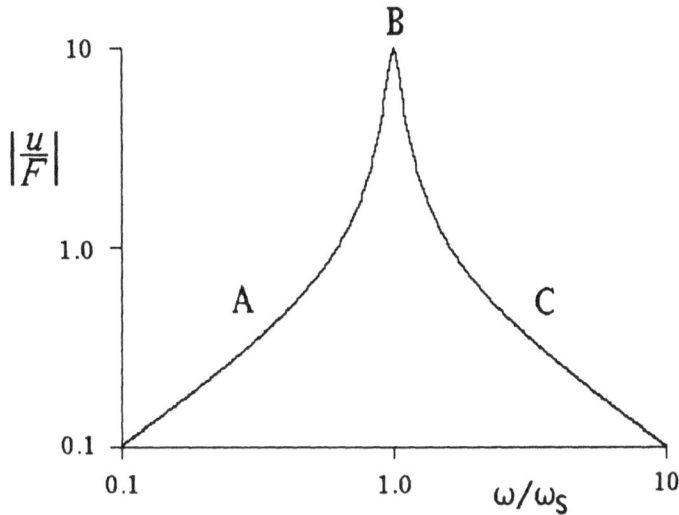

Fig 1.4 Response of resonant mechanical system as a function of frequency ($\omega_S m = 1$, $r = 0.1$).

C)   Well above resonance, if the system has only one resonance, the first term of (1.10) predominates, and $u \simeq F/j\omega m$. This is the "mass-controlled" region, in which $u \propto F/\omega$.

These regions are illustrated by the curve in *Fig 1.4*, which shows the variation of u/F (from equation (1.11)) against frequency for a typical system.
    If the acoustic pressure on the face of a piston-type hydrophone is p, and the piston area is A_p, then the force on the face is given by $F = pA_p$, ie $F \propto p$. For a piezoelectric ceramic transducer, displacement of the piston generates a charge (Q_c) on the ceramic which is proportional to the displacement (See Chap 3). Since the electrical impedance is predominantly capacitive, the open-circuit voltage produced by the pressure is given by $V_1 = Q_c/C$

ie, $V_1 \propto Q_c \propto x \propto u/\omega$

    Combining this relationship with equation (1.11), the open-circuit voltage of the transducer operating as a hydrophone behaves in the three frequency regions as:-

A)   Well below resonance   – $V_1 \propto p$

B)   Near resonance   – $V_1 \propto p/\omega$

C)   Well above resonance   – $V_1 \propto p/\omega^2$

The ratio of the output voltage (V_1) to the applied acoustic pressure (p) is the hydrophone pressure sensitivity, so these relationships indicate how the pressure sensitivity varies with frequency. The free–field hydrophone sensitivity differs from the pressure sensitivity by a factor which is generally between one and two, the pressure at the face of the hydrophone being doubled (compared with the pressure in the absence of the hydrophone) if the face is much larger than a wavelength. The factor tends to remain approximately constant except for frequencies where the dimensions are comparable with a wavelength. The relationships above can therefore be taken to indicate also the approximate behaviour of the free–field hydrophone sensitivity (M) with frequency.

The hydrophone sensitivity is related to the current projector sensitivity S_I through the principle of reciprocity [1.5], which shows that $\hat{M}/\hat{S}_I = $ constant$/\omega$. This allows us to deduce how $\hat{S}_I$ varies with frequency. Also, since a piezoelectric transducer is generally capacitive, the input current is related to the applied voltage by $V/i \propto 1/\omega C$. The voltage projector sensitivity is therefore related to the current sensitivity by $\hat{S}_V \propto \hat{S}_I.\omega$. We thus obtain the following dependencies on frequency for the transducer sensitivities in the three frequency regions:–

	$\hat{M}$	$\hat{S}_I$	$\hat{S}_V$
Well below resonance	Const	ω	ω^2
Near resonance	$1/\omega$	Const	ω
Well above resonance	$1/\omega^2$	$1/\omega$	Const

These relationships are applicable only to the piezoelectric transducers themselves; if a tuning inductor is connected across the transducer, it will of course introduce a further frequency dependence.

One frequency region is of particular interest and significance. If a constant (ie. independent of frequency) hydrophone response is wanted, a piezoelectric transducer used below resonance is the obvious choice, and this will be discussed further in Chapter 10. It may also be observed that a piezoelectric transducer used above resonance would have a constant projector sensitivity S_V. This is however much less frequently used, since other resonances often interfere with this ideal response.

REFS Chapter 1

References

1.1   Hunt,F.V., *Electroacoustics*, Harvard University Press and Wiley & Sons, 1954, Chapter 1. (Reprint published by Acoust Soc Am, 1982.)

1.2   Arfken,G., *Mathematical Methods for Physicists*, Academic Press, 1970, p324.

1.3   Kinsler,L.E., and Frey,A.R., *Fundamentals of Acoustics*, (2nd Ed), Wiley & Sons, 1962, p121.

1.4   Bobber,R.J., *Underwater Electroacoustic Measurements*, Naval Research Laboratory, 1970, Chapters 1 and 2.

1.5   Kinsler,L.E., and Frey,A.R., op. cit., p325

Additional Reading

1   Tucker,D.G., and Gazey,B.K., *Applied Underwater Acoustics*, Pergamon Press, 1966

2   Kinsler,L.E., and Frey,A.R., *Fundamentals of Acoustics*, Wiley & Sons, 2nd Ed 1962.

3   Hueter,T.F., and Bolt,R.H., *Sonics*, Wiley & Sons, 1955.

4   Hunt,F.V., *Electroacoustics*, op. cit.

5   Cady,W.G., *Piezoelectricity*, McGraw-Hill, 1946.

6   Hueter,T.F., "Twenty years in underwater acoustics: generation and reception," *J Acoust Soc Am*, 51, 1025-1040, (1972).

7   Sherman,C.H., "Underwater sound – a review; 1. Underwater sound transducers," *IEEE Trans on Sonics and Ultrasonics*, SU-22, 281-290, (1975).

8   Woollett,R.S., "Ultrasonic transducers: Pt 2, Underwater sound transducers," *Ultrasonics*, 3, 243-253, (1970).

Chapter 2
PRINCIPAL
CHARACTERISTICS

2-1 INTRODUCTION

It is essential that the characteristics which any transducer needs to satisfy its particular application should be agreed between the user and the designer. In this book we shall of course be concerned with designing transducers to achieve particular characteristics, but sometimes the requirement could be met by an existing design, and this is one of the first questions to be considered. The user is primarily concerned with specifying the overall characteristics of the transducer, and is less interested in the details of how these will be effected by the designer. This chapter will concentrate on the principal features of the transducer which need to be considered and agreed by both user and designer, whilst the actual design methods which can be applied will be discussed in later chapters.

Although the singular "transducer" has been used in the paragraph above, it is often necessary to use a number of transducers assembled into an array to achieve the required characteristics. This leads to some difficulties in terminology. If a number of transducers, each in their own casings, are assembled together to form an array, it is straightforward to refer to this as an array. It is however quite common to fit a number of transducer elements into a single housing (eg as in *Fig 2.1a*), and the whole is then often referred to as a transducer, especially for high frequency designs as illustrated in *Fig 2.1b*. Sometimes, transducer elements are assembled in a line into a common housing, and this is often called a "stave" (*Fig 2.1c*). Despite the opportunities for confusion, it is hoped that the use of the various terms will be clear enough from their context.

2-2   DIRECTIVITY REQUIREMENTS

The user is primarily concerned with the characteristics of the complete array, but one of the first decisions to be made in the design process is concerned with whether the array should be divided into separate elements, and if so how many. The decision is affected by a variety of considerations. Probably the most fundamental is the question of whether electrical beam steering of the array is required, since this will restrict the maximum inter-element separation as indicated below. In many high frequency applications (eg above about 100kHz), the array is used to form a single beam, and beam steering can be effected simply by rotating the whole array. The beam steering rate is limited by the maximum rate of rotation of the array structure, and this is often fast enough to satisfy the needs of the particular application. For lower frequencies, array sizes increase and it becomes increasingly difficult to rotate the array quickly enough, or even in some cases to rotate it at all. It is therefore often necessary to use some form of electrical beam steering, by applying phase or time delays to the array elements, which allows the direction of the acoustic beam to be steered without rotating the array itself. This may also be necessary at high frequencies if rapid steering of the beam is wanted.

The main effect of this requirement may be understood by considering the beam patterns of a line array of point sources [2.1]. These will be discussed in more detail in Chapter 6, and at this stage we need only to consider how and where the major lobes in the beam pattern are formed. This may be done in terms of either a transmitting or a receiving array, and it is probably easier to visualise the process for a receiving array. Suppose that a plane wave is incident on a line array of point receivers, as in *Fig 2.2a*, in which θ is the angle of propagation of the

(a) Elements forming planar array. Courtesy Ferranti– Thomson Sonar Systems UK Ltd.

(b) High frequency array. Courtesy Ministry of Agriculture, Fisheries & Food.

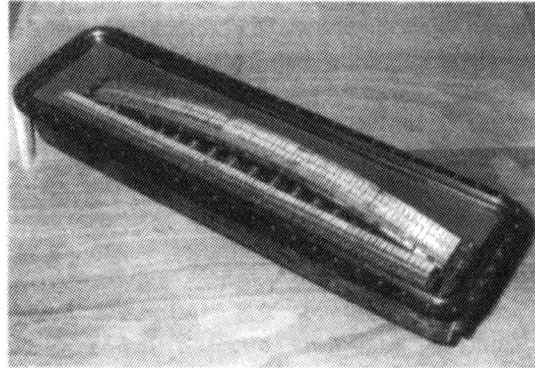

(c) Elements in stave. Courtesy Ministry of Defence.

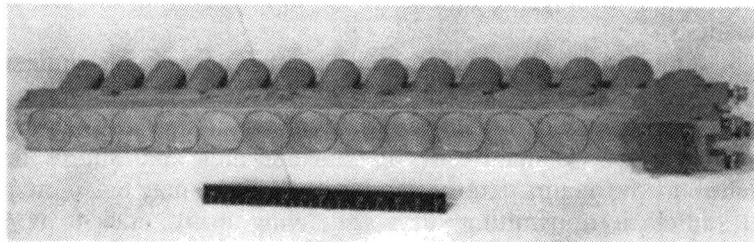

Fig 2.1   Various types of transducer

incident wave measured from **broadside** (ie perpendicular to the line of the array), and *d* is the separation between the receivers. If the receivers are numbered from 0 to $N - 1$ (for an N-element array), it is clear from the diagram that signals arriving at the m-th element are delayed by the time taken to travel a distance $md\sin\theta$ compared with the time at which the signal arrived at the 0-th receiver. By applying electrical delays to the elements to compensate for these geometric delays, it is possible to ensure that the signals arriving from the chosen steering direction produce output voltages which are all in phase, and thus give the maximum output from the array. It is in general only in this steering direction that all the signals add up in phase, and the response in other directions is then less than in the direction of the steered main beam, which is what is normally required. This is not always the case, however, and we need to establish whether other angles can exist for which the signals also add in phase and thus form further major diffraction lobes.

In order to study this question, consider two elements of the array, with phase delays applied to produce a steering direction of θ_0, as in *Fig 2.2b*. Now consider what is the resultant when a signal is incident at an angle θ to the array. The phase difference corresponding to the direction of the incident signal is $2\pi(d/\lambda)\sin\theta$ and the applied phase shift corresponding to the steering angle is given by $-2\pi(d/\lambda)\sin\theta_0$. The resultant phase delay of the output from B compared with A is therefore $2\pi(d/\lambda)(\sin\theta - \sin\theta_0)$. The phase difference is zero when $\theta = -\theta_0$, the intended steer angle. But the signals will also reinforce if the phase difference reaches 2π, – or in general whenever it is an integral number times 2π, – and the response is then as large as for the steered main beam. Directions corresponding to these values are referred to as the angles of the major diffraction lobes of the array, and it is usually desirable to

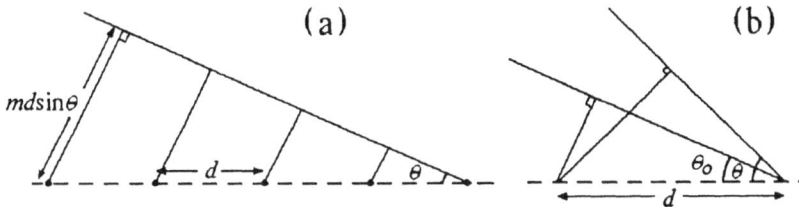

Fig 2.2 **(a) Steering of linear array of point receivers; (b) Parameters for steering of array.**

prevent their occurrence. From the expression above, the first diffraction lobe would appear when $\theta = \theta_1$, where

$$2\pi(d/\lambda)(\sin\theta_1 - \sin\theta_0) = 2\pi$$

ie when

$$\sin\theta_1 = (\lambda/d) + \sin\theta_0 \qquad (2.1)$$

For an unsteered array, when $\theta_0 = 0$, $\sin\theta_1$ will have a value in the real range (not exceeding unity) if $\lambda/d \leq 1$, ie $d \geq \lambda$. Thus, a major lobe will appear if the element spacing is as large as, or greater than, λ. It is easy to see that for $d = \lambda$ this corresponds to a wave incident along the line of the array, for which the geometric phase delay due to the element spacing would be 2π, thus giving rise to major lobes in both directions along the line of the array. The existence of these diffraction lobes can be prevented by making the element spacing less than λ. For an array which may be steered to **endfire** (90° from broadside), θ_0 may take values up to $\pi/2$, and in that case the first diffraction lobe may appear if $\lambda/d \leq 2$ ie if $d \geq \lambda/2$. The diffraction lobe appears along the line of the array in the opposite direction to the steered beam. Curves showing the angles of the first diffraction lobe, from (2.1), are given in *Fig 2.3*, with the spacing d/λ as a parameter. It is clear from these calculations

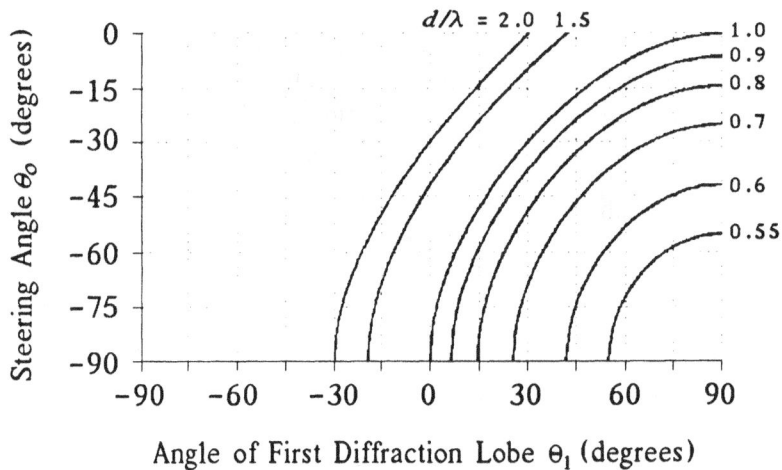

Fig 2.3   Angles of first diffraction lobe, as a function of steering angle and element separation.

that the safest way to avoid the possibility of major diffraction lobes is to make the element spacing less than $\lambda/2$. If the array is to be used only near broadside, some increase in spacing, up to nearly λ, may be allowed.

Two factors should be mentioned which complicate this simple criterion to some extent. The limits above are concerned only with the direction of the peak of the diffraction lobe. Even though the spacing may be made small enough to avoid this peak, there may nevertheless be some effect on the beam pattern from the gradual penetration into real space of the diffraction lobe because of its finite beamwidth. It is therefore desirable to keep significantly below the limits indicated by (2.1). The second factor arises from the finite size of the individual elements. The product theorem (Sect 6.4) shows that the beam pattern of a linear array of uniform elements of finite extent is given by the product of the pattern for a line array of point sources at the element centres and the pattern for a single element. The effect of this is to reduce the amplitude of the side lobes away from the broadside direction, and the diffraction lobes are therefore lower than the main beam if it is at broadside. In the limit, the elements fill the array length and the beam pattern becomes that of a continuous line array without any diffraction lobes. However, if the array is steered away from the normal, the diffraction lobes may increase relative to the main lobe, and for large steering angles the diffraction lobe may even exceed the intended main lobe. Unless a careful study of the beam patterns is to be carried out, the safest course is, as before, to make the element spacing less than $\lambda/2$. Using a spacing much less than $\lambda/2$ would, however, involve using more individual transducer elements, and this would normally result in higher costs. It is therefore a good general rule to aim for a spacing between elements of just under $\lambda/2$.

Over the main frequency band of interest, from 2 to 20kHz, this rule leads to spacings ranging from 375mm to 37.5mm. These appear reasonably practical at the higher frequencies, down to say 5kHz ($\lambda/2 = 150$mm), but below about 5kHz it becomes difficult to make piston-type transducers with radiating heads large enough to fill an array with $\lambda/2$ spacing satisfactorily. The transducer size at these low frequencies tends to be determined by the maximum size of piston which meets the acoustic requirements, is mechanically practical, and also avoids problems with flexural resonances (Sect 8.5). These considerations may well result in a piston size which is appreciably less than $\lambda/2$. Using these elements with $\lambda/2$ spacing could then leave large inactive areas between pistons, and this is usually undesirable for transmitting arrays because it reduces the radiation loading and the power handling capacity of the array. It is therefore preferable in these cases to assemble the transducers as closely as possible in the array, despite the increased costs, and accept a spacing less than $\lambda/2$. Although this can produce quite satisfactory beam patterns, problems can arise from acoustic interactions between resonant elements which are very closely spaced (Sect 6.2), and it is best to avoid spacings less than about $\lambda/4$ if possible. Interaction between non-resonant elements is generally not significant, and this condition is thus less applicable to receiving arrays operating away from resonance.

Suppose that the requirement is for a planar array which is to be steered over wide angles in both planes normal to the face of the array. Then an array of separately driven elements spaced less than half a wavelength at the operating frequency is needed. The natural choice would be to fit the elements into individual housings, each with its own electrical drive cable. The mechanical realisation of this arrangement is reasonably straightforward at the lower frequencies, but at higher frequencies the transducer sizes become rather small, and the possibility of fitting all the elements into one case would have to be considered, despite the complication of the separate electrical drives.

If steering in only one plane is wanted, it is often convenient to connect in parallel all the elements along a line in the non-steering direction. These may then be assembled into a common housing, usually referred to as a stave, as in *Fig 2.1c*. Only one electrical connection to the stave is needed, internal wiring being used to transfer the drive to the elements along the stave. This has clear advantages in reducing the number of external cables and connectors, but incurs some extra risk in that a defect in a single element will necessitate changing the whole stave. If no steering is required, it is possible to go further and fit all the elements into a single housing, as in *Fig 2.1a*. As is to be expected, this simplifies the cabling to a single connection, but may involve changing the whole array if a single element becomes faulty.

Choosing between these options is of course affected by the overall mechanical arrangement envisaged. For example, cylindrical arrays as illustrated in *Fig 2.4* are sometimes used to permit all-round radiation, and these are most conveniently assembled from a number of staves. Low-frequency transducers tend to become large and heavy, and it is then often more convenient to assemble them as individual elements than to

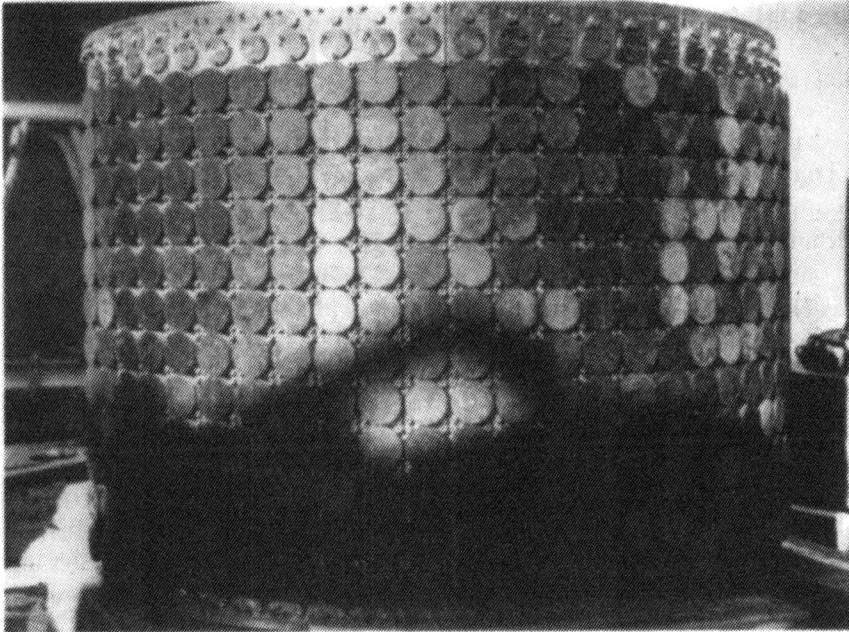

Fig 2.4 Cylindrical transmitting array. Courtesy Ferranti-Thomson Sonar Systems UK Ltd.

combine them into an even heavier array. At the other end of the spectrum, it is usually easier to assemble a high frequency (eg 100kHz or above) array into a single housing than to attempt separate housings for each element. Ease of assembly and replacement are important factors in making these choices, especially if the array cannot easily be removed from its operating environment.

Beam Width.

One of the major parameters of the array to be specified is the required beam width. Beam patterns will be discussed in more detail in Chapter 6, but it is convenient to note here some simple approximate relationships which establish the overall size and shape of the array. We shall define the beam width ($\theta(-3dB)$) as the angle between the directions where the response is 3dB less than at its maximum (the "$-3dB$ points"). This beam width depends on the dimensions of the transducer in accordance with the following expressions [2.2],

$$\theta(-3\text{dB}) \simeq 60°(\lambda/D) \text{ for a circular piston of diameter } D \quad (2.2)$$

$$\text{and } \theta(-3\text{dB}) \simeq 50°(\lambda/L) \text{ for a line source of length } L \quad (2.3)$$

Beam widths are sometimes quoted in terms of the −6dB or −10dB points, or even the angle between the first minima in the beam pattern. Approximately, $\theta(-10\text{dB}) = 1.8\theta(-3\text{dB})$, and $\theta(\text{between nulls}) = 2.3\theta(-3\text{dB})$. These expressions apply to either a single transducer or an array, *provided that the overall dimension is greater than about one wavelength of sound* in water at the frequency of interest. A rectangular source is treated as a line source, with a length equal to that of the edge of the rectangle parallel to the plane of the beam under consideration.

Expressions (2.2) and (2.3) are true only for arrays in which all the elements are driven with equal amplitude. It is sometimes desirable to vary the amplitudes of the elements across the array, in order to control the beam shape (this is often described as "shading" the array). For example, reducing the amplitudes towards the array edges may be used to reduce the side lobe levels, and in that case, the beam width is increased over the values calculated from (2.2) or (2.3). Indeed, the wider beam for the circular array arises from the geometric shading of the array compared to that of the rectangle, the effective sensitivity reducing towards the edges. As might be expected, the theoretical value of the first (and biggest) side lobe for the circular array is lower, at −18dB relative to the peak sensitivity, than that for the line array, which is −13dB. Thus, it is not particularly unusual for the beam width to exceed the values above; but only in rare circumstances can the beamwidth be made less than the value given by (2.3). Equations (2.2) and (2.3) can therefore be used to indicate the approximate minimum size of transducer needed to achieve the desired beam width.

The discussion in this section of the beam characteristics of the array has been restricted to the main features which affect the overall size of the array and how it needs to be divided into separate elements. A rather more detailed treatment of the radiation characteristics is given in Chapter 6. In order to produce the calculated beam patterns, the elements must behave as predicted. One source of deviations from this idealised performance is the mechanical coupling between elements which can arise from practical mounting arrangements, and this is an area which needs attention in considering the overall design. For example, mounting elements on a common support plate may easily cause problems, and should be adopted only with caution.

Omni-directional Beam Patterns.

The possibility of using a cylindrical array of separate elements (as in *Fig 2.4*) to produce wide beams in the plane normal to the axis of the array has been mentioned above. In the limit, such an array may be used to radiate uniformly at all angles in that plane, – ie to produce a beam pattern which is omni-directional in one plane. The directivity in the other plane is then determined by the height of the array, as for a line source. The omni-directional nature of the beam pattern is not affected by the diameter of the array, which could therefore be made as small as convenient, provided that sufficient power could be radiated to achieve the required source level. However, it is sometimes desirable to use the same array to form directional receive beams, by applying phase shifts to the array staves around a sector of the cylinder, to make the curved array behave approximately as a planar array. In that case, the beam width as a receiver is related to the effective width of the sector by (2.3), and the array diameter is likely to be determined by the required beam width of the array in its receive mode. If the diameter is then several wavelengths across, the elements may to a first approximation be treated as though they were in a planar array.

For other applications wanting an omni-directional beam pattern in one plane, it may be more suitable to allow the diameter of the array to shrink to a small value, of about one wavelength. It is then impractical to fit in separate elements around the circumference of the array, and the use of transducers in the shape of complete rings should be considered. The design of such ring transducers is discussed in Chapter 12. In general, their diameter is about one wavelength of sound in water at the resonance frequency. If a radiation pattern is required which is omni-directional in both planes, it may similarly be produced by using a large sphere made up of separate elements, or by means of a single spherical transducer, which again is likely to have a diameter of about one wavelength.

It is sometimes desirable to radiate sound from a source which is even smaller than a wavelength across. We shall see later (Chapter 6) that the radiation impedance of a source much smaller than a wavelength is such that it becomes very difficult to achieve good transducer efficiency, and this requirement is therefore only practicable for low power outputs, where efficiency may not be important. In those cases where this is practicable, a very small source (less than about $\lambda/4$) acts as a simple generator of volume displacements, and the actual shape of the source should have very little effect on the radiation pattern. It is however quite feasible to design a receiving hydrophone which is much smaller than a wavelength, because of the lesser importance of its electro-acoustic efficiency. Such a hydrophone should also in principle have an omni-directional receiving pattern, regardless of the shape of the hydrophone itself, if it is a true pressure sensor. This ideal behaviour is not always easy to achieve, since in practice it is sometimes difficult to avoid some response from the acoustic particle velocity of the water, which does introduce some directionality into the pattern. This is however a problem primarily for the transducer designer, and in principle an omni-directional receiving pattern should be achievable.

The fact that properly designed piezoelectric hydrophones respond only to pressure represents an important difference between acoustic and electromagnetic receivers. The latter essentially extract power from the radiating field, and hence cause a significant disturbance of the field. A non-resonant piezo-ceramic sensor, however, has a high acoustic impedance, and the voltage it generates may be amplified and processed without needing to extract significant power from the field. Its disturbance to the acoustic field is then due mainly to its diffraction effects, and these are generally negligible for small, high impedance obstacles. This makes it straightforward to use open arrays of widely spaced small hydrophones which are effectively acoustically transparent. Each hydrophone may be regarded as sampling the field without affecting the field at the other sensors, and beam forming may then be carried out by applying time delays to the received signals to compensate for the geometric path delays. More complex beam forming techniques may be used to satisfy particular requirements, but these will not be considered here. The main point to note is the possibility of using open, acoustically transparent arrays of receiving hydrophones, in which all of the hydrophones may be used in the beam forming. Projector arrays are usually more nearly filled, as in *Fig 2.4*, in order to handle the power; they are thus not acoustically transparent, and only the front-facing elements can be used as inputs to the beam former.

As noted above, it is common for hydrophones to be intended as pressure sensors in which the omni-directional response is due to the small size of the hydrophone dimensions compared with a wavelength at the frequencies of interest. Sometimes, however, it would be desirable for the sensor to have some directivity even at these low frequencies without having to increase the size of the hydrophone. This can be achieved in some circumstances by deliberately making the receiver respond to the acoustic particle velocity, or pressure gradient, which has a cosinusoidal dependence on angle relative to the direction of propagation. Techniques for doing this are discussed in Chapter 10. They are subject to a number of practical difficulties, and should not be specified without careful consideration of the likely problems described in that chapter.

Directivity Index.

The characteristics required of a transducer may often be best described in terms of the beam patterns, as above, but sometimes it is appropriate to use a single factor known as the directivity index to represent the beam forming performance. We firstly recall that the acoustic axis of a transducer is defined as the direction in which the acoustic response has its maximum value. The **directivity factor** (DF) for a projector is then defined as the ratio of the transmitted acoustic intensity along the acoustic axis (I_0) to the intensity which would have resulted from radiating the same power uniformly in all directions (I_{ref}), both measured at the same distance from the source. It thus represents the overall concentration of power along the acoustic axis achieved by the beam forming of the transducer or array, without explicitly involving the details of the beam shape or side lobes. For a receiver, the analogous definition is that the directivity factor is the ratio of the output electrical power generated by an acoustic signal incident along the direction of maximum response to that generated by a uniformly distributed noise field having the same mean acoustic power. The **directivity index** (DI) is then the logarithmic expression of the directivity factor, ie DI = 10logDF. Because the beam patterns depend on frequency, the directivity index is also a function of frequency.

For an omni-directional source, the directivity factor has the value unity. If a small source is in a rigid **baffle**, so that it radiates only on one side of the baffle, the directivity factor is two, and as the radiation becomes more directional the directivity factor increases. For a circular piston of area A in a rigid baffle, it can be shown (eg [2.2]) that for piston diameters greater than $\lambda/2$ the directivity index is given approximately by

$$DI \simeq 10 \ \log(4\pi A/\lambda^2) \qquad\qquad (2.4)$$

The same expression can be applied to transducers with faces of other shapes, provided that the dimensions are at least $\lambda/2$ in both directions, and if the dimensions exceed 2λ the expression is applicable whether the array is in a baffle or not. For a continuous line source of length L, provided $L \gg \lambda$, the DI is given [2.2] by,

$$DI = 10 \ \log(2L/\lambda) \qquad\qquad (2.5)$$

For an array of N point sources separated by at least $\lambda/2$, the DI is approximately equal to $10\log N$.

Because of the link between DI and beam shapes, it is possible to relate the DI to the **beam widths** of a transducer. This is particularly useful when the specified beam widths require an array with a length much longer than its width, or when we wish to analyse the behaviour of a transducer with known beam widths. *Fig 2.5* gives curves (from [2.3]) showing the relationships between beam widths and directivity index for elliptical and rectangular transducer faces. The difference between the sets of curves represents the effects of the different side lobe levels calculated for the two cases; if the measured side lobes differ significantly from theoretical values, the curves should be used only with caution. A typical value of DI may be deduced by considering a circular plane array producing a −3dB beamwidth of 12° (ie ±6°); − eg this would be generated at 15kHz by an array of diameter 50cm (from (2.3)), and would have a directivity index of 24dB.

2-3   SOURCE LEVEL

The **source level** of a transmitting transducer or array is a parameter representing its strength as an acoustic source; it is normally expressed in logarithmic terms, since this facilitates its use in calculations of sonar performance. The source level is defined as the intensity of the radiated acoustic wave in a specified direction, in decibels relative to the intensity of a plane wave of rms pressure 1µPa, referred to a point 1 metre from the acoustic centre of the source. Although it is possible to refer the source level to any chosen direction relative to the

MINIMUM BEAM WIDTH AT
10 DB DOWN (deg)

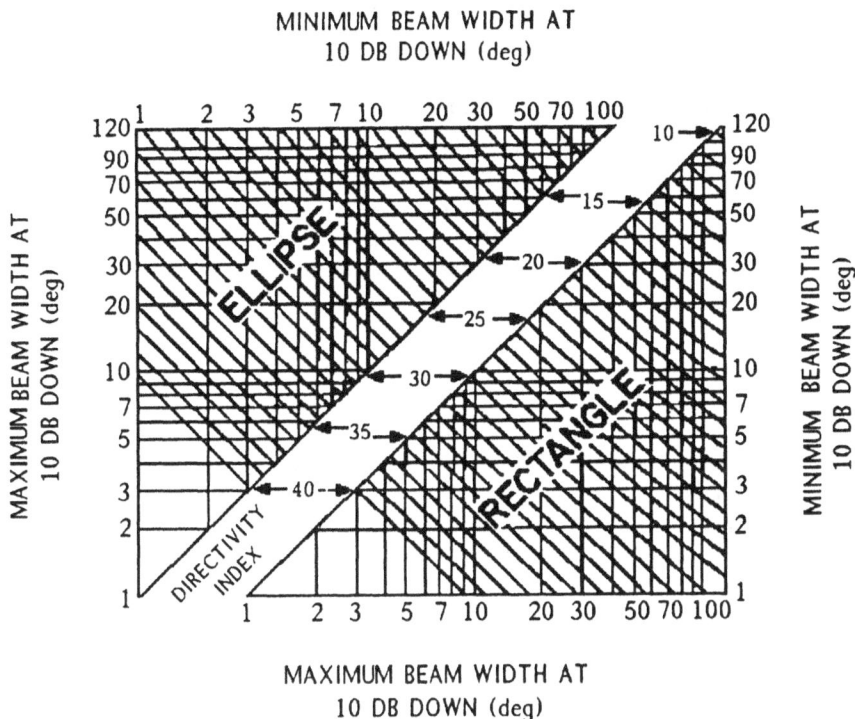

MAXIMUM BEAM WIDTH AT
10 DB DOWN (deg)

Fig 2.5 Relationship between directivity and beamwidth between −10dB points for elliptical and rectangular radiating surfaces. ($\theta(-10\text{dB}) \simeq 1.8\theta(-3\text{dB})$). From Albers, *Underwater Acoustics Handbook − II*, Pennsylvania State Press, 1965, Fig 21.6, p292.

transducer, it is in practice almost always referred to the acoustic axis, since that is usually pointed in the direction of interest, and this will be assumed to be the case in this section.

It may appear from the above definition that the source level could be determined simply by measuring the acoustic pressure at a distance of 1 metre from the source, but that is generally not practicable because the sound field near a projector does not fall off smoothly as it would near a point source. These near−field effects arise mainly from the finite size of the transducer, and extend for a distance of about $r_0 = L^2/\lambda$, where L is the length of the transducer face. Within this range r_0, often referred to as the Fresnel range, the sound field fluctuates rapidly with position, and it is very difficult to make reliable measurements. The source level is therefore generally determined from measurements made at a range r beyond r_0 and referred back to 1 metre by assuming spherical spreading, ie by adding 20 log r.

Suppose that a projector radiates a total acoustic power of W_a watts uniformly in all directions. Let the acoustic intensity at a large range r from the centre of the source be denoted by I_r. This intensity is given by the total power divided by the area of the spherical surface, ie $I_r = W_a/4\pi r^2$. For large r, the wave front over a small area approximates to a plane wave, and the intensity is then related to the rms acoustic pressure (p) by $I_r = p^2/\rho c$. Thus,

$$p^2 = \rho c I_r = \rho c W_a / 4\pi r^2 \qquad (2.6)$$

When $r = 1$, this gives the effective rms acoustic pressure at the reference distance of 1 metre, ie the source level for an omni-directional source. Substituting $\rho c = 1.54 \times 10^6$ kgm^{-2}s^{-1} for sea water, this gives the expression for the source level (SL) of an omni-directional source as

$$SL = 10 \ \log(\rho c W_a/4\pi)$$

$$= 50.9 + 10 \ \log W_a \qquad \text{(dB re 1Pa at 1m)}$$

$$= 170.9 + 10 \ \log W_a \qquad \text{(dB re 1}\mu\text{Pa at 1m)}$$

For a source with a transmitting directivity index DI, we thus obtain the general expression,

$$SL = 170.9 + 10 \ \log W_a + DI \qquad \text{(dB re 1 Pa at 1m)} \qquad (2.7)$$

Thus, one watt radiated omni-directionally produces a source level of 170.9 dB re 1μPa at 1m. If the source is directional, the source level on the acoustic axis is increased according to equation (2.7), which shows the relationship between source level, power, and directivity. If the beamwidth is specified for its particular application, this effectively determines the directivity index, and (2.7) then tells us what acoustic power would be needed to produce a specified source level. Note that W_a represents the *acoustic* output power, which is lower than the input electrical power by a factor corresponding to the electrical-acoustic efficiency. (See chapter 4.) If the input electrical power is denoted by W_e, and the electro-acoustic efficiency by η_{ea}, then $W_a = \eta_{ea}W_e$, and (2.7) may be expressed in terms of the input power as

$$SL = 170.9 + 10 \log W_e + 10 \log \eta_{ea} + DI \qquad (2.8)$$

Typical values of transmitting DI for directional systems range from 10 to 30, and efficiencies for resonant transducers are commonly between 20 and 90 percent. As an example, a system with an electrical input of 2kW, efficiency of 50 percent, and DI of 20dB, would produce a source level of about 221dB re 1μPa at 1m.

Cavitation.

Equation (2.7) thus determines what output power is needed to produce a given source level, whilst the minimum dimensions of the transducer are given by the beam shape requirements (using eg (2.2) or (2.3)). The question that now arises is whether the required power can be produced from the calculated transducer area. This question has two aspects. Firstly, can we design transducers in the available space which are capable of handling the power without internal failure? Secondly, are there limitations arising from the actual radiation of acoustic power from the transducer into the medium? The first aspect will be dealt with as part of the treatment of transducer design in Chapter 7. Radiation of acoustic power into water may cause cavitation, and the limitations arising from this effect are discussed in this section.

The intensity in an acoustic plane wave of peak pressure p_0 is given by $I = p_0^2/2\rho c$. As the intensity is increased, the peak acoustic pressure may thus reach a value equal to the hydrostatic pressure. If the intensity is increased still further, the acoustic pressure peaks will exceed the hydrostatic pressure, and the instantaneous pressure at various positions in the field will then be negative. At these points, there will be a stress tending to rupture the water, and if sufficient suitable nuclei exist bubbles will form in the water during these negative pressure peaks. This phenomenon is called "cavitation", and gives rise to various undesirable effects [2.4]. Firstly, it is associated with additional power dissipation due to the creation of the bubbles, and hence causes a reduction in the electro-acoustic efficiency. Secondly, it generates a bubble screen in the field, and thus reduces the radiation through it. And thirdly it causes non-linearity in the radiated pressure wave. The bubbles are often formed predominately at the surface of the transducer, and if the power is maintained at its high level for some time it may also cause erosion of the face of the transducer due to the collapse of the bubbles.

The theoretical **"cavitation threshold"** at which cavitation may start near the sea surface is readily found by calculating the intensity for which the peak acoustic pressure is equal to one atmosphere (approx 10^5Pa). This gives a level of 0.3 watts/cm^2, or 3000 watts/m^2, as the intensity near the surface at which cavitation may start to appear. In practice, its onset is fairly gradual, starting with some distortion of the pressure waveform and progressively reducing the conversion efficiency as the power is increased, so that the ratio of output to input powers decreases. Any non–uniformities in the field, giving rise to regions of high acoustic pressure, tend to reduce the average threshold, and it is difficult to quote a precise value for the cavitation threshold which is to be expected for any particular conditions. Making some allowance for the fact that many transducer arrays are composed of separate elements (often with circular radiating faces), so that the radiating face is not completely filled, and for the gradual onset of the effects, a level of 1 watt/cm^2 (10 kW/m^2) of total area may be taken as the upper limit for the cavitation threshold near the water surface. The difference in pressure levels corresponding to the difference between the theoretical value of 0.3 and 1W/cm^2 is sometimes interpreted as the "tensile strength" of the water. For very pure water free of nuclei, this tensile strength can be very high [2.5], but in sea water there are usually many nuclei and the increase in level is more plausibly associated with the gradual onset of the effects.

Because the formation of bubbles is a nucleation process taking some time to develop, and the periods of negative pressure in the acoustic wave become shorter as the frequency is increased, the cavitation threshold rises as the frequency is increased [2.4]. The increase is small up to 10 kHz, but may be up to a factor of about 30 (in power level) as the frequency is increased from 10 to 100kHz. An increase in threshold may also be observed if the pulse length is reduced below about 5 ms, in agreement with the concept of a nucleation process. For example, shortening the pulse to 0.5 ms may approximately double the power output before cavitation becomes serious. Both of these effects are rather difficult to quantify precisely, because of the uncertainties in defining the onset of cavitation. In practice, a pragmatic approach is usually advisable, with the cavitation level being determined experimentally, taking into account the penalties which would arise from the various effects.

A further, and generally more important, influence on the

cavitation threshold is that of the ambient pressure. If the transducer is immersed to a depth (h), the ambient pressure increases from one atmosphere at the surface to a value of approximately $(1 + h/10)$ atmospheres (for h in metres). The intensity for the onset of cavitation is then proportional to the square of the total effective pressure, including the tensile strength of the water, and Urick [2.5] thus obtains an expression for the cavitation threshold (I_c) of long pulse systems as

$$I_c = 0.3\gamma\{P_c(0) + h/10\}^2 \text{ watts/cm}^2$$

where h is in metres, $P_c(0)$ is the cavitation threshold in terms of pressure at the surface, and γ is a factor allowing for any pressure "hot spots" arising from non-uniformities in the field. Schofield [2.6] quotes cavitation levels for low frequency systems as

$$I_c = 0.3\{1.8 + h/10\}^2 \text{ watts/cm}^2 \quad \text{(for } h \text{ in metres)} \quad (2.9a)$$

$$= 0.3\{1.8 + h/33\}^2 \text{ watts/cm}^2 \quad \text{(for } h \text{ in feet)} \quad (2.9b)$$

These expressions give a threshold of nearly 1 watt/cm^2 at the surface, and thus represent about the upper limit for working levels for low frequency systems. Extra caution should be exercised if there is any reason to suspect particularly high "hot spots" in the field. For example, free-flooding ring transducers are likely to generate high fields at the centre of the ring, and cavitation can therefore be expected to occur there at levels below that calculated from the surface area of the rings themselves. If a transducer is to be used for long periods at high power, the potential danger of damage due to cavitation erosion must also be taken into account.

Equation (2.7) gave the relationship between source level, power output, and directivity index. The DI depends on the beam shape required, and this effectively determines the area of the transducer (or array) face, at least for a planar array. The expressions (2.9) can then be used to establish whether the radiating area is large enough to transmit the necessary power without exceeding the cavitation threshold, and thus establish whether the requirements for source level and beam shape are compatible. If the required power density turns out to be greater than the cavitation threshold for the operating conditions, the possibility of increasing the radiating area has to be considered. This would normally reduce the beam width below that specified, but it may be possible to curve the array, or apply phase delays to generate a curved wavefront, in order artificially to broaden the beam to the required width. These techniques, though feasible, add complexity to the array and are best avoided if possible. The relationships above may instead be used to establish the combination of source level and DI which can be achieved without recourse to such techniques. The problem is more likely to arise for transducers with wide beams operating near the surface. Sources which are required to radiate omni-directionally in one plane are particularly prone to cavitation limitations, but these can sometimes be tackled by using larger diameter cylindrical arrays, if these are compatible with the required operating frequencies.

2-4   BANDWIDTH

The final major acoustic parameter to be considered is the frequency band over which the transducer is required to operate. The definition of the term "bandwidth" is discussed in Chapter 5, and the effects of the bandwidth requirements on the transducer design are considered in Chapter 7. For the moment, the main point to note is that achieving very narrow or very broad transmitting bands can cause difficulties. Even if the transducer is to be used only over a narrow frequency band, it is not usually necessary to demand a narrow bandwidth (ie a high Q-factor) for the transducer itself, and such a requirement could pose difficulties arising from the need for accurate control of the transducer parameters. At the other extreme, some applications need a transducer which can transmit over a wide band. The important factor for the transducer design is the ratio of the required bandwidth to the centre frequency, and it is shown in Sect 5.3 that this fractional bandwidth for transmitting transducers is generally limited to about the value of the effective coupling coefficient of the electro-acoustic material used in the element. For a piezoelectric ceramic transducer, this gives a maximum practical bandwidth of about half the resonance frequency. Achieving this limiting frequency does however have a variety of practical problems, and it is probably not too demanding to ask for a bandwidth of one-third of the centre frequency for frequencies below 10 kHz, reducing to a quarter or less as the frequency is increased to 20-30 kHz. Fractional bandwidths of 0.1-0.2 are acceptable for many applications, and should not generally prove difficult to achieve.

These limitations relate to the bandwidth of a transmitting transducer where good system efficiency is needed. The bandwidth of a receiving hydrophone is defined in quite different terms, since efficiency is not of major importance, and much wider bandwidths are readily achievable.

2-5   GENERAL REQUIREMENTS

There are various other aspects of the transducer which need to be agreed between designer and user, but which may be difficult to define quantitatively, and may be to some extent matters of individual judgement. The first is the question of whether the same transducer should be used for both transmitting and receiving. In many active sonar systems it is both convenient mechanically and economical financially to use the same transducer for both functions. The transmitting and receiving bands for such cases are generally nearly the same, and the effects of the variation of hydrophone sensitivity around resonance are usually acceptable in these applications. If the desired receiving beam pattern is also the same as the transmitting beam pattern, the use of one transducer for both functions would appear satisfactory and generally desirable from size, weight, and cost aspects. However, in some applications where, for example, a much narrower receive beam is wanted than the transmit beam, the choice is not so straightforward. Assuming that the source level and beam shaping could be met by a transmit array which is much smaller than the receive array, a choice has to be made between using two separate arrays, each just large enough to satisfy its own specification, or to use a single array to satisfy both. In the first case, the receive array can be assembled from hydrophone elements, which are usually much smaller and cheaper than the projector elements. In the second case, it would be usual for the whole array to be made using the same elements, – ie the more expensive projector elements. The choice thus depends on how much different the transmit and receive array sizes need to be, what the difference in size of the two types of elements is, and what constraints exist for the overall size, weight and cost of the combined system. No universal answer to the choice can be given, but it is an important decision to be resolved at an early stage. The question only arises, of course, for systems which need both a transmitter and a receiver; this does not apply to passive systems, for which only a receiver is needed.

Another factor which needs to be specified is the range of depths of operation. The most important of these is the maximum depth, since this determines the strength of the housing, and some of the internal features of the transducer. If the transducer were only required to operate at a single depth, however, it might be possible to use a pressure balancing system to permit a reduction of the housing strength, so the full range of operating pressures must be known. Other factors arise from any limitations on access to the transducer or array mounting, or any restrictions in materials arising from the dangers of electrolytic corrosion between the transducer and the mounting to which it is connected. The operating life required of the transducer must also be specified, since it affects not only the detailed design but also the testing of the transducer.

Size and weight have already been mentioned as factors in the choice above. These, together with any restrictions on the permissible shape, are important for all applications, and need early clarification. For example, weight is sometimes of critical importance, and this suggests that great attention must be paid to minimising the size of the transducer, or that its density must be made low. The latter course may be tackled by using aluminium or plastic housings, both of which have their own particular design problems and implications for the rest of the system. The costs of the system are also important factors influencing the design. These may involve the costs associated with the design, or the unit production cost, or the costs of maintenance over its life [2.7]. The penalties arising from a failure of the transducer must also be taken into account; in some cases a failure may necessitate docking a ship to effect a change or repair of the transducer, whilst in others it may involve only an insignificant delay in an experiment. Again, these questions are not susceptible to a rigid analytical treatment, and their resolution depends primarily on scientific or engineering judgement, based on a full understanding of the requirement and its application.

A different range of possibilities is opened up by the use of **parametric arrays**, in which a single array is used to radiate two primary frequencies (f_1 and f_2). Non−linearity in the water causes conversion of some of the power to the difference frequency ($f_2 - f_1$), and the far field radiation is primarily of this difference frequency. As a result, narrow beams can be produced by relatively small arrays, which can be a major advantage for some applications. The parameters of such an array obey quite different rules from those for conventional arrays, and are discussed in Section 6.5.

It is essential for the user and the designer to discuss the application in sufficient detail to understand and agree the major features required of the transducer. In some cases, the requirements are minimal and can be met easily and probably cheaply. In other cases, the requirement could be met readily if the specification were modified slightly to ease a particular difficulty, or to make an existing design suitable. But in some cases the application genuinely needs characteristics which are not readily achievable. This book deals primarily with these latter occasions, where the specification is demanding and can be met only by a careful and systematic design. It is important for both user and designer to recognise what category of difficulty applies to any particular request for a transducer, in order to understand the approach needed and the effort required.

REFS Chapter 2

References.

2.1   Tucker,D.G., and Gazey,B.K., *Applied Underwater Acoustics*, Pergamon Press, 1966, Chapter 6.
2.2   Urick,R.J., *Principles of Underwater Sound*, McGraw-Hill, 3rd Ed, 1983, Chapter 3.
2.3   Albers,V.M.(Ed), *Underwater Acoustics Handbook - II*, Pennsylvania State University Press, 1965, Section 21.5.
2.4   Urick, op. cit., Sect 4.2.
2.5   Flynn,H.G., "*Physics of Acoustic Cavitation in Liquids,*" in: Mason W.P.(Ed), *Physical Acoustics*, Vol 1, Part B, Chap 9, Academic Press, 1964.
2.6   Schofield,D., "Transducers," in Albers,V.M., *Underwater Acoustics*, Plenum Press, 1963
2.7   Massa,F., "Sonar transducer developments during the period of World War II and beyond," *Proc I.O.A.*, 9, Part 2, *Sonar Transducers Past, Present and Future*, 1-22, (1987)

Additional Reading.

1   Urick,R.J., *Principles of Underwater Sound*, McGraw-Hill, 3rd Ed, 1983, Chapters 3,4.
2   Tucker,D.G., and Gazey,B.K., *Applied Underwater Acoustics*, Pergamon Press, 1966, Chapter 6.
3   Albers,V.M.(Ed), *Underwater Acoustics Handbook - II*, Pennsylvania State University Press, 1965, Part 3.
4   Stansfield,D., "Reliability in transducer systems," *Proc IoA*, 6, Pt 3, 1-7, (1984).
5   Gale,R.J., "Some British sonar transducers of the 1950s," *Proc IoA*, 9, Pt 2, 70-78, (1987).

Chapter 3
MATERIALS

At the heart of any electro-acoustic transducer is some mechanism for converting electrical to acoustic energy, and vice versa. In this book, we are concentrating on piezoelectric transducers, and therefore on piezoelectric materials as the basic mechanism for converting energy from one form to the other. There are numerous references describing the various characteristics of these materials, some of the more important being listed at the end of this chapter. The aim of this chapter is to describe briefly the main types and properties of these materials as they affect transducer design. Since a range of materials with slightly differing properties are produced by the various manufacturers, no attempt will be made to give exact figures for any particular variety. Before using any selected material for a particular design, the designer would need to establish its relevant parameters to the necessary accuracy.

Most of this chapter is therefore concerned with the characteristics of piezoelectric ceramics, but some of the most significant properties of other materials are also described.

3-1   PIEZOELECTRICITY

The history of piezoelectricity goes back to 1880, when the effect was first reported by the Curie brothers, and subsequent developments are well recorded by various authors (eg [3.1],[3.2]). The original discovery, that an applied pressure could generate an electrical voltage in certain crystals, is known as the direct **piezoelectric effect**. The converse effect, that the application of an electric field could produce a mechanical strain, was soon demonstrated, and measurements confirmed that the effects were linear and reciprocal. Although a number of crystalline materials showed these effects, the most important was quartz, which was later used as the basis for many piezoelectric transducers. Indeed, quartz played a vital role in the production of many sonar transducers throughout the second world war. However, quartz crystals of any appreciable size could only be obtained from natural geological sources, and efforts were made to find alternative materials. It was found possible to grow other piezoelectric crystals such as Rochelle salt, ammonium dihydrogen phosphate (ADP), and lithium sulphate, amongst others, and these also found use for some applications. One disadvantage of these crystals is that they are soluble in water, so that very careful precautions have to be taken to avoid the presence of even small amounts of moisture. They may nevertheless still be used for particular applications where precise characteristics are wanted, or in the case of lithium sulphate to obtain a good hydrostatic sensitivity. However, these crystalline materials are generally expensive to cut and grind to the required shape, are restricted in size, and are limited in their piezoelectric properties, so that their use for underwater transducers is nowadays rare, although they are still widely used for other applications such as higher frequency filters in which their stability is important. A comprehensive description of design techniques for their use in underwater transducers is given in Ref [3.3].

The next major development was that of **magnetostrictive** materials, which depended on the interaction beween mechanical stress and magnetic (rather than electric) field in some metallic materials. Magnetostrictive transducers thus avoided some of the restrictions inherent in the piezoelectric crystals, and were relatively robust. They were commonly used through the 1940s and into the 1950s, but were largely superseded by piezoelectric ceramics, except for some particular applications. In recent years some improved magnetostrictive materials have been developed, and have led to a revival of interest in magnetostrictive designs; they are discussed briefly in Sections 3.6 and 12.2.

Piezoelectric ceramics were introduced in the early 1950s, and soon became the dominant materials for underwater transducers because of their good piezoelectric properties and their ease of manufacture in a variety of shapes and sizes, making use of well developed ceramic technology. The first piezoceramic in general use was barium titanate, and that was followed during the 1960s by lead zirconate titanate compositions, which are now the most commonly employed. Although these ceramics are widely used for underwater transducers, they have various other applications, which tend to dominate the market because of their volume requirements, despite less demanding specifications.

The piezoelectric ceramics were originally investigated because of their **ferro-electric** properties. A ferro-electric material has a spontaneous electric polarisation, due to a lack of symmetry in its ionic structure, which can be reversed by an applied electric field. As a consequence, the material has a high dielectric constant, and exhibits dielectric hysteresis. This D/E behaviour is analogous to the B/H behaviour of ferro-magnetic materials, and hence gave rise to their being called ferro-electric. The ferro-electric properties exist up to a critical temperature known as the **Curie temperature**, above which the dielectric constant decreases progressively. The spontaneous polarisation of the material may be represented by small dipoles associated with the cell structure, and these dipoles tend to line up parallel with each other to form domains [3.4]. A crystal contains several of these domains, generally oriented head to tail (and at other angles determined by the crystal structure) to minimise the total potential energy by avoiding free charges. A ceramic consists of many small crystallites sintered together by firing at a high temperature, the crystallites being oriented at random throughout the ceramic. Thus, although the properties of each individual crystallite depend on direction, the macroscopic characteristics of the fired ceramic average out to be isotropic.

Each crystallite contains one or more domains, depending on its size, and application of an electric field has two effects, not easily separable. Firstly, it causes a piezoelectric strain within each crystallite; and secondly it causes some domains to grow and others to shrink, thus effectively producing a domain rotation towards the field direction. The effect of the electric field is thus to disturb the random orientation of the domains within the material, by encouraging domain growth in preferred directions. The length of a unit cell of the structure along the dipole direction is slightly longer than in the orthogonal directions, and rotation of the domains is therefore accompanied by an overall change in dimensions, which reinforces the action within the crystals themselves. Application of an electric field thus causes an expansion of the material along the direction of the field. Application of a field in the opposite direction also causes an expansion of the material, and the curve of mechanical strain against applied field is approximately a square law curve. If a mechanical stress is applied, the resulting domain rotations generally balance each other out, and virtually no net electric field is produced. These ceramics do not inherently behave as piezoelectric materials, therefore, since their responses are not linear. They are generally described as **"electrostrictive"** materials. They may however be converted to a more useful form, in which they do show a linear response, by a treatment known as **"poling"**. In this process, a high field is applied, usually at a high temperature, which has the effect of rotating the domains permanently into a preferred direction, and thus introducing asymmetry into the properties of the material. A ceramic which has been poled by this means then shows linear behaviour over a reasonable range of inputs, and is usually referred to as piezoelectric.

The two types of piezoelectric ceramic most commonly used for transducer applications are those based on **barium titanate** and **lead zirconate titanate** [3.5]. One of the major advantages of the ceramics is that their properties can be modified by introducing various chemical additives during their manufacture, and the use of the term "barium titanate" covers a range of compositions based on that material, and similarly "lead zirconate titanate" includes an even wider range of compositions. The characteristics of some of the more important varieties will be described later. The corresponding disadvantage of this susceptibilty to additions is that it may be difficult to control the characteristics of the ceramic precisely during manufacture, and the effects of variations in their parameters need to be taken into account in any transducer design. When it is necessary to control the tolerances of the ceramic more closely than normal, there may well be a significant increase in the costs of production. Nevertheless, the piezoelectric ceramics can be made in a great variety of shapes, and with good characteristics which are generally controllable to adequate accuracy, and they are much the most commonly used active materials for underwater transducers.

The processing of piezoelectric ceramics is described by Jaffe et al in [3.5]. Much care is necessary to achieve consistency in the properties of the ceramic, and some thought should be given to what tolerances are needed for any particular transducer design, since specifying tighter tolerances than normal can be a cause of considerable extra expense and delays. For example, shrinkage of the ceramic occurs during its firing, and it can therefore be difficult to obtain very accurate dimensions of the pieces without an extra grinding operation. One of the other sources of difficulty is the attachment of electrodes to the ceramic. These need to make an intimate and strong bond to the ceramic, and this is often achieved by some form of metallic silver, usually fired on to the ceramic surfaces. In some applications, it is necessary to solder connections to the electrodes, and this can be difficult unless the electrode is strongly attached and well matched to the solder. Air-drying silver electrodes are seldom satisfactory except for temporary experimental purposes.

The poling process needed to convert the ceramic to its piezoelectric form involves the application of a strong electric field, and this is probably the most unreliable process in the manufacturing sequence. Poling of barium titanate is usually carried out by maintaining a field of about 10kV/cm on the sample whilst cooling it through its Curie point. Unfortunately, the application of such a high field is likely to cause dielectric breakdown, especially if the ceramic contains internal flaws. Because the probability of flaws increases with the thickness of the specimen, poling is more difficult for thick pieces than for thin, and thicknesses in excess of about 10mm are therefore unusual. Even higher poling fields are common for lead zirconate titanate compositions, and this accentuates the need for high quality in the ceramic and great care in the poling process. Poling fields in fact depend greatly on composition, and trade-offs are possible between temperature, field, and length of time of application of the field. The Curie temperatures of typical lead zirconate titanate ceramics are about 350°C, and it is impractical to pole these ceramics above these temperatures, so the range of variations of these poling conditions is large, and the final choice of conditions for production is determined from careful study by the manufacturers. Poling is a skilled operation, and should not be undertaken without a full understanding of the difficulties involved in the process.

One of the important features of ceramic materials is their brittleness. Although generally strong in compression, they are weak in tension, and have a tensile strength which can vary considerably and unpredictably. They are therefore unreliable when subjected to any tensile stress, and this has important consequences in the transducer design, where precautions have to be taken to avoid any appreciable tensile stress on the ceramic. On the other hand, their characteristics do vary to some extent with applied stress, and this is one of the features which can be controlled by appropriate additions during manufacture. If the applied stress becomes very high, the ceramic will be partially depoled, and suffer irreversible changes in parameters. Depoling can also be caused if the electrical stress is raised to too high a value, either directly because of the electric field itself or as a result of the heating of the ceramic caused by the dielectric loss. Depoling is likely to occur if the temperature of the ceramic approaches the Curie point, since that permits readier relaxation of the domains to a more random arrangement. Limiting values of the stresses depend on the particular compositions used, and some guidance is given below. Even at normal temperatures, some relaxation of the internal stresses in the ceramic occurs, and this gives rise to gradual changes in the properties, known as "aging" (or sometimes "ageing").

Because of these effects, there are two main classes of ceramics. For those which are intended for use in projectors, the compositions are selected to have good properties even when high stresses are applied, either mechanical or electrical. Achieving these good characteristics at high stress involves some compromises in other parameters, and the other main class of ceramics has compositions which produce rather better performance for low level applications, though being unsuitable for high power projectors. Within these two main classes, compositions have been devised to improve characteristics in particular areas, such as stability against temperature variations, aging, or susceptibility to the effects of mechanical stress.

3-2   PARAMETERS OF PIEZOELECTRIC MATERIALS

In this section, we shall discuss the low field characteristics of the piezoelectric materials which have the greatest significance for low frequency transducers. The effects of applying higher fields will be considered in the next section.

The overall behaviour of the materials may be described by their equations of state, expressed in matrix notation [3.2, 3.4, 3.6]. The observable parameters are

Stress T (N/m^2)
Strain S (a dimensionless numerical ratio)
Electric field strength E (V/m)
Electric displacement D (C/m^2)

These quantities are in general tensors, but may be greatly simplified when considering simple ceramic constructions. For a normal (non–piezoelectric) material, the ratio of stress/strain is given by the elastic stiffness coefficients. For piezoelectric materials, there is an extra term to represent the effect of the electric field generated in the material. Similarly, in the ratio of D/E for the electrical conditions, there is an extra term to represent the charge generated by the mechanical strain. In most cases, it is satisfactory to assume adiabatic conditions, and the equations of state may then be written as

$$T = c^E S - e_t E \qquad\qquad (3.1)$$

$$D = eS + \varepsilon^S E \qquad\qquad (3.2)$$

where

    c is the elastic stiffness coefficient (N/m^2)
    ε is the dielectric constant (absolute) (F/m)
    e is a piezoelectric stress constant (C/m^2 or N/Vm).

The superscripts indicate the quantities held constant during the measurement of the particular parameter, and the subscript t denotes the transpose of the matrix – ie with the rows and columns interchanged.

Alternatively, treating the stress and electric field as the independent variables, the equations may be written as

$$S = s^E T + d_t E \qquad\qquad (3.3)$$

$$D = dT + \varepsilon^T E \qquad\qquad (3.4)$$

in which

    s is the elastic compliance coefficient (m^2/N) (and $s = 1/c$)
    d is a piezoelectric strain constant (C/N or m/V).

Two further pairs of equations may be written, taking other combinations of quantities as the independent variables, and giving two further piezoelectric coefficients, g and h. The four piezoelectric constants are thus given by

$$
\begin{aligned}
d &= \partial D/\partial T \ \ (E \text{ constant}) = \partial S/\partial E \ \ (T \text{ constant})\\
e &= \partial D/\partial S \ \ (E \text{ constant}) = -\partial T/\partial E \ \ (S \text{ constant})\\
g &= -\partial E/\partial T \ \ (D \text{ constant}) = \partial S/\partial D \ \ (T \text{ constant})\\
h &= -\partial E/\partial S \ \ (D \text{ constant}) = -\partial T/\partial D \ \ (S \text{ constant})
\end{aligned}
$$

where the coefficients strictly refer to the ratios of infinitesimal changes in the variables. In practice, the condition that E is constant usually corresponds to the short circuited case, whilst D constant is usually the open circuited condition, T constant corresponds to having no force resisting the displacement, and S constant implies perfect clamping of any motion, a condition which is difficult to achieve.

The coefficients which are most generally useful are d, which expresses the strain produced by an applied field, or the charge generated by an applied stress, and g, which expresses the field generated by an applied pressure. Thus d relates primarily to operation as a projector, and g to use as a hydrophone. For crystalline materials, the expressions can become very complex because of the different crystal axes, but for the piezoelectric ceramics which are our main concern in this book the position is considerably simplified, since the direction of poling is the only axis of symmetry. The coefficients are therefore referred to the poling axis, and this is done by a system of suffixes. These are based on the three orthogonal axes, labelled 1,2,3, with the poling direction taken as the 3-axis. The axis which is perpendicular to the electrodes attached to the ceramic (ie the direction of the applied field) is indicated by the first suffix, and this is normally the same as the poling axis, 3. The second suffix indicates the direction of the relevant stress or strain. Thus, d_{33} is the piezoelectric coefficient describing the strain in the poling direction (the 3-axis) produced by a field in the same direction. The coefficient for the strain perpendicular to the poling direction produced by a field in the poling direction is d_{31}. Because of the symmetry about the poling direction, axes 1 and 2 must be equivalent, and d_{31} must therefore be equal to d_{32}. A similar convention is used for the other piezoelectric coefficients, the assumption being made that all stresses other than that relevant to the second subscript are constant. For some particular applications, the field may be applied in a direction perpendicular to the poling axis, to generate a shear stress around the 2-axis, and this is indicated by a subscript 15; - eg g_{15} indicates the shear mode hydrophone coefficient. The relative dielectric constant has subscripts with the same significance as above, and a superscript to denote constant stress or strain as in (3.1-3.4), eg $\varepsilon_{33}{}^{T}$. (Sometimes only one subscript is used, to indicate the direction of the applied field.) The d and g coefficients are related by $d = \varepsilon g$, in which ε is the absolute permittivity, and appropriate subscripts are taken together; eg $d_{33} = \varepsilon_{33}{}^{T}\varepsilon_{0}g_{33}$ (where ε_0 is the permittivity of free space, 8.85×10^{-12}F/m).

Other parameters, related to those above, are often more directly useful. For example, the stiffness constants and density may be combined to derive frequency constants for the various modes of vibration of a ceramic. The resonance frequency (f_S) for a particular mode varies inversely with the linear dimension (L) determining the resonance, so that we may write $f_S L = N$, where N is a frequency constant appropriate to the mode concerned. This is useful in directly determining the resonance frequency of a single piece of ceramic; for example, the frequency constant for a disc vibrating in the radial mode is usually denoted by N_p. When pieces are assembled into transducer stacks, the Young's modulus itself is the more significant parameter.

Another derived parameter can be used to calculate the output voltage from a sample when it is subjected to a uniform stress on all three axes, as when exposed to an applied hydrostatic or acoustic pressure. In this case, the output voltage is the sum of the voltage due to g_{33} and the (opposing) voltages due to g_{31} and g_{32}. Since $g_{31} = g_{32}$ for a poled ceramic, the resulting **"hydrostatic coefficient"** (g_h) is given by

$$g_h = g_{33} + 2g_{31} \qquad\qquad (3.5)$$

The corresponding charge coefficient is denoted by d_h.

The relative dielectric constant of the ceramic along the poling direction is denoted by ε_{33}, with a superscript to indicate whether it is measured at constant stress or constant strain in the direction of the second suffix. Since the applied field is almost always in the same direction as the poling field, the subscripts are often omitted, although ε_r may be used where appropriate to emphasise that it is the relative dielectric constant which is being used. In many of the subsequent design equations, it will be the absolute dielectric constant (or permittivity) which is relevant, and this is generally denoted by $\varepsilon = \varepsilon_r \varepsilon_0$, where ε_0 is the permittivity of free space. The capacitance of a hydrophone at low frequencies (C_{LF}) is an important characteristic, a high capacitance permitting a lower limit to the operating frequency band, for a given input resistance to an amplifier. Some trade-off is often possible between output voltage and capacitance, a higher sensitivity being achievable at the expense of a lower capacitance. This could be effected, for example, by means of an (ideal) transformer, which could step up the voltage by a factor n whilst increasing the impedance (ie reducing the effective capacitance) by a factor n^2. (More practically, it can be achieved by changing the connections between the pieces of ceramic in the hydrophone; for example, if the two halves of a hydrophone element are connected in series instead of in parallel, the voltage sensitivity is doubled, whilst the capacitance is reduced by a factor of four.) The value of the product $\hat{M}^2 C_{LF}$ thus remains constant for the basic element, and can be used to indicate the limits of such trade-offs, a high value of $\hat{M}^2 C_{LF}$ permitting a wider range of practical possibilities than a lower value. For the ceramic itself, this would be expressed by the factor $g^2 \varepsilon$ (with the appropriate suffixes), and this factor is sometimes used as a **figure of merit** for hydrophone applications. Since $d = \varepsilon g$, this may alternatively be expressed as gd.

In general, the most important parameter of a piezoelectric material is its **coupling coefficient** (k). It may be defined most straightforwardly in terms of the measured series and parallel resonance frequencies of the sample [3.5], but it may also be interpreted in physical terms as the square root of the ratio of the mechanical stored energy to the total input energy. Its importance derives from its significance in indicating the efficacy of the material in converting from electrical to acoustic energy, and in its influence on the maximum achievable bandwidth for a transducer using the material. The coupling coefficient is related to the other parameters by

$$k^2 = \frac{d^2}{\varepsilon^T s^E} \qquad (3.5a)$$

in which the necessary suffixes should be added as appropriate for the particular mode under consideration (using the same conventions as before). This may be related to the hydrophone figure of merit by using the expression $d = \varepsilon g$, which leads to the relationship

$$k^2 = \frac{\varepsilon^T g^2}{s^E} \qquad (3.5b)$$

ie, the figure of merit is given alternatively by $k^2 s^E$, again with the appropriate subscripts.

For the low frequency transducers which are the main topic for this book, it is k_{33} which is the most important parameter in determining the potential performance of the transducer (See Chapter 4). This coefficient, however, assumes that the stress is uniform throughout the sample during the measurement, and this is not easy to realise for practical measurements on individual samples. It is therefore quite common for the quality of a ceramic to be judged by measurements of k for other modes, and especially by measurements on thin discs vibrating in the radial mode. The relevant coupling coefficient is then denoted by k_p, which is related to k_{31} by

$$k_P^2 = \frac{2}{1-\sigma^E} k_{31}^2 \qquad (3.6)$$

where $\sigma^E = (-s_{12}^E/s_{11}^E)$ is Poisson's ratio, which has a value between 0.28 and 0.32 for most piezoelectric ceramics of interest. Definitions and methods of measurement are described in [3.7], which is reproduced in [3.5].

Some care is necessary in using quoted values of some ceramic parameters, especially the coupling coefficient. The most straightforward way of quoting the coupling coefficient for a particular sample is to give its value derived directly from measurements of its resonance and anti-resonance frequencies, as described in [3.7], and in Chapter 9. This will be a value reflecting the actual stress distribution in the sample for the appropriate mode of vibration. Some authors give equations to correct for this stress distribution, in order to obtain a value for a condition of uniform stress, eg to obtain k_{33}. This is sometimes useful in deriving basic ceramic properties, but can introduce uncertainties when measurements are being carried out for other purposes, such as quality control.

Stability.

The parameters show some variation with external factors such as temperature and time. These are usually quoted in manufacturers' literature for the various compositions produced. Variability is critical for applications such as piezoelectric filters, and compositions with very stable characteristics have been developed for such uses. For wide band underwater transducers, these variations do not usually present great problems, provided the ceramic is sensibly chosen to keep the temperature dependence to reasonable values. Operating temperatures in water do not vary over a very wide range, and heating in the ceramic stack itself needs to be minimised to avoid depoling during high power operation, so that overall temperature excursions are generally not too large.

The variations with time ("aging") can occasionally cause problems in applications where the performance is especially sensitive to the matching of transducer characteristics. Aging of ceramic properties is generally a linear function when plotted against the logarithm of the elapsed time, – ie the percentage change is constant if the elapsed time is multiplied by a constant factor. Aging rates are thus often quoted as changes per "decade", giving the change in characteristic when the time is increased by a factor of ten. For example, the change between 1 and 10 days is approximately the same as between 10 and 100, or 100 and 1000 days. The importance of making a correct choice of the time zero in measuring aging rates is emphasised by Jaffe et al ([3.5],p28), a suitable choice of zero usually being the latest time at which the internal stresses in the ceramic were significantly altered by external changes such as major temperature changes or applied stresses, either electrical or mechanical. Any problems which arise from aging are usually due to carrying out measurements and assembly soon after poling of the ceramic, when the rate of change is greatest. The most practical method of avoiding these problems is to obtain the ceramic early and store it for some time to allow the aging to slow down, a procedure which necessitates early decisions on the size and characteristics of the ceramic. If this storage is practicable, aging of the ceramic does not usually cause very serious problems.

Experiments have been carried out in which ceramics have been stored for a period at elevated temperatures to hasten the aging process. However, although the aging rate is increased during the time at the higher temperature, the efficacy of the method in the long term seems doubtful, as the aging rate after the treatment generally settles down to its earlier value if the time zero is chosen correctly. It is usually preferable to avoid the need for such treatments by choosing a ceramic with a low enough aging rate, and obtain the material early enough to allow storage, if possible for 100 days or more.

3-3 HIGH STRESS EFFECTS

In the previous section, the piezoelectric parameters have been assumed to be linear and have real values, which is equivalent to postulating that there are no internal electrical or mechanical losses, and is a satisfactory assumption for small stresses. In practice, however, the ceramics do not behave exactly in this ideal way, the deviations from ideal behaviour becoming important when high stresses are applied. Two types of effect are significant. Firstly, the losses within the material arising from the application of high alternating stresses need to be taken into account; and secondly, the effects of static applied stresses must be considered.

Dielectric losses.

In practice, a piezoelectric ceramic behaves as an imperfect capacitor, losing a fraction of its stored energy for each cycle of the applied field. Its internal power loss due to an applied alternating field is thus proportional to the number of applied cycles per second, as for hysteresis losses. Such losses are denoted by a factor (tan δ) which represents the losses per cycle of the applied field. This loss factor may be measured at low field, and this is often the value quoted in manufacturers' data. For projector applications, however, it is the value at a typical operating field which is more significant. Berlincourt et al [3.6] give examples of how tan δ varies with applied field for a number of ceramics, the variation being very dependent on the composition of the ceramic. Typical operating fields are about 2kV/cm, and a good projector material should have tan δ not exceeding about 0.02 at that field to avoid excessive heating of the ceramic. Keeping the losses down to this limit involves some sacrifice in other parameters, and some ceramics are specially formulated for hydrophone applications, with characteristics chosen to optimise performance as a receiver but with much higher loss factors. The loss factor shows very little variation with frequency over the probable range for transducer applications. The effect of dielectric losses in causing heating of the ceramic needs to be calculated by the methods of Chapter 7, to assess its significance, but the approximate limit of tan $\delta \leq 0.02$ quoted above is generally a fair guide.

The dielectric constant itself also shows a variation of a few percent as the field is increased to about 2kV/cm. This is rarely of any great significance, although it does indicate that the material is not perfectly linear in its characteristics. Fortunately, piezoelectric strain remains a linear function of the applied electric field to considerably higher fields than does the dielectric displacement [3.6]. Fields in excess of 5kV/cm can cause depoling of the ceramics, particularly at elevated temperatures, but such high fields are generally not achievable in operational conditions, because other factors, such as flash-over across the ceramic surfaces, limit the field to lower values (See Ch 7).

Mechanical losses.

Losses also arise from hysteresis effects associated with the mechanical vibrations, and these are again dependent on the composition and structure of the ceramic. In this case, losses are generally expressed in terms of the mechanical Q-factor of the ceramic, typical values being 500 or more for the "hard" projector materials, and 100 or less for the "soft" hydrophone ceramics. As for electrical losses, the mechanical losses increase as the alternating stress or strain is increased, the Q-factor falling to about 100 for the projector materials at alternating stresses of the order of 15MPa ($\simeq$2200psi) [3.6]. Although these losses make a contribution to the internal losses in a transducer, they are usually outweighed by the other losses in the element, such as those due to joints between the ceramic pieces, or by O-ring losses.

Effects of static mechanical stress.

The domain rotations which result from the poling process produce an expansion along the poling direction. If a high compressive stress is applied along this direction, it tends to rotate the domains away from their preferred orientations, and thus causes some depoling of the ceramic. This is associated with changes in several of the ceramic parameters, especially the dielectric constant and coupling coefficient, some of these changes being reversible and some non-reversible. They are particularly significant with the "soft" lead zirconate titanate compositions [3.5, 3.6]. The effects of mechanical stress are not easy to describe concisely, and they are not usually included in manufacturers' literature, but some guidance can be obtained from references such as those listed at the end of this chapter [3.8-3.14].

Changes in the dielectric constant of the ceramic under high stress conditions cause changes in the capacitance of the transducer, and these introduce deviations from the optimum matching between the transducer and its drive amplifier. This applies to both reversible and irreversible changes, but in addition the irreversible changes in parameters reveal a permanent degradation of the material which is clearly to be avoided if possible. The nature of the effects may be indicated by considering what happens if a compressive stress is applied parallel to the poling direction in a projector ceramic such as PZT-4 (a US Navy Type I composition, see Sect 3.5). For such conditions, Nishi [3.10] and Krueger [3.11] measured a marked increase in dielectric constant. For example, a stress of 10,000psi ($\simeq$70MPa) caused an increase of about 50% in ε_{33}, most of which was recovered when the pressure was released. This arises because domains are reorientated away from the poling direction, and the dielectric constant normal to the c-axis of the domain is higher than along the axis. Not quite all of the increase is recovered because not all of the domains revert to their original poled orientation. Changes in the piezoelectric d_{33} constant, and tan δ, also occur. The changes depend to some extent on time after the application (or release) of pressure, and differ slightly for subsequent pressure cycles. Some stabilisation occurs with successive cycles, and Krueger [3.11] concludes that the changes, although significant, are generally not catastrophic in their effects for stresses up to 140MPa (20kpsi). Above that, substantial depoling of the ceramic can occur. Treatments of the ceramic have been devised [3.12] to improve the resistance to stress, resulting in the production of materials such as PZT-8. However, Krueger suggests a practical upper limit of 15kpsi for the applied stress for either PZT-4 or PZT-8, if prolonged stress or numbers of cycles are involved, and even this stress may involve considerable changes in characteristics under pressure. Although the dielectric constant of a "soft" piezo-ceramic such as PZT-5 (a US Navy Type II material) changes less than PZT-4 under an applied stress, it suffers drastic and irreversible degradation due to depoling at stresses above 40MPa (8kpsi) [3.11], and this appears to be a direct consequence of the easy domain rotation which gives the composition its high sensitivity as a hydrophone material. It is therefore recommended that the applied stress should not exceed about 35MPa (5kpsi) for this ceramic. For a barium titanate ceramic formulated for projector use, stresses up to about 10kpsi appear tolerable without causing unacceptable degradation. Typical changes in dielectric constant up to this pressure are less than 15% [3.10].

Studies have also been made of the behaviour of these piezoelectric ceramics when subjected to stresses perpendicular to the poling direction, and to hydrostatic pressure [3.9, 3.13]. Only small variations in parameters were found to result from applying hydrostatic pressures up to 70MPa, – eg not more than 7% change in dielectric constant or the piezoelectric strain constants. The result of applying a stress normal to the poling axis is somewhat more complicated, since it disturbs the symmetry around the poling direction, so that d_{31} is no longer equal to d_{32}. However, the effects of the transverse stresses are generally much less than for those parallel to the poling axis; for example, for a "hard" material such as PZT-4, the change in ε_{33} is less than 5% for an applied transverse stress of 70MPa. For the "soft" PZT-5 ceramic, the change in ε_{33} up to 70MPa is less than 10%. Brown [3.14] studied the effects of applying two-dimensional stresses normal to the direction of poling, as occurs when an air-filled sphere or cylinder is subjected to external hydrostatic pressure. These transverse stresses cause a decrease in dielectric constant, instead of the increase caused by a stress parallel to the poling axis. The effects are again relatively small – typically less than 5% change in ε_{33} for combined stresses up to 70MPa for the "hard" materials, and about 10% for PZT-5. However, it should be noted that these transverse stress conditions are most likely to arise for spherical or cylindrical hydrophones, in which the stresses in the ceramic shell are appreciably higher than the external pressure, and it is then necessary to take into account the stress magnification in the ceramic.

The comments above are only a brief summary and simplification of a very complex subject, in which even the measurement techniques present considerable difficulties, as noted for example in [3.13]. The main point to be noted is the possibility of changes in characteristics, or even permanent depoling of the ceramic, being caused when the ceramic is subjected to high mechanical stresses. If the stresses are kept below about 70MPa (10kpsi) for the "hard" ceramics (Types I and III), problems should not be severe, whilst for "soft" ceramics (Type II) the stress should not exceed 35MPa (5kpsi). But these are not precise limits, and the suitability of any particular design for its required operating conditions should be confirmed experimentally.

3-4 BARIUM TITANATE

Barium titanate was the first piezoelectric ceramic to be developed commercially, and came into wide use during the 1950s because of its high coupling coefficient and ease of manufacture. Small single crystals can be grown, and these have been used to study the underlying mechanisms of the basic ferroelectric and piezoelectric processes in the material. At temperatures between 130° and 1460°C, barium titanate has a cubic perovskite structure [3.5]. Below 130°, the unit cell becomes elongated along one edge, thus becoming tetragonal, and this distortion from cubic symmetry gives rise to a spontaneous polarisation along the longer (c) axis which is the source of the ferroelectric nature of the material. Another transition occurs at about 0°C, below which the structure is orthorhombic, with the cell elongated along a face diagonal, and a further transition occurs at about –90°C to a rhombohedral form.

It is the tetragonal form, existing around room temperature, which is of interest, originally because of its ferroelectric behaviour. The relative dielectric constant measured along the longer c axis has a value of about 300, whilst along a perpendicular (a) axis it is about 4000. These values show peaks at the transition temperatures, the upper limit of the tetragonal region being the Curie temperature, above which the material loses its ferroelectric properties. In an unpoled ceramic, the relative dielectric constant is some average of these values, and its variation with temperature also exhibits peaks at the transition temperatures. Thus, in addition to the peak at the Curie temperature of 130°C, there is another peak, though a smaller one, at about 0°C. This comes inconveniently within the operating range for many transducer applications, and it was found that this transition temperature could be lowered by the addition of calcium (Ca^{+2}) ions to replace some of the barium (Ba^{+2}), without significantly affecting the Curie temperature [3.5]. Other additions may also be made to achieve suitable characteristics for particular applications. One example is the addition of a small amount of cobalt with the calcium to reduce the high field losses for projector applications.

Major characteristics of a typical barium titanate ceramic composition formulated for projector applications are given in Table 3.1, together with data for some lead zirconate titanate ceramics. Barium titanate compositions optimised for hydrophone use have been developed, but have few advantages over the projector material, and have now been almost entirely superseded by lead zirconate titanate compositions. The most noteworthy differences in the properties of the barium titanate ceramic compared with lead zirconate titanate are its lower coupling coefficients, its higher Young's modulus, and its lower density. Because of the difference in coupling coefficients, barium titanate ceramics have largely been replaced by lead zirconate titanate in most commercial applications, but they are still used for some underwater transducers. We shall see in Chapter 5 that an optimum value of coupling coefficient may be derived for some transducer applications, and where this matches the coupling achievable with barium titanate it may be simpler and cheaper to use this ceramic rather than lead zirconate titanate. It is arguable also that barium titanate ceramic is less susceptible to the depoling effects of applied mechanical stress.

Typical variations of properties with temperature are indicated by the following values for the changes between 0° and 30°C of a projector-type ceramic:-

$$\varepsilon_{33}^T, \; -5\%; \qquad k_p, \; -12\%; \qquad N_p, \; +4\%$$

3-5 LEAD ZIRCONATE TITANATE

The ferroelectric properties of ceramics of the lead zirconate titanate type were reported by Japanese workers in the 1950s, and this was soon followed by studies of their strong piezoelectric activity by the US National Bureau of Standards. Since that time, many compositions have been investigated and many papers published on their characteristics, a small selection being quoted at the end of this chapter. The book by Jaffe et al [3.5] gives a very useful summary of the work on many piezoelectric ceramics, including both barium titanate and lead zirconate titanate.

The lead zirconate titanate ceramic compositions are solid solutions of the $Pb(Ti,Zr)O_3$ type, which can range from pure lead zirconate to pure lead titanate. Within this range, the structure adopts various crystalline forms which depend on composition and temperature and can be delineated by means of a phase diagram. Most of the compositions of interest for transducer applications lie near the boundary between tetragonal and rhombohedral phases, which occurs for compositions having about 48 mol% of $PbTiO_3$. In addition to the range of compositions near this boundary, the effects of numerous additives have been studied, and these investigations have resulted in a number of different compositions being available from manufacturers. Some of the earliest were produced by the Clevite Corporation with the trade name of PZT, and this is still often used as a common name for these ceramics. However, to preserve generality, the abbreviation LZT will be used in this book to indicate ceramics of the lead zirconate titanate family.

Because of the wide variety of possible LZT compositions, their properties also show considerable variation, but they generally fall into two main categories. One group is intended for high power transmitter applications, with low dielectric loss maintained up to high fields, and low internal mechanical losses. The original Clevite composition, known as PZT-4, was reproduced by several manufacturers (often with a 4 in the type number), and became incorporated in a US Navy Mil STD 1376 (Ships) as Navy Type I. Typical characteristics are listed in

TABLE 3.1
Properties of piezoelectric ceramics

Property	Units	$BaTiO_3$
Density	$10^3 Kg/m^3$	5.55
Young's Mod, ($1/s_{33}^E$)	$10^{10} Pa$	11.9
Poisson's ratio		0.30
Curie Temp.	°C	115
k_{33}		0.46
k_{31}		0.19
k_{15}		0.47
k_p		0.31
d_{33}	$10^{-12} m/V$	150
d_{31}	$10^{-12} m/V$	−58.5
d_{15}	$10^{-12} m/V$	245
g_{33}	$10^{-3} V.m/N$	14.3
g_{31}	$10^{-3} V.m/N$	−5.6
g_{15}	$10^{-3} V.m/N$	20.1
g_h	$10^{-3} V.m/N$	3.1
Freq. constants for discs		
N_{3t}	kHz.m	2.87
N_p	kHz.m	3.18
(N_{3t} = thickness × f_r in thickness mode,		
$\varepsilon_{33}{}^T$		1220
Tan δ		
at low field		0.005
at 2kV/cm		0.015
Change in ε_{33} to 2kV/cm	% change	5
Mech. Q_M		600
Aging	% per time decade	
$\varepsilon_{33}{}^T$		−1.6
k_p		−1.4
N_{33}		+0.6
$g_{33}d_{33}$	$10^{-12} m^2/N$	2.1

Properties (*contd*)

LZT I	LZT III	LZT II	LZT IIH
7.6	7.55	7.7	7.3
6.3	7.1	5.8	5.6
0.29	0.29	0.30	0.30
320	300	370	210
0.68	0.67	0.70	0.72
0.35	0.32	0.32	0.35
0.71	0.60	0.68	0.65
0.58	0.55	0.58	0.60
295	245	420	520
−125	−105	−180	−240
500	390	600	680
24.5	25.0	28.0	18.5
−10.5	−11.0	−12.0	−8.5
39.5	31.5	40.0	30
3.5	3.0	4.0	1.5
2.08	2.03	1.94	1.86
2.21	2.30	2.00	2.20

N_p = diameter $\times f_r$ in radial mode)

LZT I	LZT III	LZT II	LZT IIH
1300	1100	1700	3200
0.005	0.004	0.018	0.02
0.015	0.007	NA	NA
5	2	NA	NA
500	1000	70	60
−4.2	−5.0	−1.6	−0.7
−2.1	−2.0	−0.4	−2.5
+1.0	+1.0	+0.2	+0.4
7.2	6.1	11.8	9.6

Table 3.1 under the heading LZT I. This group includes
compositions such as Channel 5400, EDO EC-64, Quartz &
Silice P7-62, and Unilator PC4A [3.15]. The application of
repeated compressive stress cycles in excess of about 70MPa
(10kpsi) to this type of material may cause some depoling, and
studies to reduce this degradation led to the development of a
PZT-8 ceramic with improved stability under high electric fields
and high operating pressures (Section 3.3). Compositions of this
type form a second sub-group within the projector family,
indicated by the LZT III column in Table 3.1. These
compositions, corresponding to US Navy Type III, include (for
example) Channel 5800 and EDO EC-69. Their reduced losses
under electrical or mechanical stress are associated with lower
mobility of the domain walls, and this leads also to lower aging
rates. The price to be paid for this improved stability is a slight
reduction in piezoelectric properties.

The other main category of LZT ceramics is intended for
use in hydrophones, the compositions being formulated to give
high permittivity and sensitivity at the expense of some
degradation in high field characteristics. A considerable spread
in characteristics is possible, and there are again sub-groups
within this category. The main sub-group, indicated by LZT II
in Table 3.1, originated from PZT-5, and now includes
ceramics such as Channel 5500, EDO EC-65, Quartz & Silice
P1-60, and Unilator PC5 (generally conforming to US Navy
Type II). They achieve high relative dielectric constants, of
about 1700, without sacrificing sensitivity (ie the
g-coefficients). The values of hydrophone figure-of-merit
indicated by $g_{33}d_{33}$ in Table 3.1 illustrate the advantages of
these compositions for hydrophone applications. Their losses
under mechanical or electrical stress are however relatively
high, and increase rapidly with increasing stress, so that they
are not suitable for high power use. Typical variations of
characteristics with temperature change from 0° to 30°C are:-

	ε_{33}^{T}	k_p	N_p
LZT I	-2%	-2.5%	1.3%
LZT II	10%	3%	-0.7%

Other compositions are produced with advantages for
particular applications (which may have little connection with
underwater transducers), including for example Channel 5700
and Unilator PC5H, typical characteristics of which are given
under LZT IIH in Table 3.1. These ceramics have dielectric

constants of 3200, which may be advantageous if high capacitance is needed in a small hydrophone, although the figure of merit is less than that of the LZT II compositions. The table illustrates the wide range of properties available in these materials. The advantages of the LZT II and IIH compositions for hydrophones are evident from the values of $g_{33}d_{33}$ (or $g_{33}{}^2\varepsilon_{33}{}^T$), the figure of merit, and these materials are therefore commonly used when only receiving sensitivity is important. When higher power operation is needed, the LZT I and III types are more appropriate. It is generally desirable to use the same type of ceramic for most designs, to establish a sound basis of experience, rather than seeking to optimise the material for each particular application.

It is of interest to use these parameters to calculate some of the piezoelectric effects. For example, suppose that a voltage V is applied across a ceramic of thickness t, and produces a change in thickness of Δt. Then the strain is related to the field by the d_{33} coefficient, ie,

$$d_{33} = \frac{\Delta t/t}{V/t} \qquad (3.7)$$

$$= \Delta t/V$$

Thus, taking $d_{33} = 295 \times 10^{-12}$m/V for LZT I, an applied voltage of 1kV would produce an expansion of $\Delta t = 295 \times 10^{-12} \times 10^3$m $= 0.295 \times 10^{-3}$mm. If the ceramic thickness is 5mm, the field is 2kV/cm and the strain is 0.59×10^{-4}. If a stack of n discs is assembled and connected so that the mechanical deformations are in series, the overall change in length is $n\Delta t$. A stack of 20 discs would thus produce an expansion of 5.9×10^{-3}mm (ie 5.9µm) in a total length of 100mm for an applied voltage of 1kV.

Alternatively, the d-coefficient expresses the ratio of the charge density generated by an applied stress, ie,

$$d_{33} = \frac{Q_c/A}{F/A} \qquad (3.8)$$

$$= Q_c/F$$

where Q_c denotes the generated charge, A the electroded area of the disc, over which the stress is applied, and F the total force. For the material used in the example above, a force of 1 newton would thus generate a charge of 295×10^{-12}coulomb,

which is independent of the area of the disc. Using the
g–coefficient, we can derive the voltage generated by this
stress, by writing

$$g_{33} = \frac{V/t}{F/A} \qquad (3.9)$$

Hence,

$$V_{33} = g_{33}tF/A \qquad (3.10)$$

For example, a stress of 10MPa ($\simeq$1,400psi) applied to a
LZT I sample 5mm thick would generate a voltage of
$24.5\times10^{-3}\times5\times10^{-3}\times10^{7}$ = 1225 volts. This is an example of
the large voltages which can be produced by these ceramics if a
high mechanical stress is applied. This is made use of in
devices such as piezoelectric ignition systems, but it represents
a danger to be avoided in acoustic transducers. Since these
large stresses are generally applied relatively slowly, it is
usually possible to allow the generated charge to leak away
through a parallel resistance which, with the ceramic
capacitance, constitutes a high pass filter; the cut–off frequency
can generally be adequately below the acoustic frequencies of
interest, but high enough to permit dissipation of any charge
generated by large external stress changes.

For large voltages may also be generated in these ceramics
through the **pyroelectric effect**, which has the same physical
basis as the piezoelectric effect in the spontaneous polarisation
of the material. The pyroelectric coefficient for a "soft" LZT II
type of ceramic is of order $0.2\times10^{-3}C/m^{2}{}^{\circ}C$ at 20°C, rising to
about 0.8×10^{-3} at 150°C ([3.5],p170). If this charge cannot
leak away, it can generate a large voltage on the ceramic,
which may cause problems with pre–amplifiers, or even with
simple handling of large pieces. For example, if the area of one
electroded surface of a sample is $10cm^{2}$ (= $10^{-3}m^{2}$), and the
temperature is changed by 1°C (from 20°C), a charge of about
$0.2\times10^{-6}C$ is generated. If the capacitance of the sample is
$10^{4}pF$, the resulting voltage across the electrodes, in the
absence of any leakage, would be approximately 2×10^{4} volts.
In order to avoid electric shocks during handling, it is advisable
for large piezoelectric ceramic pieces to be supplied and
handled with their electrodes shorted by a piece of thin foil.

When piezoelectric ceramic rings are assembled into stacks,
they are usually bonded together by using an epoxy resin which
is cured at an elevated temperature. On cooling, some stress
may be introduced into the bonds by the differential thermal
expansion of the various component parts of the stack, which

should ideally be as well matched as possible. The thermal expansion coefficient of LZT ceramics is approximately $2.5 \times 10^{-6}/°C$ up to about 100°C, but above that temperature the behaviour can become complicated because of domain re-orientation during the first heating cycle. The thermal expansion coefficients for LZT up to 100°C are approximately equal along and perpendicular to the poling axis, whilst for barium titanate ceramic the coefficient is about $4.5 \times 10^{-6}/°C$ along the poling axis, and $7.5 \times 10^{-6}°C$ in the perpendicular direction. The thermal conductivity of LZT ceramics is quoted [3.6] as 1.25W/m°C, and 2.5W/m°C for barium titanate, and these values are useful in assessing the rise in temperature which may be caused by the power dissipation in a transducer element during operation.

3-6   MAGNETOSTRICTIVE MATERIALS

Magnetostrictive materials have constituted the main alternative to piezoelectric ceramics as the basis for medium frequency electro-acoustic transducers. They were developed primarily to overcome the limitations associated with designs using piezoelectric crystals, but were generally displaced during the 1950s by the more versatile piezoelectric ceramics. This section gives a brief description of the main characteristics of the conventional magnetostrictive materials, to show their essential differences from piezoelectric ceramics. Unfortunately, the properties of magnetostrictive materials depend on many factors, which make it difficult to describe them concisely. In the interests of brevity, and since magnetostrictive transducers are not the main subject of this book, this section will give only a very simplified description of their main characteristics. For more information, the reader is referred to more comprehensive treatments, such as those given in the references at the end of the chapter.

Magnetostrictive materials, sometimes called **piezomagnetic** materials, generate a mechanical strain when a magnetic field is applied. They are thus to some extent analogous to piezoelectric materials, in which the strain is produced by an electric field, and their behaviour can be described by equations of state similar to those of the piezoelectrics, except that the electric field is replaced by a magnetic field. Piezomagnetic coefficients can thus be defined and measured for materials of interest, but it is again the coupling coefficient which is of greatest significance in assessing the potential performance of the materials for transducer applications. As for the electrostrictive ceramics, it is necessary to provide a biassing field in order to achieve a linear response, but in this case there is not so convenient a technique as the poling of the electrostrictive ceramics. The remanent field which remains after removing a strong magnetic field from a magnetostrictive material is generally rather small and permits only low power operation. For higher power applications, it is necessary to apply a biassing field, which is normally provided either by permanent magnets or by a coil carrying direct current wound around the element. The magnetic permeability (μ) and the piezomagnetic coefficients depend on where they are measured on the magnetic (B/H) curve, and in specifying the characteristics of a material they must be related to the value of the biassing (or polarising) field.

The first magnetostrictive material in common use was nickel, which was widely used during the 1940s and 1950s, and various metallic alloys with good magnetostrictive properties were also developed for use during that era [3.16]. In addition to the dependence on applied field referred to above, the properties of these materials generally depend on their thermal and mechanical treatment, such as any rolling and annealing applied to them. It is therefore often necessary to quote a range of values for any particular characteristic. Typical values of coupling coefficient for these materials range from 0.15 to 0.35; they can thus exceed that of quartz and ADP, but are rather lower than that of barium titanate. Because of their finite resistivities, they suffer from eddy current losses within the metal. This is minimised, as it is for transformers, by dividing the element into thin laminations, the thickness of which is determined by the "penetration depth" of the magnetic fields at the operating frequency. The laminations are then stacked and cemented together to form a resonant element. Electrical contact between adjacent laminations is prevented by forming an oxide film on their surfaces, or by incorporating an insulating filler in the bonding resin to give adequate separation between the individual laminations.

2V-Permendur, an alloy of cobalt and iron with a small vanadium addition, has a moderate remanent field, and has been used at remanence for modest power outputs. It also has good power handling capability when suitably biassed, There is however no clear choice of material amongst the various alloys available. Indeed, their use is nowadays confined only to special applications, such as for toroidal rings, as the properties of the piezoelectric ceramics are generally superior. Properties of typical magnetostrictive materials are listed in Table 3.2. It is worth repeating that the properties of these materials are very dependent on their treatment and precise composition, and the values quoted in Table 3.2 should only be taken as a guide; more accurate values should be obtained from specialist sources if necessary. The penetration depth of the field into the material is an indication of the eddy current losses, and the frequency at which the penetration depth becomes equal to $\sqrt{2}$ times the lamination thickness (t) is defined as the characteristic frequency (F_c). The product $F_c t^2$ is then given [3.16] by

$$F_c t^2 = \frac{10^{10}\rho_e}{2\pi^2\mu_r} \qquad (3.11)$$

in which ρ_e is the electrical resistivity, μ_r is the blocked or reversible permeability, F_C is in kHz, and t is in mm. Values of $F_C t^2$ are listed in Table 3.2. The higher the value of this product, the greater is the allowable lamination thickness, and hence the simpler the transducer construction. In practice, eddy current losses make these metallic materials generally unsuitable for use above about 100kHz. Typical efficiencies at resonance for these types of transducer are in the region of 20% for elements resonant between 10 and 80kHz.

Investigations into means of improving the properties of magnetostrictive materials led during the 1950s to the development of a range of **magnetostrictive ferrites** [3.17]. Their main advantages compared with the alloys are their much higher electrical resistivities, thus reducing eddy current losses, and the ease of producing them by ceramic techniques. The reduced eddy current losses result in considerably higher electro-acoustic efficiencies for transducers based on ferrites than for those using nickel, but the mechanical unreliability of the ceramic under stress can introduce difficulties at high powers. Some of the characteristics of Ferroxcube 7A1 (a nickel-copper-cobalt-ferrous ferrite) are given in Table 3.2, as an example of this type of material. The coupling coefficient is about 0.3, which is less than that of barium titanate ceramic, and the ferrite materials never achieved great popularity for underwater transducer applications.

Recent researches in magnetostrictive materials have concentrated on the rare earth-iron alloys. US workers discovered in the early 1970s that binary alloys of some of the rare earth elements and iron showed magnetically generated strains which were 10 to 100 times greater than that of nickel. Since that time, many other alloys have been investigated, to optimise the properties for particular applications, and some of the characteristics of a ternary alloy of terbium-dysprosium-iron ("Terfenol") are

TABLE 3.2
Properties of Magnetostrictive materials.

Property	Units	Nickel
k_{33}opt		0.15–0.31
d_{33}opt	10^{-9}Wb/N	~–3.1
$\mu_{33}{}^S$opt		22
H_{opt}	10^2A/m	7–10
Saturation magnetostrain	10^{-6}	–33
Resistivity	ohm.m	7×10^{-8}
Young's mod.	10^{10}Pa	21
Density	10^3kg/m^3	8.8
Curie temp.	°C	358
$F_c t^2$	kHz.mm^2	1.7

Notes.

Values for Nickel from Berlincourt et al [3.6]

2V–Permendur (Composition 2%V–49%Co–49%Fe) [3.16],[3.17]

Ferroxcube 7A1 (Nickel–copper–cobalt–ferrous ferrite) [3.17]

TbDyFe ($Tb_{0.27}Dy_{0.73}Fe_{1.95}$ alloy) (Terfenol) [3.18]

These values are approximate and subject to considerable variation as a function of thermal and mechanical treatment.

Properties (*contd*)

2V– Permendur	Ferroxcube 7A1	TbDyFe
0.20–0.37	0.25–0.30	0.60
14–30	−2.8 to −4.4	
80	15–25	4
25	15–24	160
70	−27	1100
30×10^{-8}	>10	60×10^{-8}
22	16.6	5
8.2	5.3	9.2
980	530	375
2.4	$>10^6$	110

k_{33}opt = optimum value of coupling coefficient along bias direction.
d_{33}opt = optimum value of magnetostrictive d–coefficient.
$\mu_{33}{}^S$opt = relative permeability at constant strain, at optimum bias.
H_{opt} = bias magnetic field for optimum k_{33}.
Young's modulus (= $1/s_{33}{}^B$, where s^B = elastic compliance coefficient at constant magnetic flux density).

The 33-mode has occasionally been referred to as the 'thickness mode'; to agree with IEEE standards, this should more formally be described as the 'extensional mode'.

TABLE 3.3

Properties of other useful materials

	Tensile strength MPa	Yield stress MPa	Fatigue limit MPa
Mild steel (En 2)	390	230	170
High tensile steel (En 30A)	1500	1310	620
Stainless steel (En 57)	880	700	370
Brass	420	150	140
Al bronze	690	460	300
Ph bronze (extruded)	360	120	110
Monel	590	280	300
Be Cu	1300	1030	290
Aluminium (DTD5064)	540	460	150
Titanium	930	770	390
Uranium			
Alumina	140 −200		
PTFE	17 −40		

Other materials (*contd*)

Young's mod. 10^3MPa	Density 10^3kg/m^3	Poissons ratio	Thermal ex coef 10^{-6}/°C
207	7.85	0.29	11
207	7.85	0.29	11
210	7.75	0.31	11
95	8.50	0.33	19
124	7.70	0.33	16
96	8.90	0.33	17
169	8.84	0.31	14
127	8.30	0.29	17
70	2.75	0.34	23
117	4.50	0.32	9
165	19.1	0.21	
240 -340	3.3 -3.8	0.22	7.7
0.34 -0.62	2.1 -2.3		100

shown in Table 3.2. The coupling coefficient of this alloy may be as high as 0.6, but its main potential advantage is its very high magneto-strain, which is over twice the maximum strain for LZT and some 20 times that of nickel [3.18]. The power handling capacities of these alloys are therefore much higher than for nickel, and somewhat better than for the LZT ceramics. However, these high strains are obtained only by combining very large magnetic fields with high biassing fields, and this presents some difficulties in practice. Experimental transducers have been made using these alloys in the form of rods, in which configuration eddy current losses have been significant. Developments are continuing on these alloys, and their future usefulness remains to be demonstrated. Further details of these developments are given in the references and in Chapter 12 (Section 12.2).

3-7   OTHER MATERIALS

The practical construction of transducers involves the use of a number of other materials in addition to piezoelectric ceramic. Data on the relevant characteristics of some of these other materials are listed in Table 3.3. Some of these properties are again very dependent on the thermal and mechanical treatment of the materials, and the values should be taken as a guide – especially the values for tensile strength, yield, and fatigue limit.

REFS Chapter 3

References

3.1   Hunt,F.V., *Electroacoustics*, Harvard University Press and Wiley & Sons, 1954. (Reprint published by Acoust Soc Am, 1982.)

3.2   Brown,C.S., Kell,R.C., Taylor,R., and Thomas,L.A., "Piezoelectric materials: A review of progress." *Proc IEE*, 109B, 99–114, (1962) (Paper No 3798)

3.3   NDRC Summary Technical Report, Division 6, Vol 12, (1946). *Design and Construction of Crystal Transducers.*

3.4   Katz,H.W., *Solid State Magnetic and Dielectric Devices*, Wiley & Sons, 1959, Ch 2.

3.5   Jaffe,B., Cook,W.R.Jr., and Jaffe,H., *Piezoelectric Ceramics*, Academic Press, 1971.

3.6   Berlincourt,D.A., Curran,D.R., and Jaffe,H., "Piezoelectric and Piezomagnetic Materials and their Function in Transducers," in Mason,W.P.(Ed), *Physical Acoustics*, Vol 1 – Part A, Academic Press, 1964, p170.

3.7   IRE Standards on Piezoelectric Crystals: Measurements of Piezoelectric Ceramics, 1961. (IEEE Std 179-1961.) (Reprinted in Jaffe et al [3.5].)

3.8   Krueger,H.H.A., and Berlincourt,D., "Effect of high static stress on the piezoelectric properties of transducer materials." *J Acoust Soc Am*, 33, 1339-1344, (1961).

3.9   Nishi,R.Y., and Brown,R.F., "Behaviour of piezoceramic projector materials under hydrostatic pressure." *J Acoust Soc Am*, 36, 1292-1296, (1964).

3.10   Nishi,R.Y., "Effects of one-dimensional pressure on the properties of several transducer ceramics." *J Acoust Soc Am*, 40, 486-495, (1966).

3.11   Krueger,H.H.A., "Stress sensitivity of piezoelectric ceramics: Part 1. Sensitivity to compressive stress parallel to the polar axis." *J Acoust Soc Am*, 42, 636-645, (1967).

3.12   Krueger,H.H.A., "Stress sensitivity of piezoelectric ceramics: Part 2. Heat treatment." *J Acoust Soc Am*, 43, 576-582, (1968).

3.13   Krueger,H.H.A., "Stress sensitivity of piezoelectric ceramics: Part 3. Sensivity to compressive stress perpendicular to the polar axis." *J Acoust Soc Am*, 43, 583-591, (1968).

3.14   Brown,R.F., "Effect of two-dimensional mechanical stress on the dielectric properties of poled ceramic barium titanate and lead zirconate titanate." *Can J Phys*, 39, 741-753, (1961).

3.15   Manufacturers' Piezoelectric Ceramic brochures, eg Unilator Technical Ceramics, EDO Corporation, Channel Industries Inc, Quartz & Silice.

3.16   Davies,C.M.Jr., "Properties of conventional magnetostrictive materials for use in underwater transducers", in *Proceedings of 25-26 Feb 1976 Workshop on Magnetostrictive Materials*. (NRL Report 8137, June 1977), p39.

3.17   van der Burgt,C.M., "Piezomagnetic ferrites." *Elec Tech*, 37, 330-341, (1960),

3.18   Clark,A.E., "Introduction to highly magnetostrictive rare-earth materials", in *Proceedings of 1976 Workshop on Magnetostrictive Materials*, op cit, p109

Additional Reading

Some of the above references provide much general information, including especially [3.1, 3.2, 3.3, 3.5, 3.6]. The following references also give good general treatments:-

1   Mason,W.P., "Piezoelectricity, its history and applications." *J Acoust Soc Am*, 70, 1561-1566, (1981).

2   Berlincourt,D., "Piezoelectric ceramics: Characteristics and applications." *J Acoust Soc Am*, 70, 1586-1595, (1981).

3   *Proceedings of the 25-26 Feb 1976 Workshop on Magnetostrictive Materials*, NRL Report 8137, June 1977. (Timme, R.W.,(Ed), Naval Research Laboratory, Orlando, Fla.)

4   Neppiras,E.A., "New magnetostrictive materials and transducers," Part I, *J Sound Vib*, 8, 408-430, (1968): Part II, *J Sound Vib*, 8, 431-456, (1968).

5   NDRC Summary Technical Report, Division 6, Vol 13, (1946), *Magnetostrictive Transducers*.

6   IEEE Standard on Piezoelectricity, *IEEE Trans on Sonics and Ultrasonics*, SU-31, (No 2), March, 1984.

Chapter 4
EQUIVALENT CIRCUITS

4-1 INTRODUCTION

From the point of view of the associated electrical circuit, the transducer appears as an electrical load, and in this and the following chapter we shall concentrate on the transducer as an electrical component of the system. We therefore need first of all to consider the electrical characteristics of the transducer, and what they can tell us about its parameters. After this we shall be able to consider how the parameters can best be selected in order to achieve the performance needed for its particular application. All of this can be treated entirely as an electrical problem, and for the time being this is the approach that will be adopted, without any consideration of what method is used for effecting the conversion from electrical to acoustic energy.

It may nevertheless be useful to make a short diversion to consider in a simple way how there may exist a relationship between the mechanical or acoustic behaviour of a transducer and its electrical characteristics. We saw in Chapter 1 that the equation for a simple mechanical system of a mass and spring as shown in *Fig 1.3* was

$$ m \frac{\mathrm{d}u}{\mathrm{d}t} + ru + K \int u \, \mathrm{d}t = F_0 \exp(j\omega t) \qquad (1.9) $$

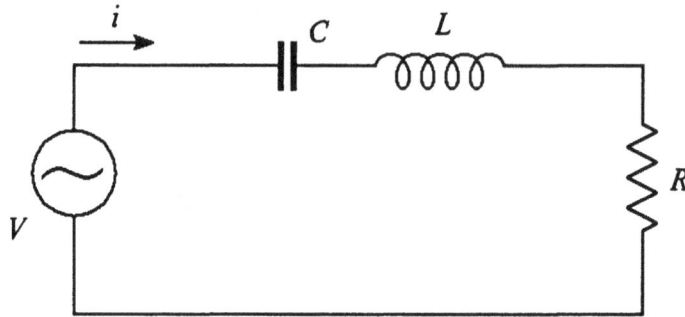

Fig 4.1   Electrical series resonant circuit.

where u is the velocity of m, K is the stiffness of the spring, and r represents the mechanical damping. Now consider the series-resonant electrical circuit shown in *Fig 4.1*. If a sinusoidal voltage represented by $V_0 \exp(j\omega t)$ is applied to this circuit we obtain the equation

$$L \frac{di}{dt} + Ri + \frac{1}{C} \int i \, dt = V_0 \exp(j\omega t) \qquad (4.1)$$

The similarity in form between these two equations is an indication that there is an analogy between the electrical and mechanical systems. It is therefore reasonable to expect that a mechanical system such as a transducer may be represented by an equivalent electrical circuit, and we shall find that this is in fact extremely useful in analysing and understanding the performance of these systems. In particular, it is clear that a resonant mechanical device such as a transducer might be analysed in terms of a resonant electrical circuit, and that the behaviour of the complete system might then be interpreted in electrical terms. Although this example is of a simple resonant circuit, the analogy may be confirmed also for more complex systems.

In the example given above the applied force was analogous to the voltage, the current to the velocity, the capacitance to 1/stiffness (ie to the compliance), and so on, as indicated in the following table:-

MECHANICAL	ELECTRICAL
Mass	Inductance
Compliance	Capacitance
Resistance	Resistance
Velocity	Current
Force	Voltage
Displacement	Charge
Impedance	Impedance

Other systems of analogues have been devised, but this one, generally known as the impedance analogue, has the advantage of being well established and will be the one used in this text.

We can now return to considering the behaviour of a transducer solely as an electrical load. Piezo-electric transducers generally have a high impedance – considerably greater than that of most voltage sources. It is thus natural to expect practical sources of electrical power to behave approximately as low impedance voltage sources rather than as high impedance current sources. The input current into the transducer is given by the source voltage multiplied by the admittance of the load, and it is then convenient to consider the transducer in terms of its admittance rather than its impedance.

4-2 ADMITTANCE LOOPS

It is well known that the electrical impedance of a device can be decomposed into its real and imaginary parts, its resistance and its reactance. Similarly the admittance is composed of a real part, its conductance, and an imaginary part, its susceptance.

i.e. $$Y = G + jB \qquad (4.2)$$

where Y denotes the admittance, G the conductance, and B the susceptance. Suppose that we measure the admittance of a transducer over a band of frequencies around its resonance, and plot the variation of its conductance as a function of frequency. A typical plot for a well-behaved transducer with a clearly-defined resonance is shown in *Fig 4.2a*. The main features to note are the peak rising to a maximum value of

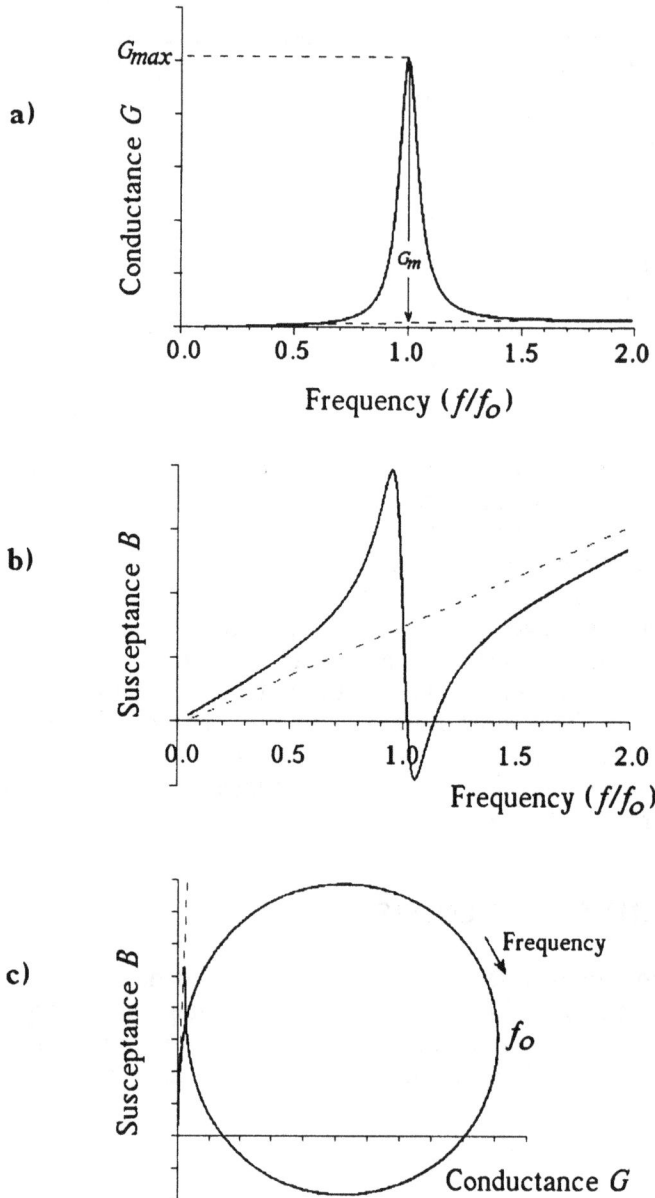

Fig 4.2 Electrical admittance of simple piezoelectric transducer around resonance: (a) conductance versus frequency; (b) susceptance versus frequency; (c) admittance loop.

G_{max} at the frequency f_0 superimposed on a steadily rising base line. The base line can be interpolated, and the "motional conductance" G_m derived, as shown in the figure. The power into a device is given by V^2G, and the variation of G thus indicates how the power into the transducer varies with frequency when a constant voltage source is applied. The increase in power shown by the peak corresponds to the resonance of the transducer, whilst the baseline represents the power which is dissipated in the transducer independently of its resonance. It is common to find that this baseline shows an approximately linear increase, as in the diagram, for reasons that we can ignore for the present.

The variation of the susceptance is typically as shown in *Fig 4.2b*. In this case the resonance is associated with the pair of peaks, one above and one below a line (dotted) indicating an approximately linear increase which is independent of the resonance. The frequency at which the curve crosses the dotted line is approximately f_0, the same as that for the maximum conductance. These two figures may be combined into a single diagram, the admittance loop, as shown in *Fig 4.2c*. This diagram, although it may appear at first glance less easy to interpret, is in fact a most instructive tool in analysing transducer behaviour, and well merits some attention to understand its use.

In this presentation the admittance is plotted as a vector in the susceptance–conductance plane, with frequency increasing along the loop. The admittance loop shown in *Fig 4.2c* would again be typical for a well–behaved transducer with a clearly defined resonance, for frequencies around resonance. The curve starts from the (0,0) point at zero frequency, describes an approximately circular loop in the region of f_0, and then becomes asymptotic to the dotted line which again represents the behaviour of the device in the absence of its resonance. This dotted line is thus said to represent the "clamped" characteristics of the transducer. It is perhaps worth noting at this point that deviations from this idealised curve are all too often observed, for example because of the effects of other resonances within the frequency band plotted, but we will for the moment continue to examine this simple case. Such a well–behaved loop is more likely to be measured for a highly resonant system such as a lightly damped transducer, for which the loop will generally be larger and the range of frequencies to be covered relatively small. When the loading is very small the loop approximates to a circle; in that case, the susceptance may become negative for a band of frequencies just above resonance.

4-3   DERIVATION OF EQUIVALENT CIRCUIT

The analysis of a transducer's performance in an electrical system is best understood through the use of equivalent circuits, so we now need to consider how to convert the admittance measurements into the elements of an electrical equivalent circuit. In doing this we shall derive also the main transducer parameters themselves. The admittance curve shown in *Fig 4.3* is for a lightly loaded piezoelectric vibrator, for which the response is dominated by the resonance and the loss represented by the clamped line is not significant. For practical transducers, the clamped loss may well be significant, and the treatment is much simplified if the effects of the mechanical resonance can be isolated, by subtracting the clamped conductance from the total conductance to derive the motional admittance loop as in *Fig 4.3*. The frequencies of importance for the analysis are indicated on the diagram, in accordance with the IRE Standards for measurements on piezoelectric vibrators [4.1], viz:–

f_S = motional (series) resonance frequency
　　 (ie frequency of maximum conductance)

f_m = frequency of maximum admittance

f_p = parallel resonance frequency

f_n = frequency of minimum admittance

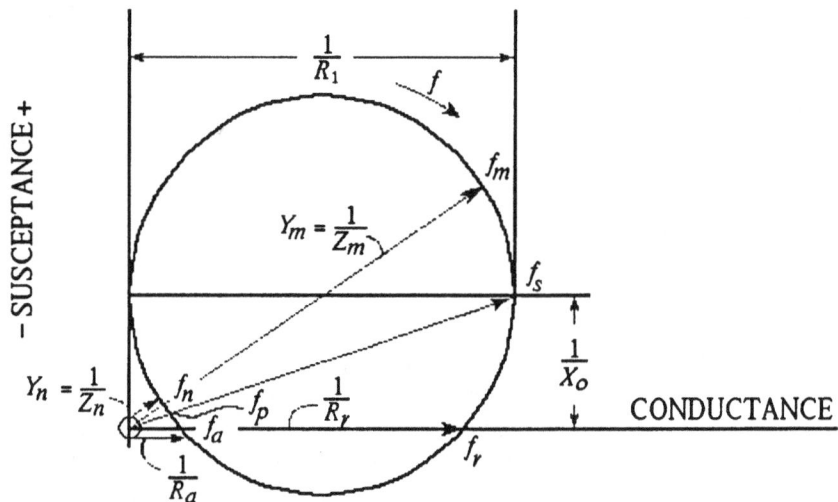

Fig 4.3   **Motional admittance loop for simple resonator; from *Proc IRE*, March 1957, 353-358, IRE Standards on Piezoelectric Crystals – The Piezoelectric Vibrator, Fig.3.**

The series resonance frequency f_S is that at which the motional conductance is a maximum, and is the frequency at which the motional power is a maximum for a constant applied voltage. The parallel resonance frequency f_p is that at which the motional resistance is a maximum, and lies on the line from the origin to f_S, as indicated in the diagram. The frequencies at which the conductance is half the value G_{mS} at resonance are denoted by f_1 and f_2 ($f_2 > f_1$). The mechanical Q-factor Q_M is then defined by:-

$$Q_M = \frac{f_S}{f_2 - f_1} \qquad (4.3)$$

In many cases, the clamped conductance is small compared with the motional conductance, and its effect is then often ignored by taking the resonance frequency as that at which the *total* input conductance is a maximum; the error involved in this approximation is usually small. More care has to be taken in determining f_1 and f_2, and it is generally safer to subtract the clamped conductance from the total measured value before determining f_1 and f_2.

At resonance the admittance Y_S is composed of a conductance G_{mS} in parallel with a susceptance B_S. A parameter which is sometimes useful is the electrical Q-factor (Q_E) at resonance, defined by

$$Q_E = B_S/G_{mS} \qquad (4.4)$$

The input susceptance may be interpreted in a useful way by noting that the susceptance of a parallel capacitance is given by

$$B = \omega C$$

where $\omega = 2\pi f$. The input susceptance may therefore be treated as a parallel capacitance given by

$$C = B/\omega$$

The variation of B with frequency shown in *Fig 4.2b* may thus be converted to represent the variation of input capacitance, with a typical result as shown in *Fig 4.4*. At frequencies well below resonance, the capacitance is effectively constant, its value being denoted by C_{LF}. At high frequencies the capacitance for an ideal transducer should also become asymptotic to a constant value, which turns out to be equal to the capacitance at

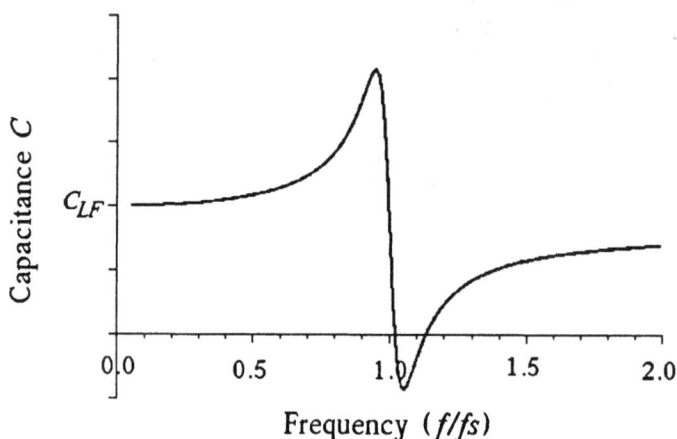

Fig 4.4 Relationship between frequency and capacitance, for piezoelectric transducer with single resonance.

resonance, but this simple behaviour is often disturbed by the effects of other resonances. The behaviour at low frequencies is however generally much nearer to the ideal, and C_{LF} is a useful and easily measured parameter. The dielectric loss factor (tanδ) can be determined at the same time as the capacitance measurement at low frequency.

These electrical measurements thus permit the determination of the main parameters of the transducer, f_S, Q_M, C_{LF}, and tan δ. There remains just one further parameter of importance, whose relevance is not quite so immediately apparent. This factor, which together with Q_M controls the size of the admittance loop, is known as the "coupling coefficient" (k), and as will become apparent later the coupling coefficient does in fact have a crucial role in determining the performance of any transducer. It can be determined from the ideal admittance loop by using the frequencies quoted above, thus:–

$$k^2 = 1 - (f_S/f_p)^2 \qquad (4.5)$$

As in this equation, it is common for the coupling coefficient to appear as its square. The relationship above may be used for measuring k^2 when the resonance is sufficiently clearly defined, but other methods will be described later which are more accurate when the admittance loop is smaller.

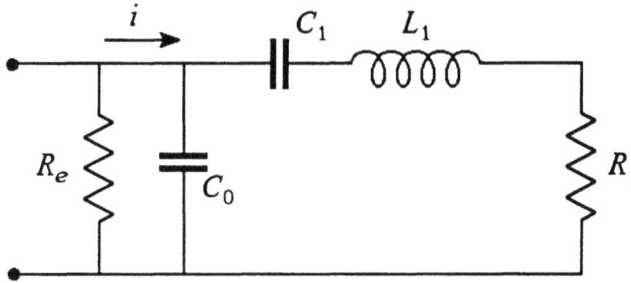

Fig 4.5   Equivalent circuit representing electrical behaviour of piezoelectric transducer around resonance.

This completes the list of parameters derived directly from the electrical measurements. We now move on to consider whether the variation of the admittance with frequency could also be represented by an electrical circuit which would show the same characteristics. Such a representation is useful only if the circuit elements needed to produce a reasonably accurate approximation to the transducer behaviour over a significant frequency range are independent of frequency. This is generally true to sufficient accuracy, and for piezoceramic transducers having a single well-defined resonance, a circuit such as that shown in *Fig 4.5* provides a good approximation to the variation of input admittance for frequencies around resonance. *Fig 4.5* is thus a typical "equivalent circuit" for a low frequency piezo-electric ceramic transducer.

In this circuit the series $L_1 C_1 R_1$ arm represents the mechanical resonance, and C_0 and R_e the dielectric clamped capacitance and loss. At low frequencies the circuit reduces to the capacitance C_0 in parallel with C_1 and R_e. Thus the low frequency capacitance is given by

$$C_{LF} = C_0 + C_1 \qquad (4.6)$$

At resonance the reactance of the series arm vanishes and the circuit appears as C_0 in parallel with R_1 and R_e. Thus the motional conductance at resonance (G_{mS}) is given by

$$G_{mS} = 1/R_1 \qquad (4.7)$$

and

$$G_{max} = 1/R_1 + 1/R_e \qquad (4.8)$$

The dielectric loss is measured at low frequency, so

$$\tan \delta = \frac{1}{\omega C_{LF} R_e} \qquad (4.9)$$

Note that in this case the resistance R_e needs to vary inversely with frequency, since $\tan \delta$ is assumed to be constant; however, this variation is predictable and usually has only a small effect, because R_e is generally much larger than R_1.

The resonance of the series LCR arm is at f_S;

i.e. $\qquad \omega_S^2 = \dfrac{1}{L_1 C_1} \qquad$ (where $\omega_S = 2\pi f_S$) $\qquad (4.10)$

And the mechanical Q-factor is given by

$$Q_M = \frac{\omega_S L_1}{R_1} = \frac{1}{\omega_S C_1 R_1} \qquad (4.11)$$

It will be shown later that the coupling coefficient is related to the circuit components by

$$k^2 = \frac{C_1}{C_0 + C_1} \qquad (4.12)$$

The relationships above may be inverted, to give equations by which the components of the equivalent circuit may be derived from the measured parameters of the transducer, i.e:-

$$C_1 = k^2 C_{LF} \qquad (4.13)$$

$$L_1 = \frac{1}{\omega_S^2 C_1} \qquad (4.14)$$

$$R_1 = \frac{1}{\omega_S C_1 Q_M} = \frac{1}{\omega_S k^2 C_{LF} Q_M} \qquad (4.15)$$

$$C_0 = (1 - k^2) C_{LF} \qquad (4.16)$$

$$R_e = \frac{1}{\omega C_{LF} \tan \delta} \qquad (4.17)$$

These relationships thus provide a means of deriving an electrical equivalent circuit which reproduces the input admittance of the transducer around resonance.

4-4 INPUT ADMITTANCE

The input admittance of the circuit of *Fig 4.5* represents that of the transducer, and its variation with frequency may readily be calculated. The impedance of the series *LCR* arm is given by

$$Z_1 = R_1 + j\omega L_1 + \frac{1}{j\omega C_1}$$

$$= R_1 + j\omega L_1 \left(1 - \frac{\omega_S^2}{\omega^2}\right) \quad \text{where } \omega_S^2 = \frac{1}{L_1 C_1}$$

$$= R_1 + j\omega_S L_1 \left(\frac{\omega}{\omega_S} - \frac{\omega_S}{\omega}\right)$$

$$= R_1(1 + jQ_M\Omega) \quad \text{where } \Omega = \frac{\omega}{\omega_S} - \frac{\omega_S}{\omega} \quad (4.18)$$

$$\text{and using } Q_M = \frac{\omega_S L_1}{R_1}$$

The input impedance (Z_{in}) is that due to the parallel combination of R_e, C_0, and Z_1, and the input admittance (Y_{in}) is thus given by

$$Y_{in} = \frac{1}{Z_{in}} = \frac{1}{R_e} + j\omega C_0 + \frac{1}{R_1(1+jQ_M\Omega)}$$

$$= \frac{1}{R_e} + \frac{\{j\omega C_0 R_1(1+jQ_M\Omega) + 1\}}{R_1(1+Q_M^2\Omega^2)}(1-jQ_M\Omega)$$

$$= \frac{1}{R_e} + \frac{1}{R_1(1+Q_M^2\Omega^2)} + j\omega \left\{C_0 - \frac{\omega_S C_1 Q_M^2\Omega}{\omega(1+Q_M^2\Omega^2)}\right\} \quad (4.19)$$

The second and third terms on the right hand sides of this equation represent the motional conductance and susceptance of the transducer, whilst the first term represents the clamped conductance. At resonance, $\omega = \omega_S$ $\Omega = 0$, and $G_m = 1/R_1$, $B = \omega_S C_0$ as would be expected. The capacitance is given as a function of frequency by:–

$$C = \frac{B}{\omega} = C_0 - \frac{\omega_S C_1 Q_M^2 \Omega}{\omega(1+Q_M^2\Omega^2)} \qquad (4.20)$$

At low frequencies, when $\omega \to 0$, $\Omega \to -\omega_S/\omega$, and $C_{LF} \to C_0+C_1$. At frequencies well above resonance, $\Omega \to \omega/\omega_S$, and $C_{HF} \to C_0$.

All of these expressions should strictly be used only if the equivalent circuit is an accurate representation of the transducer over the whole frequency band of interest; i.e. if the components of the circuit remain independent of frequency. This assumption is often accurate enough for frequencies at and below resonance, but is less reliable above resonance, where other resonances often have a significant effect. Thus the capacitance and admittance are usually quite well represented below resonance by these equations, and the value of C_{LF} can be used with confidence if it is measured sufficiently far below f_S.

The **resonance frequency** f_S occurs when L_1 resonates with C_1; the **anti-resonant** frequency f_p occurs when L_1 resonates with C_1 in series with C_0. Thus,

$$\omega_p^2 = (2\pi f_p)^2 = \frac{1}{L_1} \cdot \frac{(C_0+C_1)}{C_0 C_1}$$

$$= \omega_S^2 \frac{(C_0+C_1)}{C_0}$$

$$= \omega_S^2 \left(\frac{1}{1-k^2}\right) \qquad \text{(using (4.12))}$$

Hence, the coupling coefficient k is given by

$$k^2 = 1 - \frac{\omega_S^2}{\omega_p^2}$$

$$= 1 - \frac{f_S^2}{f_P^2} \qquad\qquad (4.5)$$

$$= \frac{C_1}{C_0 + C_1} \qquad\qquad (4.12)$$

Here the coupling coefficient is defined by equation (4.5) in terms of the measurements which can be carried on the device, and by (4.12) in terms of the elements of the equivalent circuit. The expressions above demonstrate the essential equivalence of the two approaches, at least for an ideal transducer of high Q_M. We shall return again to the significance of the coupling coefficient later.

4-5 EFFICIENCY

The series LCR arm represents the mechanical resonance of the transducer, and it is therefore often called the "motional arm" of the equivalent circuit. The power dissipated in this arm is the electrical equivalent of the motional power dissipated by the transducer. The parallel C_0 and R_e represent the (clamped) electrical admittance of the transducer, the losses in R_e being those due directly to applied field. This may be used to extend the analysis based on the equivalent circuit treatment to give some further insight into the efficiency of the device.

Many low frequency transducers use a radiating piston sealed against water entry by some form of watertight seal such as a rubber O-seal. When the piston is made to oscillate, some of the power is dissipated in the O-seal and thus represents a loss of efficiency. It is reasonable to assume that the O-ring loss for a given amplitude of vibration does not depend to any significant extent on whether the transducer is in water or in air. Measurements of the admittance in air can thus be used to determine the O-seal loss, and hence to estimate the efficiency of the transducer when it is operated in water. Although the O-seal loss has been quoted in this example, the same argument applies to any internal mechanical loss which is independent of whether the transducer is in water or in air. Suppose that the admittance of the transducer is measured when it is in air. The peak motional conductance (denoted by G_{air}) gives a measure of the resistance in the motional arm due to the internal mechanical losses. If the admittance is now measured with the transducer immersed in water, the change in characteristics represents the effects of the water loading. In particular we expect to see an increase in the values of both L_1 and R_1 in the motional arm. Measurements of the resonance frequency and maximum motional conductance (G_{mS}) in both air and water thus permit derivation of the equivalent circuit shown in *Fig 4.6*, in which R_i represents the internal mechanical losses, R_r the radiation resistance and X_r the radiation reactance of the water. The power dissipated in R_r thus represents the acoustically radiated power. Denoting the appropriate values of G_{mS} by G_{air} and G_{wat}, the value of R_r is obtained from the relationships:-

$$G_{wat} = \frac{1}{R_i + R_r}$$

and
$$G_{air} = \frac{1}{R_i}$$

Thus, the mechanical-acoustic efficiency η_{ma} is given by:-

$$\eta_{ma} = \frac{R_r}{R_i + R_r} = \frac{G_{air} - G_{wat}}{G_{air}} \qquad (4.21)$$

This expression assumes that the value of R_i is not affected by the change in resonance frequency on immersion in water, an assumption which is usually sufficiently accurate.

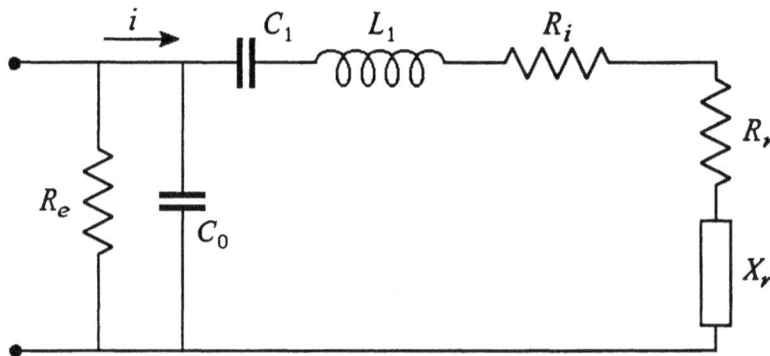

Fig 4.6   Equivalent circuit showing internal losses.

In addition to the internal mechanical losses, the losses due to the dielectric loss in the ceramic should also be taken into account. This is done by noting that at resonance the equivalent circuit reduces to C_0 with R_e in parallel with $R_r + R_i$. The powers in the motional and electric arms are then inversely proportional to their resistances, and the electrical-mechanical efficiency is given by:-

$$\eta_{em} = \frac{G_{wat}}{G_{max}} \qquad (4.22)$$

The overall electrical-acoustic efficiency of the transducer is then given by the product of these two factors, i.e:

$$\eta_{ea} = \eta_{em} \cdot \eta_{ma} = \frac{G_{wat}}{G_{max}} \cdot \frac{G_{air} - G_{wat}}{G_{air}} \qquad (4.23)$$

This expression gives the efficiency at the resonance frequency itself. The variation of efficiency with frequency may then be calculated from the characteristics of the equivalent circuit. The motional power dissipated by the device is represented by the power in the series LCR arm, and is thus given by $V^2 G_m$. The electrical-to-motional efficiency η_{em} is therefore:-

$$\eta_{em} = \frac{G_m}{G_m + 1/R_e}$$

$$= \frac{1}{1 + \omega C_{LF} \tan\delta R_1 (1 + Q_M^2 \Omega^2)} \qquad \text{using } \tan\delta = \frac{1}{\omega C_{LF} R_e}$$

$$= \left\{ 1 + \frac{\omega \tan\delta}{\omega_S k^2 Q_M} (1 + Q_M^2 \Omega^2) \right\}^{-1} \qquad (4.24)$$

where we have used the expressions:-

$$Q_M = \frac{1}{\omega_S C_1 R_1}$$

and $$k^2 = \frac{C_1}{C_{LF}}$$

Equation (4.24) can thus be used to calculate the variation with frequency of η_{em}. *Fig 4.7* shows curves for η_{em} for typical values of the parameters for a transducer operating in water. The overall electrical–acoustic efficiency η_{ea} is the product of η_{em} and η_{ma}. In many cases it is reasonable to assume that the motional–acoustic efficiency η_{ma} is independent of frequency and η_{ea} can then be derived by simply multiplying the curves in *Fig 4.7* by a constant value of η_{ma}. The curves show that the efficiency η_{em} is higher for a higher value of coupling coefficient, an example of the importance of this factor. It is also evident that the value of η_{em} at resonance is higher for a transducer of high Q_M in water than for one with a lower Q_M, but that the efficiency falls off faster on either side of resonance. The inference that the overall efficiency at resonance is higher for a higher Q_M must however be treated with caution, since a high Q_M may often be the result of a low value of the radiation resistance, ie a high value of G_{wat}. Equation (4.21) shows that η_{ma} is reduced by a high value of G_{wat}, and the value of the overall electrical–acoustic efficiency may well be lower for a high Q_M transducer than for one with a lower Q_M. The converse situation, in which a low Q_M is the result of a design having large internal losses (ie a low G_{air}) rather than because of the radiation loading, is clearly to be avoided on physical grounds. As a general rule, the overall

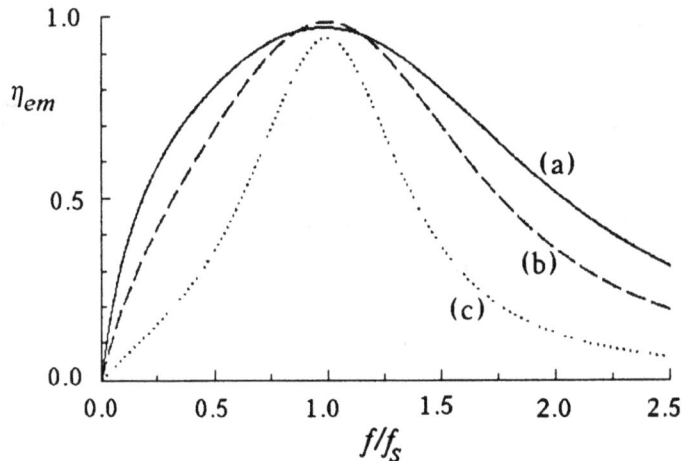

Fig 4.7   Variation of electrical-mechanical efficiency η_{em} with frequency, for tanδ=0.02. (a) k=0.5, Q_M=2.5; (b) k=0.5, Q_M=5; (c) k=0.25, Q_M=5.

electrical–acoustic efficiency is likely to be higher for a low Q_M transducer than for a higher Q_M design, provided that the low value of Q_M is due predominantly to the radiation resistance rather than internal losses.

4-6   APPLICATION TO PRACTICAL TRANSDUCERS

The method described above for measuring the coupling coefficient applies with good accuracy for transducers which show a large and well–defined loop in the admittance diagram, with no other resonances in the band. More precisely, this requires that the circle diameter is large in comparison with the change in the susceptance ωC_0 over the resonance range. The errors which arise when this condition is not satisfied have been discussed by Woollett [4.2] and in the IRE Standards on Piezoelectric Crystals [4.3]. From their treatments, a more accurate expression for the coupling coefficient than that in equation (4.5) is given by:-

$$k^2 = \{(1 - (f_S/f_p)^2\}(1 + 4Q_E^2)^{-1/2} \qquad (4.25)$$

The correction factor involves the parameter Q_E already defined in (4.4). It is useful to note the relationship between Q_E and Q_M; thus

$$Q_E = \omega_S C_0 / G_{mS}$$

and
$$Q_M = \frac{G_{mS}}{\omega_S C_1} \qquad \text{(From 4.11)}$$

Therefore
$$Q_E Q_M = \frac{C_0}{C_1} = \frac{(1 - k^2)}{k^2} \qquad (4.26)$$

For most engineering purposes the simpler equation (4.5) without the correction factor gives a sufficiently accurate value of k, provided $Q_E^2 < 1/25$, which is approximately equivalent to the condition $k^2 Q_M > 5$. Typical values of k^2 usually lie in the range 0.1–0.3, so the condition above corresponds to the condition that $Q_M > 20$–50. This would usually be satisfied by a well–designed transducer in air, though not by one operating in water. It is therefore normal to measure the coupling coefficient of a transducer in air: this is considerably simpler and more direct than carrying out the measurement in water, although alternative methods for performing these measurements for low Q_M transducers will be described in Chapter 9. It is however sometimes useful to derive a value of k from the admittance loop in water by using the expression

$$Q_M = \frac{G_{mS}}{\omega_S C_1} = \frac{G_{mS}}{\omega_S k^2 C_{LF}}$$

Hence
$$k^2 = \frac{G_{mS}}{\omega_S Q_M C_{LF}} \qquad (4.27)$$

Although this method is not particularly accurate as a means of measuring k, because it depends rather critically on the ideal behaviour of the transducer, it does sometimes serve as a useful check.

It is easy to show that the losses due to the clamped conductance are usually small compared to the motional losses at resonance. The motional conductance at resonance is (re-arranging (4.27)):-

$$G_{mS} = \omega_S k^2 Q_M C_{LF}$$

and the clamped conductance is represented by a parallel resistor R_e which varies with frequency to give a constant loss per cycle according to the relationship:-

$$1/R_e = \omega C_{LF} \tan\delta \qquad (4.17)$$

Thus the ratio of the motional to the clamped conductance at resonance is given by:-

$$\frac{G_{mS}}{1/R_e} = \frac{\omega_S k^2 Q_M C_{LF}}{\omega_S C_{LF} \tan\delta}$$

$$= \frac{k^2 Q_M}{\tan\delta} \qquad (4.28)$$

Typical values of k^2 are likely to be 0.1-0.3, and of Q_M 5-15 for a transducer in water. Assuming that the clamped losses are due to the dielectric loss in the piezoelectric ceramic, a typical value of $\tan\delta$ would be 0.02. Then the ratio of motional to clamped conductance would be of the order of $(0.2)(10)/(0.02) = 100$. In this case, small errors in the interpolation to subtract the clamped conductance would not introduce any important effect on the calculation of the motional conductance. Actual values for a particular transducer may be less favourable than this example, but subtraction of the clamped conductance does not usually pose any serious difficulties for piezoelectric designs.

If other resonances occur close to the resonance of interest, so that the admittance curve does not exhibit a single loop, the simple methods described above fail. The most serious problems arise when the other resonances are so close that the loop in air is affected. When the transducer is in water, the bandwidth needed to complete the loop is larger, and the danger of interference from other resonances is correspondingly greater. However, it is then sometimes possible to derive the equivalent circuit with reasonable accuracy from the unaffected loop in air, together with the value of admittance only at the resonance frequency in water. Although efforts can be made to derive the equivalent circuit elements where multiple resonances are present, the best procedure (if possible) is to abandon the design, and concentrate instead on creating one which will have an admittance variation which is nearer to the ideal and will therefore be more amenable to systematic analysis.

4-7 RELATIONSHIP TO MOTIONAL PARAMETERS

Except for noting that the series arm of the equivalent circuit represents the motional resonance of the transducer, the treatment so far has been entirely based on the electrical characteristics of the circuit, without reference to its motional or acoustic behaviour. The relationship of the circuit to the actual design of the transducer has not yet entered into the considerations, and will not do so until much later. All that has been required is that the admittance loop can be represented by a circuit like that in *Fig 4.6*. However, we shall find it very useful to introduce some simple relationships between electrical and mechanical parameters which have general applicability rather than depending on the details of any particular design.

To obtain these, we note that the motional output power can be expressed in either electrical or mechanical terms, and that these must be equivalent. Thus

$$W_m = i_m^2 R_1 = u^2 r_m \qquad (4.29)$$

where W_m is the motional power dissipated, i_m the rms current into the resistor R_1 in the motional arm of the equivalent circuit, u the rms velocity of the displacement of the transducer's radiating piston, and r_m the associated motional resistance in mechanical terms.

A second useful relationship is derived similarly by expressing the motional Q–factor Q_M in both electrical and mechanical terms.

$$Q_M = \frac{\omega_S L_1}{R_1} = \frac{\omega_S M_e}{r_m} \qquad (4.30)$$

where M_e is the effective vibrating mass. The evaluation of M_e and r_m will be discussed later. For the moment we merely note that these expressions allow us to make some link between the electrical and motional characteristics of the transducer without needing to consider any details of the design.

The significance of the coupling coefficient k in electrical terms is brought out by noting that the admittance of the series (motional) arm compared with that of the low frequency capacitance of the transducer is determined by the ratio C_1/C_{LF}, ie by the factor k^2. If a voltage of amplitude V is applied to the device, the electrical power stored is proportional to $V^2 C_{LF}$; the power in the motional arm is proportional to its admittance, ie to the value of $V^2 C_1$. The proportion of the total input power which is stored in the motional arm is thus given by the factor $C_1/C_{LF} = k^2$, which therefore represents the coupling between the electrical and mechanical modes of the device.

4-8   GENERAL THEORY

It is appropriate at this stage to consider the more general theory of equivalent circuits, although this section may be omitted at first reading as it is not critical to an understanding of the treatment in the next chapter. A piezoelectric transducer can be represented in general terms by a 4–terminal network as shown in *Fig 4.8*, with two electrical input terminals (1,2), and two mechanical output terminals (3,4) [4.4]. If an alternating voltage V is applied to the input terminals, the piezoelectric action of the transducer generates an alternating force F at the output and a velocity u of the radiating face, when it is radiating into an impedance Z_r. The box denoting the transducer thus represents an idealised transformer which converts between electrical and mechanical quantities with a transformation ratio φ. By analogy with the more usual electrical transformers, φ is

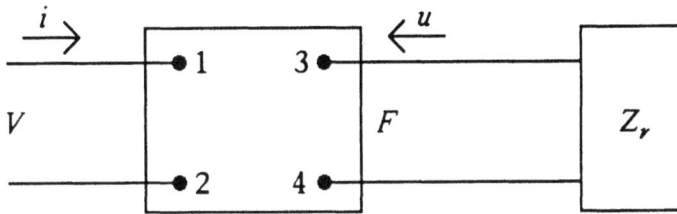

Fig 4.8 General representation of transducer as 4-terminal network.

the factor relating the piezoelectrically generated force F to the applied voltage V, when the motion of the face is blocked (ie $u=0$). Alternatively, φ is the factor relating the piezoelectrically generated current (i) between the electric terminals when they are short-circuited (ie $V=0$) to an applied velocity u at the mechanical side. When the motion is blocked, the ratio of input current i to the applied voltage V is called the blocked electrical input admittance Y_e. Note that all of these quantities are in general complex.

In the general case when the transducer radiates into an impedance Z_r, as in *Fig 4.8*, the input current is related to the applied voltage by

$$i = Y_e V - \varphi u \qquad (4.31)$$

in which the second term represents the reaction at the input due to the piezoelectric effect. Alternatively, if an alternating force F is applied to the mechanical terminals,

$$F = \varphi V + Z_m u \qquad (4.32)$$

where Z_m is the mechanical impedance at the output terminals when the electrical terminals are short-circuited (ie $V = 0$). Also, considering the energy radiated outwards into the load Z_r, the mechanical force acting on the transducer face must be related to the particle velocity v of the adjacent fluid by

$$Z_r = \frac{F}{-u} \qquad (4.33)$$

Then
$$F = \varphi V - Z_m \cdot \frac{F}{Z_r}$$

And hence
$$F = \frac{\varphi V Z_r}{Z_m + Z_r} \qquad (4.34)$$

Substituting in (4.31), and using (4.33), gives

$$i = Y_e V + \frac{\varphi^2 V}{Z_m + Z_r}$$

The input admittance Y_{in} (= i/V) is therefore

$$Y_{in} = Y_e + \frac{\varphi^2 V}{Z_m + Z_r} \qquad (4.35)$$

This corresponds to the input admittance of the circuit shown in *Fig 4.9*, in which Y_e represents the "blocked" electrical admittance, and the impedances of the mechanical elements Z_m and Z_r are reflected across to the input side of an ideal transformer of ratio φ:1. For a piezoelectric transducer, the blocked admittance is that of a dielectric capacitor, represented by its "clamped capacitance" C_0 in parallel with a resistor R_e to account for the dielectric loss. The mechanical impedance Z_m of a resonant transducer can be represented by a mechanical system with lumped elements such that

$$Z_m = r_m + j(\omega m - K/\omega) \qquad (1.10)$$

where m is the effective vibrating mass, K the effective stiffness, and r_m a resistive component associated with the mechanical losses in the transducer itself.

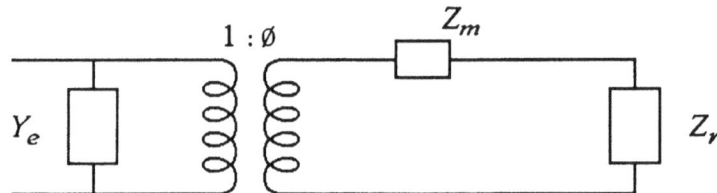

Fig 4.9   Generalised equivalent circuit of piezoelectric transducer.

The mechanical impedances are converted to their electrical equivalents by the hypothetical transformer in *Fig 4.9*, for which the impedance transformation ratio φ^2 includes conversion from mechanical to electrical units. The representation of a transducer working into its radiation load in water may thus be converted to its electrical equivalent circuit, giving a circuit as in *Fig 4.6*, with the relationships

$$R_1 = r_m/\varphi^2 \qquad L_1 = m/\varphi^2 \qquad C_1 = \varphi^2/s$$

$$R_r = r_r/\varphi^2 \qquad X_r = x_r/\varphi^2 \qquad\qquad \left.\right\} \quad (4.36)$$

where the radiation impedance in mechanical terms was expressed as $Z_r = r_r + jx_r$. The derivation of the ratio φ from a knowledge of the transducer design will be discussed in Chapter 7.

4-9   EQUIVALENT COMPONENTS

The treatment in the preceding paragraphs has again involved no knowledge of the transducer construction; in deriving the relationships (4.36) we have only used the assumption that the mechanical resonance of the element can be represented by an electrical series resonant circuit. Indeed, the expressions would apply to a variety of basic transducer designs. We will now, however, consider how a particular design of transducer element may be represented by its equivalent circuit. At this stage, the treatment will be restricted to designs satisfying the **lumped-mass** approximation. In this approximation, the masses are assumed to be effectively rigid, so that all parts of the mass move with the same displacement, and any springs have zero mass, so that the force is equal at all positions along the spring. This is equivalent to assuming that the lengths of the masses and springs are small compared with the wavelength of longitudinal vibrations at the frequency of interest in the materials of the element. This assumption is usually valid for the low frequency devices which are the main topic of this book, and will generally be adopted since it simplifies the analysis and leads to results which are most readily understood for establishing general principles. The extension to higher frequencies will be discussed later (Chap 11).

Thus, adopting this approximation, assume that two rigid masses m_1 and m_2 are mounted on a linear and massless spring, as in *Fig 4.10*, and that the system is allowed to

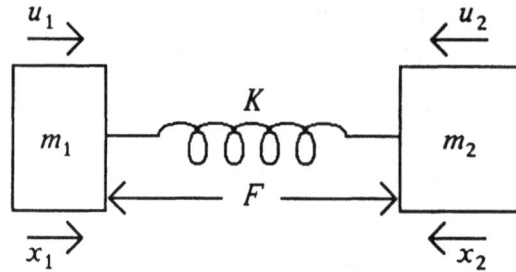

Fig 4.10   Simple mechanical system with two masses and a spring.

oscillate along the line of the spring. Displacements of the masses are denoted by x_1 and x_2, and their velocities by u_1 and u_2. Note that positive values of x or u correspond to compression of the spring. The equations of motion for the system may then be written as

$$m_1 \dot{u}_1 = -F$$

$$m_2 \dot{u}_2 = -F$$

where F is the force in the spring (of stiffness K), and $\dot{u}$ indicates the differential with respect to time (ie $\dot{u}_1 = du_1/dt$ etc). By Hooke's law, the force in the spring is given by

$$F = K(x_1 + x_2)$$

$$= K\left(\int u_1 dt + \int u_2 dt\right)$$

Thus,
$$\frac{-m_1}{K} \dot{u}_1 = \int u_1 dt + \int u_2 dt \qquad (4.37a)$$

and
$$\frac{-m_2}{K} \dot{u}_2 = \int u_1 dt + \int u_2 dt \qquad (4.37b)$$

Now consider an electrical circuit as shown in *Fig 4.11*, in which alternating currents i_1 and i_2 can flow through the two inductors L_1 and L_2, in parallel with the capacitor C. In this case, positive currents are in the direction to make the upper end of C have a positive voltage. Then,

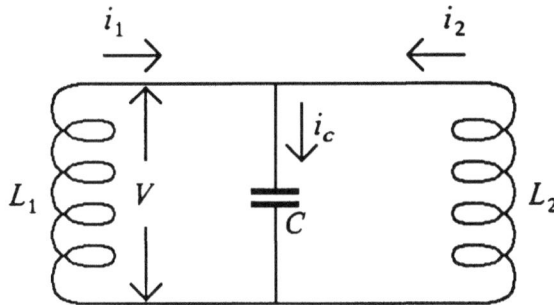

Fig 4.11 Electrical equivalent of *Fig 4.10*.

$$-L_1\dot{i}_1 = V$$

$$-L_2\dot{i}_2 = V$$

and
$$V = \frac{1}{C} \int i_c dt$$

where $i_c = i_1 + i_2$

Thus
$$-CL_1\dot{i}_1 = \int i_1 dt + \int i_2 dt \qquad (4.38a)$$

and
$$-CL_2\dot{i}_2 = \int i_1 dt + \int i_2 dt \qquad (4.38b)$$

The pair of equations (4.38) is clearly equivalent to the pair (4.37) if electrical currents correspond to velocities, inductances to masses, and capacitances to compliances (ie to the reciprocal of stiffnesses). This set of equivalences is known as the impedance system of analogues (Sect 4.1), and is the one adopted in this text, although various authors (eg [4.5, 4.6]) argue the virtues of a different system, known as the mobility analogue, in which velocity is equivalent to voltage.

The circuit of *Fig 4.11* is thus equivalent in its response to the simple mechanical system of *Fig 4.10*. However, in order to analyse the behaviour of more realistic systems, it is clearly necessary to be able to extend the method to more complex mechanical arrangements. As a simple extension, consider the system shown in *Fig 4.12a*, where two springs are in parallel in

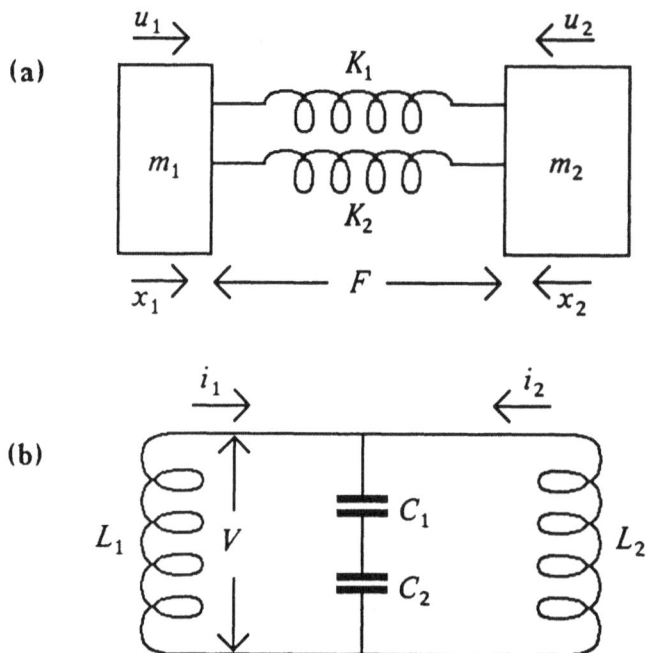

Fig 4.12   (a) Mechanical system with two springs in parallel; (b) Electrical equivalent of (a).

place of the single spring in *Fig 4.10*, and assume for simplicity that the combined spring rate is unchanged from that for *Fig 4.10*. Thus, if the stiffnesses of the two springs are K_1 and K_2, the effective stiffness will be $K = K_1 + K_2$. However, in transforming to electrical equivalent quantities, it is the mechanical compliance which is equivalent to capacitance, and we therefore expect the electrical equivalent to involve two capacitances satisfying a relationship in the form

$$1/C = 1/C_1 + 1/C_2$$

This would be the relationship for two capacitors in series, and the equivalent circuit would then be as shown in *Fig 4.12b*. It is easy to show, by comparing equations as above, that this circuit is in fact the equivalent of the mechanical system with two springs. Such an arrangement, in which the same current passes through each capacitor, corresponds to the mechanical

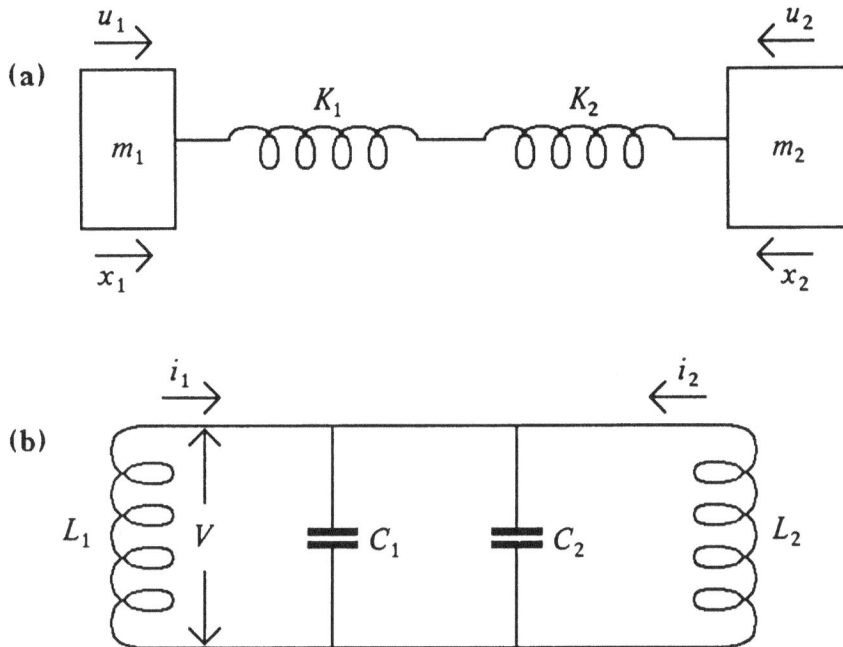

Fig 4.13 (a) Mechanical system with two springs in series; (b) Electrical equivalent of (a).

system in which the velocities are the same at the ends of each spring. We thus see that the equivalent for a mechanical system which has springs in parallel is an electrical circuit with capacitors in series.

Now consider the case of two springs in series, as in *Fig 4.13a*. The equations of motion for the two masses are as for *Fig 4.10*. The force F is constant along both springs (since they are assumed to be massless), and we may therefore write

$$x_1 - x_3 = F/K_1$$

and

$$x_2 + x_3 = F/K_2$$

Hence

$$x_1 + x_2 = F(1/K_1 + 1/K_2)$$

This is equivalent to the equation for the simple system of *Fig 4.10* if

$$\frac{1}{K} = \frac{1}{K_1} + \frac{1}{K_2}$$

ie the effective compliance is the sum of the individual compliances. For the electrical equivalent, this corresponds to the sum of the individual capacitors representing each spring, and we thus see that mechanical springs in series are equivalent to capacitances in parallel, as in *Fig 4.13b*. This may also be deduced from the equivalence of force and voltage; the equality of force through the springs corresponds to the equality of voltage applied to capacitors in parallel.

Similar methods may be used to derive further equivalent circuit elements according to this impedance system of analogues, and *Fig 4.14* lists the equivalence between mechanical and electrical components for both the impedance and mobility systems of analogues. The alternative (mobility) system of analogues is based on a correspondence between velocity and voltage, and has the advantage that the electrical elements are arranged in a topologically similar way to those of the mechanical device. In an instructive discussion of the derivation of equivalent circuits, Beranek [4.5] explains how the mobility analogue may be used as an intermediate step in deriving the impedance analogue. Full use of the mobility analogues does however require the introduction of the concept of a "gyrator", and it is largely to avoid this complication that the impedance system is preferred in this book. Careful application of the methods above permits the construction of equivalent circuits for any particular mechanical design, and this approach is used by many authors to analyse in detail the behaviour of transducers. However, considerable care is necessary to ensure accuracy of the circuit and the results for other than the simplest of designs, and it is often more straightforward and reliable to tackle any particular problem directly, for example by solving the equations of motion. Where an equivalent circuit is not too complicated, it may be useful in giving an insight into the response of the device, and especially into its variation with frequency, and these techniques will be used where appropriate in succeeding chapters. One point to note from *Fig 4.14* is that there is no mechanical equivalent (in the impedance system) to an inductor connected between a live rail and earth.

So far, this section has been concerned solely with the relationship between passive mechanical systems and their

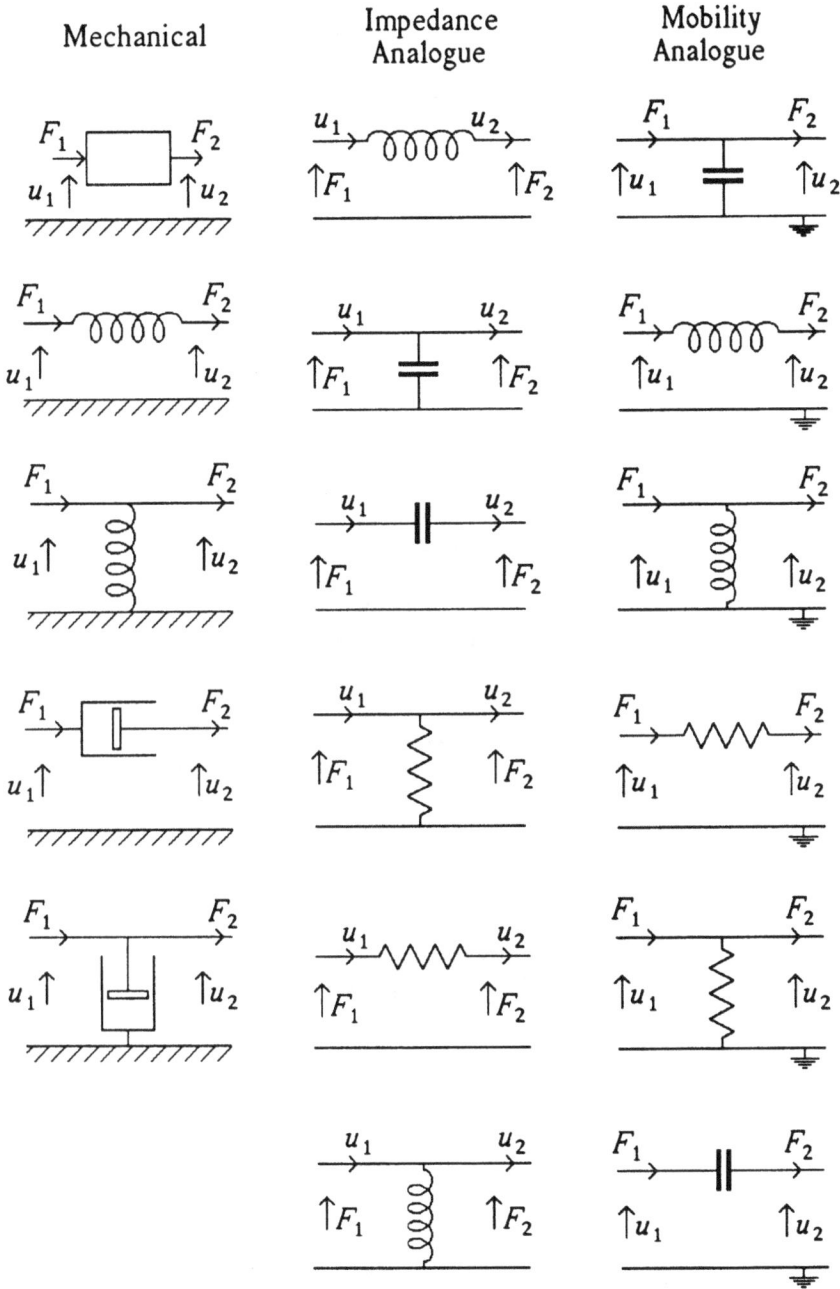

Fig 4.14 Equivalent mechanical and electrical components, for impedance and mobility systems of analogues.

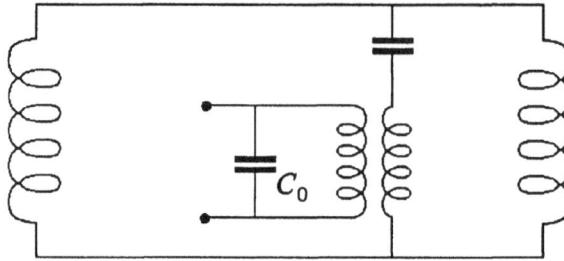

Fig 4.15   Equivalent circuit showing added piezoelectric effect.

electrical equivalent circuits. The piezoelectric effect is now introduced by noting that an applied field generates a force in the stack in addition to the elastic force due to its purely mechanical displacement. It is therefore represented by a voltage in series with C_1, but since this voltage is derived from an electrical input it is usual to apply it by means of an idealised electromechanical transformer as in *Fig 4.15*, which corresponds to the simple resonator of *Fig 4.11*. The capacitor C_0 represents the clamped (dielectric) capacitance of the piezoelectric ceramic in the stack.

The concepts above can be used to establish the relationship between the internal construction of a transducer and its equivalent circuit. This can be useful in predicting the behaviour of a particular transducer, or in deriving what construction would correspond to a desired equivalent circuit, and we shall return to these aspects in a later chapter. This relationship is not however of critical importance in using its equivalent circuit to analyse the electrical behaviour of a transducer as part of an overall system. It is to this aspect of the system design that we return in the next chapter.

4-10   SUMMARY

The characteristics of an electro-acoustic transducer can be represented by means of a lumped-component electrical equivalent circuit, the components of which may be derived solely from electrical measurements on the transducer, provided that only one resonance occurs in the frequency range of interest (Eqns 4.13–4.17). This equivalent circuit can be used to analyse the variation of electrical input admittance of the transducer over a band of frequencies around its resonance (Sect 4.4). The efficiency of the transducer, and its dependence on frequency, can be derived from measurements in air and water (Sect 4.5). No knowledge of the detailed internal construction of the transducer is involved, but a link with the main mechanical or acoustic output parameters is given by equations (4.29) and (4.30).

A more detailed equivalent circuit which is associated with the internal construction of the transducer can be obtained from the analogues shown in *Fig 4.14*. This equivalent circuit, calculated from a knowledge of the transducer design, is used mainly for predicting its characteristics, whilst that derived from measurements on the transducer is used primarily for analysing its performance.

REFS Chapter 4

References

4.1   IRE Standards on Piezoelectric Crystals: The Piezoelectric Vibrator – Definitions and Methods of Measurement, 1957. *Proc IRE*, 45, 353–358, (1957). (IEEE Std 177-1966).
4.2   Woollett,R.S., "Effective coupling factor of single-degree-of-freedom transducers." *J Acoust Soc Am*, 40, 1112–1123, (1966).
4.3   IRE Standards on Piezoelectric Crystals: Determination of the Elastic, Piezoelectric, and Dielectric Constants – The Electromechanical Coupling Factor, 1958. *Proc IRE*, 46, 764–778, (1958). (IEEE Std 178-1985).
4.4   Kinsler,L.E., and Frey,A.R., *Fundamentals of Acoustics*, Wiley & Sons, 2nd Ed, 1962, Chapter 12.
4.5   Beranek,L.L., "Some remarks on electro-mechano-acoustical circuits." *J Acoust Soc Am*, 77, 1309–1313, (1985).
4.6   Skudrzyk,E., *Simple and Complex Vibratory Systems*, Pennsylvania State University Press, 1968, Chap 1.

Additional Reading

1   Beranek,L.L., *Acoustics*, McGraw-Hill, 1954, Chapter 3. (Revised edition published by Acoust Soc Am, 1988.)
2   Redwood,M., and Rodrigo.G.C., *Analysis and Design of Piezoelectric Sonar Transducers*, Thesis, Queen Mary College, London, 1970.

Chapter 5
BANDWIDTH

5-1 DEFINITIONS

The electrical to motional efficiency of a transducer has a peak around the resonance frequency, and transmitting transducers are therefore usually operated within a band of frequencies around resonance. The width of this operating band depends not only on the transducer itself, but also on the associated drive amplifier and the nature of the application, and as a result various definitions of the concept of bandwidth have been employed. In this chapter we shall discuss these definitions and argue that one based on overall system considerations should generally be adopted. This concept then leads to methods for determining the required electrical characteristics of a transducer.

In discussing the equivalent circuit in the previous chapter we have used the frequencies (f_1, f_2) at which the motional conductance is half its peak value to evaluate the motional Q-factor. The frequency difference $f_1 - f_2$ may be used as the basis for one definition of the bandwidth, in which it represents the range within which the motional output power is greater than 50% of its value at resonance. Assuming the motional to acoustic efficiency to be constant over this range, this corresponds to the band within which the output *acoustic* power is within 3dB of its peak value, when a constant voltage (independent of frequency) is applied to the transducer. This definition involves the calculation of the motional conductance, and an alternative definition may be adopted which is based on the frequencies at which the *total* input conductance is half of its peak value. This definition avoids the need to subtract the clamped component in determining the motional conductance, and is therefore easier to calculate, although its basis does not appear to be so well founded. Either of these definitions can be used, even if the admittance loop is distorted by the presence of other resonances, although it may not be simple to derive an equivalent circuit in that case.

These definitions refer to the output power without reference to its directional distribution. However, it is sometimes more important to base the definition on the acoustic intensity in a particular direction, normally the acoustic axis. This leads to a definition of bandwidth in terms of the frequency band within which the source level lies within 3dB of its maximum value. It is usual to assume that the transducer is driven by a source of constant voltage, and to normalise this to one volt, to obtain the projector sensitivity per volt, S_V. The bandwidth is thus interpreted as the frequency range within which S_V is within 3dB of its maximum. The bandwidth obtained by this method will differ to some extent from that defined in terms of the total acoustic power because of the increase in directivity as the frequency increases, although the difference is not usually very great. Some authors have used similar definitions but based on a 6dB (or even 10dB) drop from the peak value. It is also sometimes more appropriate to use the projector sensitivity per ampere (S_I), for a constant current source, and define the bandwidth in terms of the frequencies for a 3dB (or 6 or 10dB) drop in S_I. An equalising network may also be applied to the input of the amplifier to compensate for the variation in projector sensitivity, and thus make the output power more constant with frequency. The "bandwidth" may then be defined in terms of the frequency range in which the source level for the combined system of amplifier (including equalising network) and transducer is within 3dB of its peak value.

These definitions assume that it is feasible to apply a constant voltage (or constant current in appropriate cases) to the transducer over the frequency range of interest. However, the transducer will have a significant variation of admittance over the same frequency range, so the question will arise of whether this constant voltage can be maintained in practice. The answer to that question will depend on the nature of the application. If the requirement is only for a low power output, it may well be quite practicable to use a driver amplifier with sufficient power reserve to cope with the excursions of the load impedance and maintain the voltage constant to a good enough approximation. The definitions above may then be useful, and the choice of which one to use is likely to depend mainly on personal preference or ease of measurement. However, when such reserves of power are not available, it may be more appropriate to consider efficiency as a criterion, and define bandwidth as the range of frequencies within which the efficiency exceeds some chosen value.

It is clear that several distinct definitions or interpretations of bandwidth are possible, and have been used by various authors. Although it is usually not made explicit, these definitions have limited applicability unless ample reserves of power are available. For more demanding applications, it is necessary to consider the transducer as an element in the overall system and consider how power is transferred from the source of electrical power to be dissipated as radiated acoustic power. The bandwidth is then more meaningfully defined in terms of the overall performance of the system, and this will be the topic for the rest of this chapter.

5-2 ANALYSIS AS A FILTER

A piezoelectric transducer at resonance has an admittance represented by a resistance in parallel with the blocked capacitance. In order to allow the maximum power to be transferred from the source to the transducer, it is normal to tune out its capacitance by means of a parallel inductance, so that the load at resonance appears purely resistive and can be matched to the output resistance of the source. In this section, we shall therefore assume the transducer to be fitted with a parallel tuning coil, and derive the equations describing the electrical input admittance and its dependence on frequency. The results of the matching between this admittance and the source impedance will be discussed in Section 5.4.

The effect of a tuning inductor L_0 in parallel with the capacitance C_0 in the equivalent circuit of *Fig 4.6* is to add an additional contribution of $1/j\omega L_0$ to the admittance given by equation (4.19), so that the input susceptance becomes

$$B_{in} = \frac{-1}{\omega L_0} + \omega C_0 - \frac{\omega_S C_1 Q_M^2 \Omega}{1 + Q_M^2 \Omega^2} \qquad (5.1)$$

If L_0 and C_0 are tuned to resonate at the same frequency as the series arm, so that $\omega_S{}^2 = 1/L_0 C_0$, the expression for B_{in} becomes

$$B_{in} = \omega_S C_0 \Omega - \frac{\omega_S C_1 Q_M^2 \Omega}{1 + Q_M^2 \Omega^2} \qquad (5.2)$$

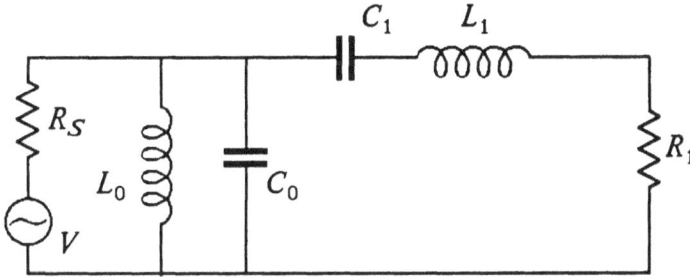

Fig 5.1   Equivalent   circuit   of   tuned   transducer   as band-pass filter network.

In this case where the parallel combination resonates at the same frequency as the series arm, the circuit resembles a half-section band-pass filter, as in *Fig. 5.1*, in which the dielectric loss resistor R_e has been omitted for simplicity since its effect on the electrical behaviour is usually small. The driving amplifier is represented by a voltage source V with its source resistance R_S. This circuit may be analysed by means of established   filter   methods   (eg   [5.1])   using   normalised parameters which can be derived from those defined earlier for transducers by the expressions:-

Bandwidth Factor $\qquad W^2 = \dfrac{C_1}{C_0} = \dfrac{k^2}{1 - k^2} \qquad (5.3)$

Frequency Variable $\quad\quad y = \dfrac{1}{W}(f/f_s - f_s/f)$

$$= \dfrac{\Omega}{W} \quad\quad\quad (5.4)$$

Nominal Resistance $\quad\quad R_N = \dfrac{1}{\omega_s W C_0}$

$$= \dfrac{W}{\omega_s C_1} = \omega_s L_1 W \quad\quad (5.5)$$

Matching Factor $\quad\quad \beta = \dfrac{R_1}{R_N} = \dfrac{1}{Q_M W} \quad\quad (5.6)$

Using these parameters, the impedances of the circuit in *Fig. 5.1* may be expressed in normalised form. Thus, the impedance of the series *LCR* arm may be written as

$$Z_1 = R_1 + j\omega L_1 - \dfrac{j}{\omega C_1}$$

$$= R_1 + j\omega_s L_1 \Omega$$

$$= \beta R_N + jy R_N \quad\quad (5.7)$$

The admittance of the parallel combination of C_0, L_0 becomes

$$Y_0 = \dfrac{1}{j\omega L_0} + j\omega C_0$$

$$= j\omega_s C_0 \Omega$$

$$= j\dfrac{y}{R_N} \quad\quad (5.8)$$

The input admittance (Y_{in}) of the circuit of *Fig 5.1* is thus given in normalised form by

$$Y_{in} = G_{in} + jB_{in}$$

$$= j\frac{y}{R_N} + \frac{1}{\beta R_N + jyR_N}$$

$$= \frac{1}{R_N} \frac{\beta}{(\beta^2 + y^2)} + j\frac{y}{R_N} \frac{(\beta^2 + y^2 - 1)}{(\beta^2 + y^2)}$$

Thus

$$G_{in} = \frac{1}{R_N} \frac{\beta}{(\beta^2 + y^2)} \qquad (5.9a)$$

$$B_{in} = \frac{1}{R_N} \frac{y(\beta^2 + y^2 - 1)}{(\beta^2 + y^2)} \qquad (5.9b)$$

Normalised admittance curves can be derived by plotting $\beta R_N B_{in}$ against $\beta R_N G_{in}$ as in *Fig 5.2*. Curves are shown for values of β from 0.6 to 1.2, and illustrate how the shapes of the curves depend on the value of the matching factor β. For $\beta = 1$, the curve is in the form of a cusp; when $\beta < 1$, the curve forms a loop; and when $\beta > 1$ the loop disappears to give a monotonic variation with frequency. The frequency variable y is used as the variable along the admittance curve itself, and the variation of the normalised frequency parameter y/β is indicated by the dashed vertical lines in the figure. The phase angle of the admittance at any frequency is given by

$$\tan\theta = \frac{B_{in}}{G_{in}}$$

$$= y(\beta^2 + y^2 - 1)/\beta \qquad (5.10)$$

It is of interest at this stage to note several features of these curves, and some approximations. The shape of the curve is determined by the parameter β, ie by the value of $Q_M W$, whilst the scaling of the axes is controlled by βR_N, ie by $R_1 = 1/\omega_S k^2 C_{LF} Q_M$. At the resonance frequency, $y = 0$ and $B_{in} = 0$, as expected for a correctly tuned transducer. There are

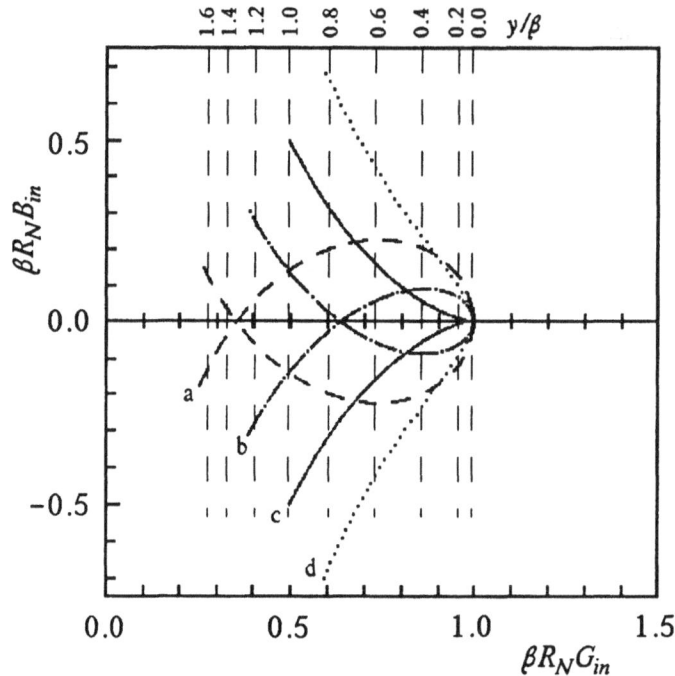

Fig 5.2   **Admittance curves for band-pass filter: (a) $\beta = 0.6$;
(b) $\beta = 0.8$; (c) $\beta = 1.0$; (d) $\beta = 1.2$.**

two further solutions for $B_{in} = 0$, provided that $\beta^2 < 1$; these occur when $\beta^2 + y^2 = 1$ and correspond to the frequencies at which the curve crosses the conductance axis. The conductance at resonance has the value $G_{in} = 1/\beta R_N$. The half-conductance points are easily shown to occur at the frequencies given by $y^2 = \beta^2$, ie $y = \pm\beta$ and hence $\Omega = \pm 1/Q_M$. The frequencies corresponding to the values $y = \pm 1$ are usually called the nominal filter cut-off frequencies in standard filter design texts. For small frequency deviations, the frequency variable Ω may be approximated by

$$\Omega = (f/f_S) - (f_S/f) \simeq 2(f - f_S)/f_S \qquad (5.11)$$

Also, if $k^2 \ll 1$, the approximation $W \simeq k$ can be made with reasonable accuracy. For example, the error is approximately 5% for $k = 0.3$, and 13% for $k = 0.5$.

Equations (5.7) and (5.8) allow us to represent the equivalent circuit in the generalised form shown in *Fig 5.3*; the

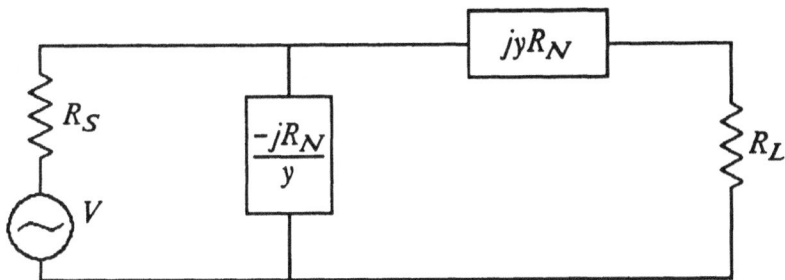

Fig 5.3 Generalised half-section band-pass filter network.

voltage source is represented by V in series with its internal source resistance R_S. The characteristics of such a filter depend on the terminating resistors, and the network shown in *Fig 5.3*, in which the resistors are related to the design resistance R_N by the factors β and $1/\beta$, is treated in standard filter texts (eg Ref 5.1). This arrangement is sometimes known as "rho" or "anti-rho" matching, with the matching factor denoted by rho, but we have avoided the use of rho here as it is to be used for other parameters. This representation of the circuit emphasises how the transducer acts as part of the complete system in converting electrical to acoustic power, and can be analysed in terms of a band-pass filter.

The power delivered into the motional arm is derived by calculating the current i_m into the series LCR arm. This is obtained from the transfer admittance, which is given by

$$\frac{i_m}{V} = \frac{1}{R_N} \frac{\beta}{1 + \beta^2 - y^2 + j2\beta y} \qquad (5.12)$$

The modulus of this transfer admittance, and hence the magnitude of i_m is given by

$$\left| \frac{i_m}{V} \right| = \frac{1}{R_N} \frac{\beta}{\left[(\beta^2 + y^2 - 1)^2 + 4\beta^2 \right]^{1/2}} \qquad (5.13a)$$

and the phase angle by

$$\tan\theta = \frac{-2\beta y}{1 + \beta^2 - y^2} \qquad (5.13b)$$

Since the circuit is linear, the principle of reciprocity holds, and the same expressions apply if the positions of i_m and V are interchanged. This can be useful when calculating the receiving response of the transducer, in which case the output current is multiplied by R_N/β to obtain the output voltage.

Fig 5.4 shows curves to illustrate the variation with frequency of the transfer admittance for values of β ranging from 0.6 to 1.2. These demonstrate how the pass band of the filter may be increased by using a value of β below unity if some ripple in the pass band is allowed. For example, $\beta = 0.8$ gives a pass band significantly wider than for a filter terminated by the design resistance R_N (ie for $\beta = 1$). In the next section we shall see that considerations of the variation in admittance also often lead to a choice of $\beta \simeq 0.8$ for the filter elements.

5-3 TRANSDUCER MATCHING

The curves in *Fig 5.2* illustrate the variation with frequency of the admittance of a transducer, and show how the shape of the admittance loop around resonance is greatly affected by the

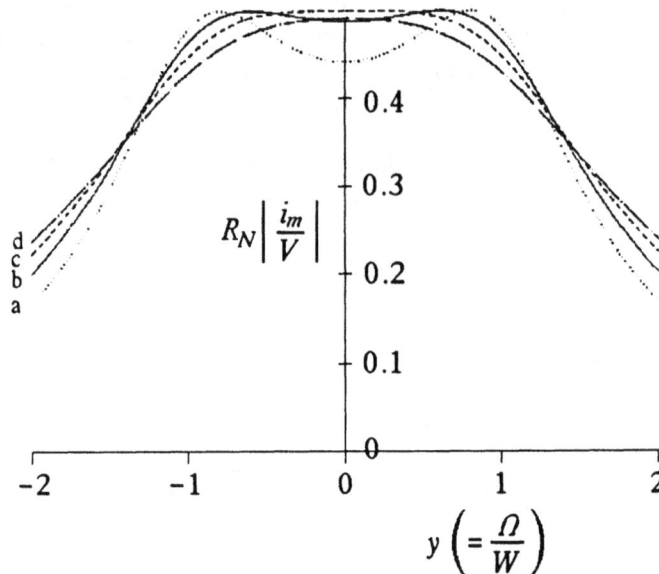

Fig 5.4 Transfer admittance of half-section band-pass filter.
(a) $\beta = 0.6$; (b) $\beta = 0.8$; (c) $\beta = 1.0$; (d) $\beta = 1.2$.

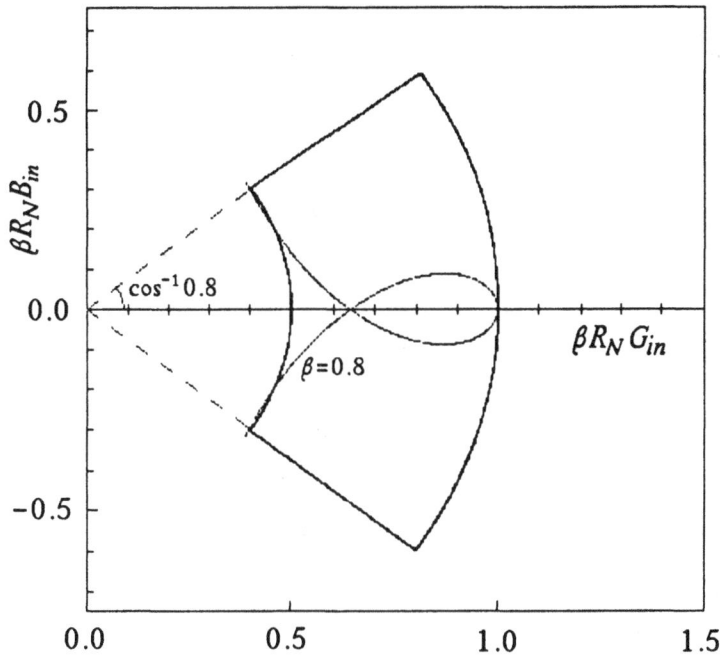

Fig 5.5   Optimum admittance curve, β = 0.8.

value of the matching factor β (= $1/WQ_M$). The transducer admittance represents the load on the driving amplifier, and the choice of matching factor for a particular application will depend on the characteristics and constraints of the amplifier design. If ample reserves of power are available, matching the load to the amplifier may not be important, but if the power output from the amplifier is to be maximised over as wide a band as possible the choice of β is critical. For these more demanding applications, the amplifier designer needs to specify the limits of the load which can be presented to the amplifier, and the frequency band over which the transducer admittance lies within these limits can then be maximised. Typical design limits are that the admittance of the load should be such that its modulus lies within a range of 2:1, and that its power factor is always greater than 0.8. Examination in more detail of curves such as those in *Fig 5.2* shows that the maximum range of frequency for which the admittance satisfies these limits is achieved by selecting a value of $\beta \simeq 0.8$. This is illustrated in *Fig 5.5*, in which the limits of modulus and power factor are shown bounding the admittance curve for $\beta = 0.8$. The filter response curves of *Fig 5.4* show that a value of $\beta = 0.8$ gives a relatively flat response over the same bandwidth, and this provides another good reason for the choice. It is common to express this optimum value in the alternative form of $WQ_M = 1.25$; expressed in terms of the coupling coefficient parameter, this gives the approximate but important relationship $kQ_M \simeq 1.2$.

This condition thus gives good matching between the transducer and the driving amplifier, and hence maintains the power output within a specified deviation from the maximum, over the widest possible band. When the optimum matching is achieved, the fractional bandwidth is approximately equal to the coupling coefficient. This result shows the fundamental importance of the coupling coefficient in determining the bandwidth of a transducer, since a wide bandwidth can only be obtained by means of a transducer with a high effective coupling coefficient. However, to achieve the optimum, it is also necessary to realise a mechanical Q-factor which matches this value of k. For example, if the effective coupling factor of a transducer is 0.3, which is typical of a barium titanate thickness-mode design, a mechanical Q_M of 4 is required, and this would result in a fractional bandwidth of about 0.3. A wider bandwidth could be achieved using lead zirconate titanate ceramic, for which an effective coupling coefficient of 0.5 might be practicable, but in this case the mechanical Q-factor would need to be reduced to 2.4. A value of Q_M below the optimum would cause large phase angles at the edges of the band, whilst too high a Q_M would cause excessive variation in the admittance modulus. This emphasises that, in addition to the need for the coupling coefficient to be high enough to meet the requirement for the bandwidth, it is also necessary to design the transducer to have the matching value of Q_M actually to achieve the potential bandwidth. It is sometimes difficult to design a transducer with a Q_M as low as 2.4, and if the minimum Q_M is greater than 2.4, then the optimum k is below 0.5, and the stack design should be adjusted to give this optimum. This may well imply deliberately reducing the coupling coefficient of an LZT stack by introducing spacers, or possibly using barium titanate instead of lead zirconate titanate.

The optimum value of kQ_M applies to the rather idealised admittance loop corresponding to a simple equivalent circuit with known parameters derived as in Chapter 4. In practice, further variations in the position and shape of the admittance curve arise because of a variety of effects such as:-

a. changes in the transducer parameters caused by changes in environmental conditions, such as temperature or pressure, and aging of the ceramic.

b. variations in the transducer parameters arising from manufacturing tolerances.

c. variations in radiation impedance caused by interactions between individual transducers in an array.

The sensitivity of piezoelectric ceramics to temperature and pressure depends on the particular ceramic used, as described in Chapter 3. Typical values for a barium titanate ceramic have been used to calculate the variations in the transducer parameters (eg C_{LF}, k, f_S) for a range of operating temperatures and depths which might be expected in service. To these effects are added a further contribution from the aging of the ceramic. Although this differs from the environmental factors in that aging could in principle be calculated and allowed for, in practice if arrays are built up from numbers of elements it is virtually impossible to know and compensate for the aging of all the elements. It is therefore convenient to include aging with the environmental changes.

Further contributions arise from the tolerances which have to be specified to allow for errors and variability during manufacture. The magnitude of these variations depends on the suitability of the design for easy production and on the effort and cost devoted to meeting tight tolerances. They are therefore to some extent controllable, but can have major implications on the cost of a large array. From the performance point of view, the tolerances need to be determined by taking into account their effect on the acoustic characteristics of the array, and care should be taken to avoid making them unnecessarily stringent.

In many applications transducers are assembled into arrays, and in those cases the acoustic pressure generated by any one transducer produces forces on the faces of all other transducers in the array. This acoustic interaction changes the radiation loading on the elements of the array, and is discussed in more detail in Chapter 6. If the elements are all connected in parallel, the interaction effects are effectively averaged in their contributions to the total load presented to the amplifier. If however the elements are driven by separate amplifiers, the variations due to these interactions need to be included with those described above, when considering the load on each amplifier. It is shown in Chapter 6 that these array interactions

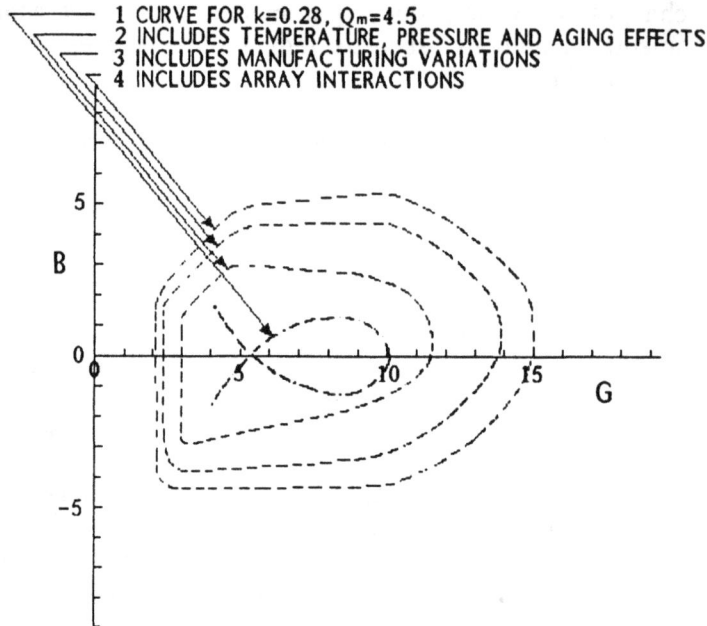

Fig 5.6   Effects of variations on admittance loop envelopes.

are generally small if the separation between neighbouring
elements exceeds about half an acoustic wavelength in water,
but for closer spacings the effects can become very important.

An example of these variations for an array of barium
titanate transducers is shown in *Fig 5.6*. It is necessary to
convert the predicted changes in parameters such as capacitance
and resonance frequency into changes in circuit components
using the equations given in Section 4.3, and then use equation
(4.19) to evaluate the admittance loop for any particular set of
environmental conditions etc. For an environmental change such
as a rise in temperature, the resulting changes in the parameters
are correlated; for this example of a barium titanate transducer,
the capacitance and coupling coefficient both decrease, whilst
the resonance frequency increases. For manufacturing variations
and array interactions, no such correlation was assumed. The
spacing between elements in this example was about half a
wavelength, and the array interaction effects then accounted for
a variation of about ±20% in radiation resistance. Manufacturing
variations were fairly tightly controlled, so that for example
resonant frequencies were within 1% and capacitances within 5%.

Fig 5.6 shows the envelopes surrounding the admittance curves, to illustrate the effects of environmental changes and the further contributions of the less predictable manufacturing and array interaction effects. The dashed curves showing the envelopes of the individual curves thus indicate the overall range of admittances which is likely to be presented to the amplifier as a result of uncontrollable variations during manufacture or in service. In this quite typical example, the modulus of the admittance may vary over a range of about 6:1, and the phase angle over ±60°. These variations are much larger than is generally appreciated, and although some techniques to reduce them can be envisaged it is in practice difficult to effect any great reduction. It is important for the design of the driving amplifier to allow for this range of possible loads, which is much larger than the nominal spread of 2:1.

To some extent this may appear to weaken the arguments above for determining an optimum value of kQ_M, but the variation in admittance for the nominal transducer is in itself a major contributor to the overall variation. It is therefore usually advisable to aim for the optimum value of $kQ_M \simeq 1.2$, as any other value will tend to increase the range of either modulus or phase angle even further. Although it is evident that it is difficult to be precise about the actual range of loads on the amplifier, it should be emphasised that the amplifier design must be able to withstand this range without failure, otherwise catastrophic breakdown of the system may occur.

This requirement to avoid breakdown of the amplifier, though essential, is only one of the criteria for safe operation of the system. If the amplifier acts as a constant voltage source, a range of 6:1 in modulus of the load would imply a similar variation in input current into the transducer, and for a correctly tuned device this would correspond to about the same range of motional currents. Since these are equivalent to piston velocities, these large variations in input current could cause mechanical breakdown of the transducer. Similarly, if the amplifier behaves as a current source, the wide range of applied voltages may cause electrical breakdown of the transducer. These variations would need to be taken into account in designing the transducer, effectively by increasing the safety factors in the construction, unless a way could be found to reduce the range of input current and voltage. One method of controlling the maximum current or voltage is to use limiting devices – such as a biassed diode to control voltage – but these have the great disadvantage that they introduce non-linearity into the circuit, and may therefore cause distortion of the output waveform. A better method is to make the source impedance resistive and comparable with the impedance of the transducer. It is well known that maximum power transfer into a resistive load is achieved by making the load resistance equal to the source resistance, and a similar condition applies if the load is complex. In that case, the reactance should be tuned out to make the load on the source appear resistive, and the value of the load made equal to the source resistance. In addition to maximising the power transfer, this also has advantages in limiting the excursions of current or voltage from the nominal values, since it is easy to show that neither current nor voltage can exceed twice the nominal value (ie the value when correctly matched) for any positive value of the load resistance. It is not usually necessary to make explicit the condition that the resistance should be positive, but we shall see later that it is possible in some circumstances for elements in an array to have negative resistive components arising from the effects of array interactions. Apart from meeting this rare eventuality, however, a safety factor of two applied to the nominal values should be adequate for the transducer design when the system has its drive impedance nominally matched to the load.

The two previous paragraphs have been concerned with preventing catastrophic failure of the amplifier or transducer. In addition, it is usually necessary to meet certain acoustic requirements over the full operating bandwidth, not merely at the resonance frequency. The "power-matched" case above, in which the impedance of the transducer is matched to the source resistance, is an effective way of limiting the deviations from the nominal behaviour, since variations in the load cause relatively small changes in the output power. The application of this in controlling array interactions will be discussed in more detail in Chapter 6. For the present, we shall consider how the power into a complex load depends on the range of its admittance values. If a generator of voltage V and source resistance $R_S = 1/G_S$ works into a complex admittance $Y = G + jB$, the power (W) delivered into the load is given by the equation

$$G^2 + B^2 + 2G_S(1 - 2W_{max}/W)G + G_S^2 = 0 \qquad (5.14)$$

where W_{max} is the maximum power obtainable from the source, ie $W_{max} = V^2 G_S/4$. For any given value of W/W_{max} this equation represents a circle having its centre at $G_S\{(2W_{max}/W) - 1\},0$ and radius $2G_S(W_{max}/W) \times (1 - W/W_{max})^{1/2}$. The power into any load with an admittance lying within the circle for a chosen value of W/W_{max} will be between W_{max} and W. For

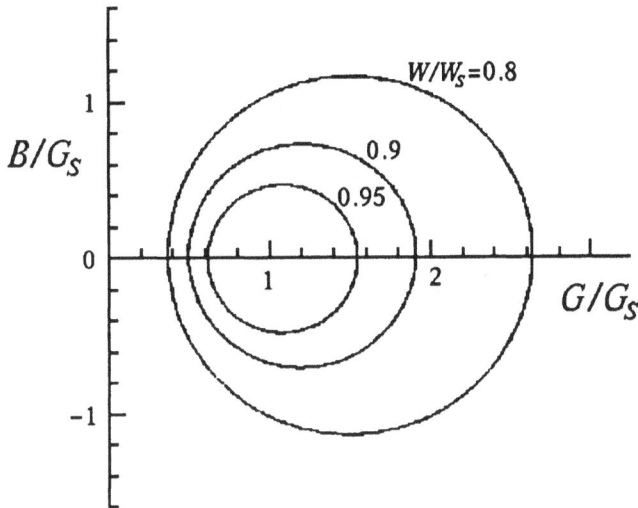

Fig 5.7   Constant power circles in admittance plane.

maximum power transfer the allowable load shrinks to a point at
$G_S,0$; as the permitted drop in power is increased, the circle
enclosing the allowable range of admittances increases in radius
and moves outwards along the G-axis, as illustrated in *Fig 5.7*.

5-4 BROAD-BAND NETWORKS

The analysis of Section 5.2 is based on the treatment of
the equivalent circuit of a tuned piezoelectric transducer as
a half section band-pass filter. It is reasonable to ask whether
the performance of the system could be improved by extending
the filter network, and in this section we shall show that some
improvement is indeed possible. For this it is convenient again
to make use of the methods of filter analysis, as described in
texts such as those listed at the end of the chapter. In
particular, we shall consider methods which can be applied to
maximise the bandwidth of a band pass-filter, a good treatment
of which is given by Green [5.1]. As in the earlier parts of this
chapter, this will be discussed initially as a network problem,
without referring specifically to transducer characteristics. This
section may be omitted by the reader who has no especial
interest in maximising the operating bandwidth of the transducer.

In this network analysis it is conventional to approach the design of a band–pass filter by first considering the design of its low–pass "prototype". The purpose of the system is to supply power from a source to a load. The source is assumed to be a voltage generator with a series resistance R_S, and the load to have an impedance Z_L. A coupling network can be connected between the source and the load to maximise the power transfer, and the fundamental problem in designing a broad band system is to devise a network which maximises the band over which the power transfer satisfies some specified limit. This coupling network is assumed to be composed of ideal components which are themselves lossless, so that no power is dissipated in the coupling network itself. The treatment makes use of a "reflection coefficient" ρ, defined by

$$\rho = \frac{Z_L - R_S}{Z_L + R_S} \tag{5.15}$$

This definition is such that $|\rho|^2$ is the fraction of the power reflected back to the source by the mismatch of the load. Thus, if the impedance matching is good, the value of $|\rho|$ approaches zero; for very poor transmission through the network (eg in the stop band), $|\rho|$ approaches unity.

The low–pass form of the load Z_L can be represented by a capacitor C in parallel with a resistor R (or an inductor in series with a resistor). For this low–pass case, Bode [5.2] showed that there was a fundamental limit to the matching which could be achieved, this being given by the expression

$$\int \ell n \frac{1}{|\rho|} d\omega \le \frac{\pi}{RC} \tag{5.16}$$

We noted above that $|\rho|$ should be small in the pass band, and tend towards unity in the stop band. Thus, $\ell n(1/|\rho|)$ should have a finite value in the pass band, and tend towards zero in the stop band. The integral in (5.16) represents the area under the curve of $\ell n(1/|\rho|)$ against ω, and shows that this area cannot exceed a value of π/RC. If $|\rho|$ is made constant, at a value of $|\rho|_{max}$, over a pass band of width ω_B, and has the ideal value of unity elsewhere, then (5.16) becomes

$$\omega_B \ell n \frac{1}{|\rho|_{max}} = \frac{\pi}{RC} \tag{5.17a}$$

Or,

$$|\rho|_{max} = \exp(-\pi/RC\omega_B) \qquad (5.17b)$$

Thus, if the bandwidth ω_B is increased, the corresponding value of $|\rho|_{max}$ must also increase — ie the matching must be less good. Or conversely, if a greater degree of mismatch is permitted, a wider bandwidth is possible. The assumption of a constant value of $|\rho|$ through the pass band and unity elsewhere represents an ideal case, rather like an ideal low-pass filter response with an infinitely sharp cut-off. In a more realistic case, there will be some variation of $|\rho|$ in the pass band. If $|\rho|_{max}$ indicates the maximum allowable reflection coefficient in the band, then variations must all be towards smaller values of $|\rho|$, ie towards larger values of $\ell n(1/|\rho|)$. These make larger contributions to the integral in (5.16), and thus reduce the pass band. To achieve a specified pass band of ω_B, we must therefore determine the corresponding value of $|\rho|_{max}$ and seek to make the actual value of $|\rho|$ as nearly equal to $|\rho|_{max}$ as possible over the band. Achieving a better match at some frequency will reduce the overall bandwidth.

The general problem of how to design a network according to these principles was treated by Fano [5.3], and is well summarised by Green [5.1]. Achieving the ideal response with a constant value of $|\rho|$ in the pass band would need an infinite number of filter sections. In practice, it is only worthwhile to use a few sections, the resulting performance becoming nearer the ideal as more sections are used. We shall consider only three or four section filter networks, with some reference to two section networks since they represent the circuits already discussed earlier in the chapter. A three section ($n = 3$) low-pass prototype is shown in *Fig 5.8a*, and a four section ($n = 4$) prototype in *5.8b*. The number of sections, and hence the value of n, is equal to the number of reactive elements in the low-pass prototype. The load is in this case represented by an *LR* series combination, since this will turn out to be more convenient later, and is normalised so that R_1 is equal to unity; L_1 is then the corresponding normalised load inductance. The basic approach is then to devise a filter having a Tchebychev response — ie a specified ripple within the pass band, and a smooth roll-off outside the band.

In order to derive the networks, Fano needed to evaluate coefficients A_1^∞, A_3^∞, A_5^∞, . . etc. (We shall omit the superscript ∞ from now on.) The first step is to calculate two further coefficients (a,b), which can be obtained from the requirement that the filter shall have a Tchebychev response, since this imposes the condition that a,b must satisfy

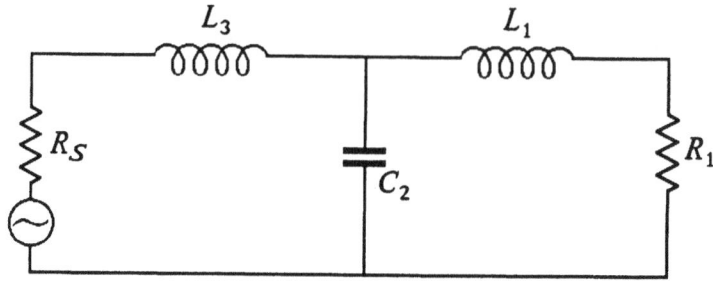

(a) $n=3$

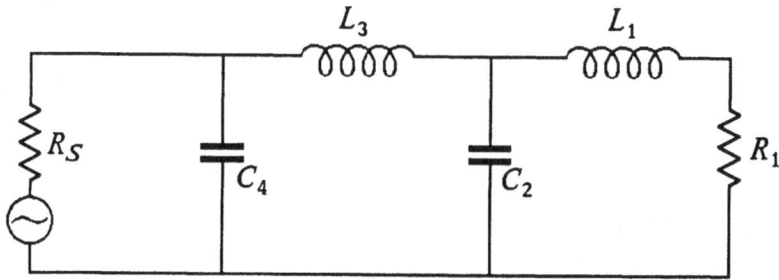

(b) $n=4$

Fig 5.8   **Low pass-filter prototypes: (a) $n = 3$; (b) $n = 4$.**

$$\omega_B \left\{ \frac{\sinh a \; - \; \sinh b}{\sin \left(\pi / 2n \right)} \right\} = A_1 = 2/L_1 \qquad (5.18a)$$

and

$$|\rho|_{max} = \frac{\cosh nb}{\cosh na} \qquad (5.18b)$$

This value of $|\rho|_{max}$ is minimised, subject to the condition given by (5.18a), with the result

$$\frac{\tanh na}{\cosh a} = \frac{\tanh nb}{\cosh b} \qquad (5.19)$$

The parameters a,b are evaluated by solving (5.18a) and (5.19), results being given in Ref [5.3]. These values can then be used to derive the coefficients A_3, A_5, . . . etc, thus:-

$$A_3 = \frac{-\omega_B^3}{2^2} \left\{ \frac{\sinh 3a - \sinh 3b}{3\sin(3\pi/2n)} + \frac{\sinh a - \sinh b}{\sin(\pi/2n)} \right\} \qquad (5.20a)$$

$$A_5 = \frac{\omega_B^5}{2^4} \left\{ \frac{\sinh 5a - \sinh 5b}{5\sin(5\pi/2n)} + \frac{\sinh 3a - \sinh 3b}{\sin(3\pi/2n)} \right.$$

$$\left. + 2\frac{\sinh a - \sinh b}{\sin(\pi/2n)} \right\} \qquad (5.20b)$$

$$A_7 = \frac{-\omega_B^7}{2^6} \left\{ \frac{\sinh 7a - \sinh 7b}{7\sin(7\pi/2n)} + \frac{\sinh 5a - \sinh 5b}{\sin(5\pi/2n)} \right.$$

$$\left. + 3\frac{\sinh 3a - \sinh 3b}{\sin(3\pi/2n)} + 5\frac{\sinh a - \sinh b}{\sin(\pi/2n)} \right\} \qquad (5.20c)$$

etc.

These coefficients are now used to calculate further parameters α_3, α_5, etc, from

$$\alpha_3 = \frac{2^2 A_3}{A_1^3} - 1/3 \qquad (5.21a)$$

$$\alpha_5 = \frac{2^4 A_5}{A_1^5} - 1/5 \qquad (5.21b)$$

$$\alpha_7 = \frac{2^6 A_7}{A_1^7} - 1/7 \qquad (5.21c)$$

The normalised values of the components of the low-pass prototypes in *Figs 5.8* are then given by

$$C_2 = \frac{-L_1}{\alpha_3} \qquad (5.22a)$$

$$L_3 = \frac{-\alpha_3 L_1}{1 + \alpha_3 - (\alpha_5/\alpha_3)} \qquad (5.22b)$$

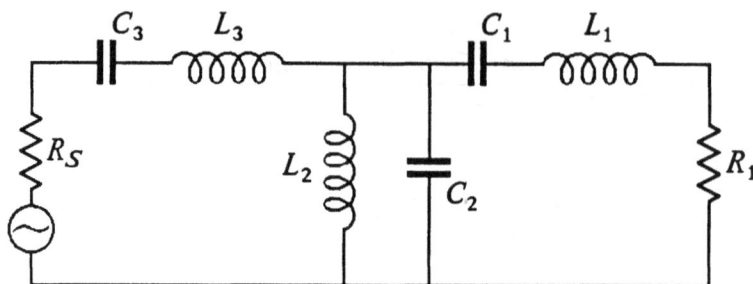

Fig 5.9   3-section band pass-filter network.

$$C_4 = \frac{\{1 + \alpha_3 - (\alpha_5/\alpha_3)\}^2 L_1}{\alpha_3\{1 + \alpha_3 - (\alpha_5/\alpha_3) + (\alpha_5/\alpha_3)^2 - (\alpha_7/\alpha_3)\}} \qquad (5.22c)$$

These equations are sufficient to design a low-pass prototype up to $n = 4$. (Note that R_1 and L_1 represent the given load, after normalisation.)

This low-pass prototype is converted to a band pass-filter by resonating all the L,C components with corresponding C,L elements (in series or parallel), as in *Fig 5.9*. The element values are chosen to make each pair resonate at the required mid-band frequency ω_S, ie:-

$$L_1 C_1 = L_2 C_2 = L_3 C_3 \cdot \cdot \cdot = 1/\omega_S^2 \qquad (5.23)$$

The width of the pass band remains the same as for the low-pass prototype. This is still a network which is normalised to a terminating resistor of unity. If the resistor has the non-normalised value R instead of 1 ohm, then all the impedances must be multiplied by a factor R - ie inductances are multiplied by R and capacitances are divided by R.

Design Procedure.

Application of this technique is simplified by using further coefficients, which have been calculated from the basic equations, giving the curves shown in *Figs 5.10* and *Fig 5.11* (based on [5.1]). These show values of coefficients α_1^2, k_{12}, k_{13}, $(k_{14},)$ and $|\varrho|_{max}$ against a parameter d_1, and can be used to derive the low-pass prototype. The parameter d_1 is defined by

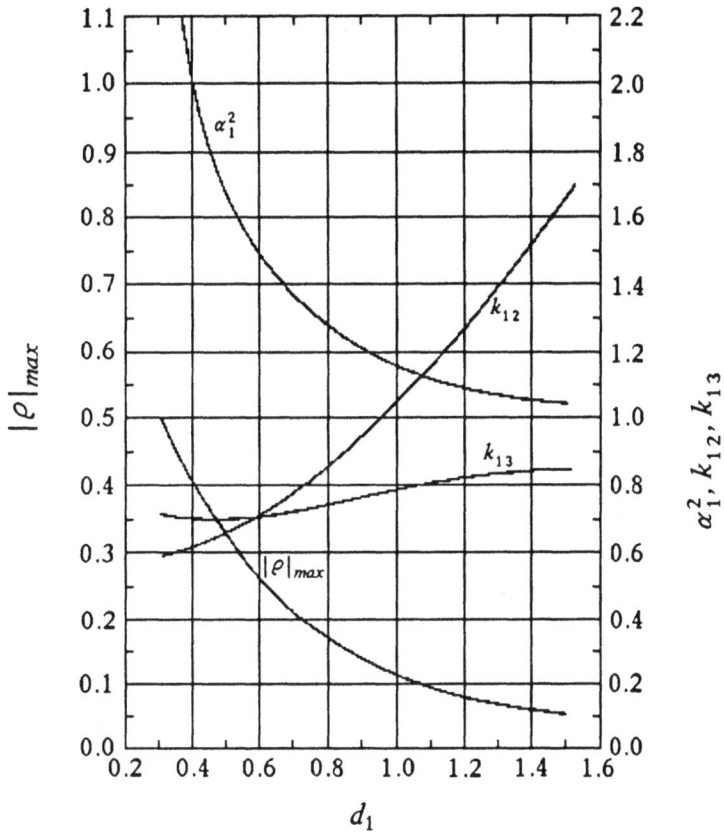

Fig 5.10   Values of parameters for *n* = 3 filter. (From [5.1])

$$d_1 = \frac{R_1}{\omega_B L_1}$$

For a given value of R_1, and a desired bandwidth ω_B, the remaining components are then obtained from the relationships:-

$$L_1 = \frac{R_1}{d_1 \omega_B} \qquad (5.24)$$

$$C_2 = \frac{d_1}{k_{12} R_1 \omega_B} \qquad (5.25)$$

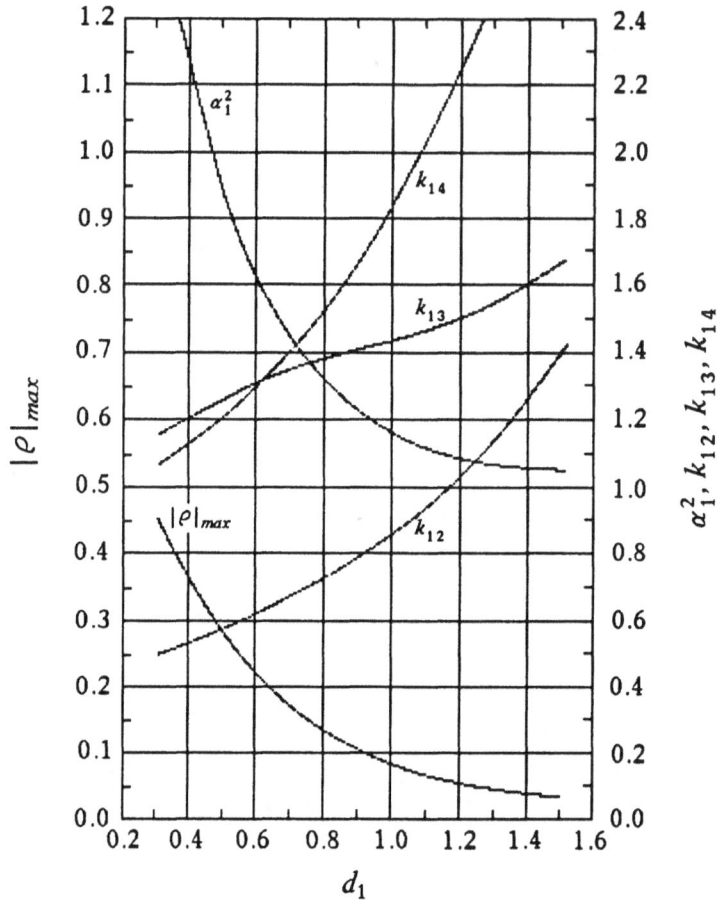

Fig 5.11   Values of parameters for $n = 4$ filter. (From [5.1])

$$L_3 = \frac{k_{13}R_1}{d_1\omega_B} \quad (= k_{13}L_1) \tag{5.26}$$

$$R_S = \alpha_1^2 R_1 \tag{5.27}$$

These are then resonated with the corresponding C,L components to give the band-pass network, with a mid-band frequency ω_S; ie:-

$$C_1 = \frac{d_1\omega_B}{\omega_S^2 R_1} \tag{5.28}$$

$$L_2 = \frac{k_{12}R_1\omega_B}{\omega_S^2 d_1} \quad (= k_{12}L_1(\omega_B/\omega_S)^2) \qquad (5.29)$$

$$C_3 = \frac{d_1\omega_B}{\omega_S^2 k_{13}R_1} \quad (= C_1/k_{13}) \qquad (5.30)$$

For a four section network, two further relationships apply, using the curves in *Fig 5.11*,

$$C_4 = \frac{d_1}{k_{14}R_1\omega_B} \qquad (5.31)$$

$$L_4 = \frac{k_{14}R_1\omega_B}{\omega_S^2 d_1} \qquad (5.32)$$

These equations relate the circuit elements to the desired bandwidth ω_B.

Application to transducers.

When the power is delivered by means of a piezoelectric transducer, the L_1, C_1, R_1 combination is represented by the motional arm of the transducer's equivalent circuit, with the clamped capacitance, C_0 in *Fig 4.7*, contributing to C_2. For the network to be realisable with an actual transducer, C_2 may be greater than C_0, but cannot be less than C_0. In general, the widest bandwidth is obtained when C_1/C_2 is as large as possible, and for a given transducer this is achieved by making C_2 equal to the clamped capacitance C_0, without adding any extra parallel capacitance, ie $C_2 = C_0$. Equations (5.28) and (5.25) can then be substituted into (5.3) to give

$$W^2 = C_1/C_2 \quad (= k^2/(1 - k^2))$$

$$= \frac{d\omega_B}{\omega_S^2 R_1} \cdot \frac{k_{12}R_1\omega_B}{d_1}$$

$$= k_{12}(\omega_B/\omega_S)^2$$

Thus the maximum fractional bandwidth for the transducer is

$$\frac{\omega_B}{\omega_S} = \frac{W}{\sqrt{k_{12}}} \qquad (5.33)$$

This expression may be interpreted as showing either the maximum bandwidth for a transducer of given coupling coefficient, or what coupling coefficient is needed to achieve a specified bandwidth.

An alternative expression may be derived by noting that the mechanical Q_M of the transducer is given by

$$Q_M = \frac{\omega_S L_1}{R_1}$$

$$= \frac{\omega_S}{\omega_B d_1}$$

Thus, an alternative expression for the maximum bandwidth in terms of the motional Q-factor, or the Q-factor needed to achieve a specified bandwidth, is

$$\frac{\omega_B}{\omega_S} = \frac{1}{Q_M d_1} \tag{5.34}$$

In considering the design of a transducer for broad band operation, it is a good starting point to estimate the maximum likely value of the coupling coefficient which can be achieved, since this will determine the greatest bandwidth. Assuming this value of coupling coefficient, and hence of W, (5.33) can be used to express equations (5.24) to (5.32) in terms of W instead of ω_B. This allows us to estimate what bandwidth can be achieved if we use a transducer with a particular value of W, rather than estimating what value of W is needed to achieve a particular bandwidth. For this purpose, the previous equations can be converted to

$$L_1 = \frac{R_1}{d_1 \omega_S} \cdot \frac{\sqrt{k_{12}}}{W} \tag{5.35}$$

$$C_1 = \frac{d_1}{R_1 \omega_S} \cdot \frac{W}{\sqrt{k_{12}}} \tag{5.36}$$

$$L_2 = \frac{k_{12} R_1}{d_1 \omega_S} \cdot \frac{W}{\sqrt{k_{12}}} \tag{5.37}$$

$$C_2 = \frac{d_1}{k_{12} R_1 \omega_S} \cdot \frac{\sqrt{k_{12}}}{W} \tag{5.38}$$

$$L_3 = \frac{k_{13}R_1}{d_1\omega_S} \cdot \frac{\sqrt{k_{12}}}{W} \qquad (5.39)$$

$$C_3 = \frac{d_1}{k_{13}R_1\omega_S} \cdot \frac{W}{\sqrt{k_{12}}} \qquad (5.40)$$

$$Q_M = \frac{\sqrt{k_{12}}}{Wd_1} \qquad (5.41)$$

and
$$R_S = \alpha_1^2 R_1 \qquad (5.27)$$

Suppose that the admittance variation is required to be within a range of 2:1. We shall show later that this corresponds to a value of $|\rho|_{max}$ = 0.172. Then the design process for a three section filter would be:–

a)   Assuming $n=3$ and the specified value of $|\rho|_{max}$, read off the corresponding value of d_1, and hence k_{12}, k_{13}, and α_1, from *Fig 5.10*.

b)   Assume a value of W^2. (Typically 0.1 for barium titanate, or 0.3 for lead zirconate titanate thickness mode designs.) Assume also the appropriate value for the required mid–band frequency.

c)   Derive the component values from equations (5.35) to (5.41) in terms of R_1.

d)   The scaling factor is in practice determined by C_2, which can be controlled through the dimensions of the piezoelectric ceramic in the transducer. From equation (4.11)

$$Q_M = 1/\omega_S C_1 R_1$$

$$= 1/\omega_S W^2 C_0 R_1$$

Hence

$$R_1 = \frac{1}{\omega_S W^2 C_0 Q_M}$$

Using (5.41) to substitute for Q_M, this may be expressed in terms of k_{12} and $d1$. Then,

$$R_1 = \frac{d_1}{\sqrt{k_{12}}\,\omega_S W C_2} \qquad (5.42)$$

This value of R_1 is now substituted in the equations obtained in (c), to give the actual component values.

If a four section network is to be used, the same procedure as above can be followed, but with coefficients determined from *Fig 5.11*. The components in the fourth section can be calculated by expressing (5.31) and (5.32) in a form similar to that for (5.35) to (5.41), ie:-

$$L_4 = \frac{k_{14}R_1}{d_1\omega_S} \cdot \frac{W}{\sqrt{k_{12}}} \qquad\qquad (5.43)$$

$$C_4 = \frac{d_1}{k_{14}R_1\omega_S} \cdot \frac{\sqrt{k_{12}}}{W} \qquad\qquad (5.44)$$

The voltage which needs to be applied to the element to produce the desired output is sometimes inconveniently high, requiring extra insulation around the drive cables and in the connectors. It is therefore sometimes desirable to scale the electrical impedances down by inserting a transformer into the circuit. This can be effected by making L_2 act as a transformer as well as a tuning inductor, and in that case the components on one side of the transformer must be multiplied by the square of the turns ratio, as usual. Although this complicates the appearance of the circuit, and the calculations, the principle of the method remains unchanged.

The procedure above is aimed at maximising the bandwidth which can be achieved with a transducer whose coupling coefficient is effectively limited by the piezoelectric material to be used, allowing for some degradation due to practical constructional effects. One of the outputs of the analysis is the required value of Q_M for the transducer (from equation (5.41)). Provided that this value of Q_M can be realised in practice, this will provide the widest bandwidth (defined in terms of the specified $|\rho|_{max}$) achievable with that particular ceramic. It is sometimes desirable to adopt a different approach; to investigate how wide a bandwidth can be achieved using a given transducer. In this case, the values of both W^2 and Q_M are given, and they will not in general match the condition expressed in (5.41) for a specified $|\rho|_{max}$. The procedure can then be modified, to find instead the lowest value of $|\rho|_{max}$ which matches the values of Q_M and W. To do this, we note that (5.41) can be rearranged as $W^2Q_M{}^2 = k_{12}/d_1{}^2$. Optimum values of $k_{12}/d_1{}^2$, derived from *Figs 5.10* and *5.11*, are plotted against d_1 in *Fig 5.12*, and these curves can be used to determine the value of d_1 which

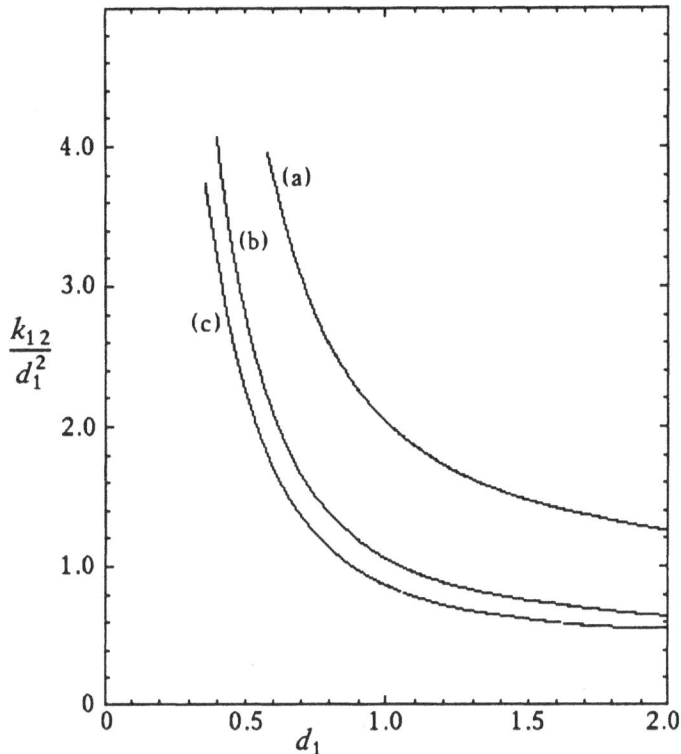

Fig 5.12   Relationship between optimum values of k_{12}/d_1^2 $(= W^2Q_M^2)$ and d_1, to maximise bandwidth for given transducer. (a) $n = 2$; (b) $n = 3$; (c) $n = 4$.

has the value of k_{12}/d_1^2 (equal to $W^2Q_M^2$) for the given transducer. *Figs 5.10* or *5.11* can then be used to evaluate the corresponding value of $|\rho|_{max}$ and the coefficients needed to calculate the remaining circuit elements. The bandwidth appropriate to this $|\rho|_{max}$ is still given by equation (5.34).

Admittance Curves.

Expressions for the admittance curves for band-pass filter networks such as that shown in *Fig 5.9* may be calculated by extending the analysis of Section 5.2 to allow for the extra section(s). For a three section filter network, the result is

$$(R_1 \varrho_M)G_{in} = \frac{(W^2-\Omega^2)(W^2-k_{13}\Omega^2)+\Omega^2\{k_{13}(W^2-\Omega^2)+W^2\}}{(1/\varrho_M)(W^2-k_{13}\Omega^2)^2+\varrho_M\Omega\{k_{13}(W^2-\Omega^2)+W^2\}^2}$$

and

$$(5.45a)$$

$$(R_1\varrho_M)B_{in}$$

$$= \frac{(\Omega/\varrho_M)(W^2-k_{13}\Omega^2)-\varrho_M\Omega(W^2-\Omega^2)\{k_{13}(W^2-\Omega^2)+W^2\}}{(1/\varrho_M)(W^2-k_{13}\Omega^2)^2+\varrho_M\Omega\{k_{13}(W^2-\Omega^2)+W^2\}^2}$$

$$(5.45b)$$

Substituting the normalising parameters used in (5.9), these become

$$G_{in} = \frac{1}{R_N}\frac{\beta}{\beta^2(1-k_{13}y^2)^2 + y^2\{k_{13}(1-y^2)+1\}^2} \qquad (5.46a)$$

$$B_{in} = \frac{y}{R_N}\frac{\beta^2(1-k_{13}y^2) - (1-y^2)\{k_{13}(1-y^2)+1\}}{\beta^2(1-k_{13}y^2)^2 + y^2\{k_{13}(1-y^2)+1\}^2} \qquad (5.46b)$$

For a four section network, it is only necessary to add the contribution to B_{in} due to the added parallel combination of L_4 and C_4. Thus, $(R_1\varrho_M)G_{in}$ is the same as in (5.45a), whilst $(R_1\varrho_M)B_{in}$ is equal to (5.45b) plus an extra contribution of $(k_{12}/k_{14})\Omega/W^2$.

It is instructive to consider the power transmitted into loads represented by these admittance loops. Equation (5.14) gave the equation of the circular locus in the admittance plane for which the power W delivered into the load was a constant fraction of the maximum power W_{max}. This can be related to the parameter $|\rho|$ by noting that $|\rho|^2$ represents the fraction of the maximum input power which is reflected back by the load. Since the network is assumed to be lossless, this implies that the power delivered into the load is $W_{max}(1-|\rho|^2)$.

In case this appears too glib, an alternative proof of the statement can be deduced by a direct calculation of the power into the load. Suppose that a generator of open–circuit voltage V and source resistance R_S is connected to a load of impedance $Z = R + jX$. The current into the load is $|i| = V/|R_S + Z|$, and the power into the load is therefore

$$W = |i|^2R = \frac{V^2R}{|R_S + Z|^2} \qquad (5.47a)$$

Now,

$$\rho = \frac{Z - R_S}{Z + R_S}$$

Therefore

$$1 - |\rho|^2 = 1 - \frac{|Z - R_S|^2}{|Z + R_S|^2}$$

$$= \frac{|Z + R_S|^2 - |Z - R_S|^2}{|Z + R_S|^2}$$

$$= \frac{(R + R_S)^2 + X^2 - (R - R_S)^2 - X^2}{|Z + R_S|^2}$$

$$= \frac{4RR_S}{|Z + R_S|^2} \qquad (5.47b)$$

Thus, from (5.47a) and (5.47b),

$$W = \frac{V^2}{4R_S}(1 - |\rho|^2)$$

The maximum value of W occurs when $Z = R_S$; ie $W_{max} = V^2/4R_S$. Therefore

$$W = W_{max}(1 - |\rho|^2) \qquad (5.47c)$$

Using this expression to substitute for W/W_{max} in (5.14) shows that the circle defining the admittance for a constant W/W_{max} (ie constant $|\rho|$) has its centre at $\{G_S(1 + |\rho|^2)/(1 - |\rho|^2), 0\}$, and its radius is $2G_S|\rho|/(1 - |\rho|^2)$, where $G_S = 1/R_S$. For maximum power transfer ($\rho = 0$), the permissible load shrinks to a point at $(G_S, 0)$. As the permissible value of $|\rho|_{max}$ is increased, the limiting circle expands and moves outwards along the G–axis. Taking the criterion of a maximum value of $|\rho|$ to define the bandwidth is equivalent to specifying that the admittance values within the band shall lie within the circle corresponding to $|\rho|_{max}$. The maximum bandwidth is achieved by making $|\rho|$ as near as possible to $|\rho|_{max}$ over the whole band – ie making the admittance curve lie as near as possible to the $|\rho|_{max}$ circle. This is illustrated in *Fig 5.13b*, which shows the admittance curve for the $n = 3$ example below. The extra filter section adds a further loop to the admittance curve of a simple tuned transducer, thus allowing it to remain close to the $|\rho|_{max}$ circle over a wider band; although not shown in *Fig 5.13b*, the effect of the series reactances in the added arm is to make the admittance curve in this case tend to the origin for very high and very low frequencies. Adding a further section in parallel, to give the $n = 4$ case, introduces yet another loop into the admittance curve, as in *Fig 5.13c*.

It is easy to show that the limiting circle intersects the G–axis at values of $G_S(1 + |\rho|)^2/(1 - |\rho|^2)$ and $G_S(1 - |\rho|)^2/(1 - |\rho|^2)$. These values correspond to the positions of maximum and minimum conductance, and it is clear from the figure that these correspond also to the maximum and minimum admittance values. The ratio of Y_{max}/Y_{min} corresponding to the circle defined by $|\rho|_{max}$ is therefore given by

$$\frac{Y_{max}}{Y_{min}} = \frac{(1 + |\rho|_{max})^2}{(1 - |\rho|_{max})^2} \qquad (5.48a)$$

Or

$$|\rho|_{max} = \frac{(Y_{max}/Y_{min})^{1/2} - 1}{(Y_{max}/Y_{min})^{1/2} + 1} \qquad (5.48b)$$

These equations give the relationship between $|\rho|_{max}$ and the maximum permissible range of admittances. For example, a range of $Y_{max}/Y_{min} = 2$ is typical of what might be required for a high power amplifier, and that is equivalent to $|\rho|_{max} = 0.172$, which corresponds to a maximum reduction of power into the load of $(0.172)^2 W_{max}$, ie 0.13dB. Similarly, an admittance range of 4:1 would correspond to $|\rho|_{max} = 0.33$, equivalent to a maximum loss of 0.5dB.

It should perhaps be observed that the limits of the pass band have been defined in this section rather differently from those in section 5.3, where we used a sector of an annulus instead of a circle to specify the limits in the admittance plane. This arises because the filter analysis techniques used here are more amenable to interpretation in terms of the circles corresponding to $|\rho|$ values, whilst the limits in section 5.3 represented assumptions about the restrictions appropriate for amplifier loads. In practice, the limits are generally not critical, and the difference between the two approaches is not very significant. The important point of this section is that the bandwidth of a transducer may be increased over that of the simple tuned transducer by extending the filter network.

(a)

(b)

(c)

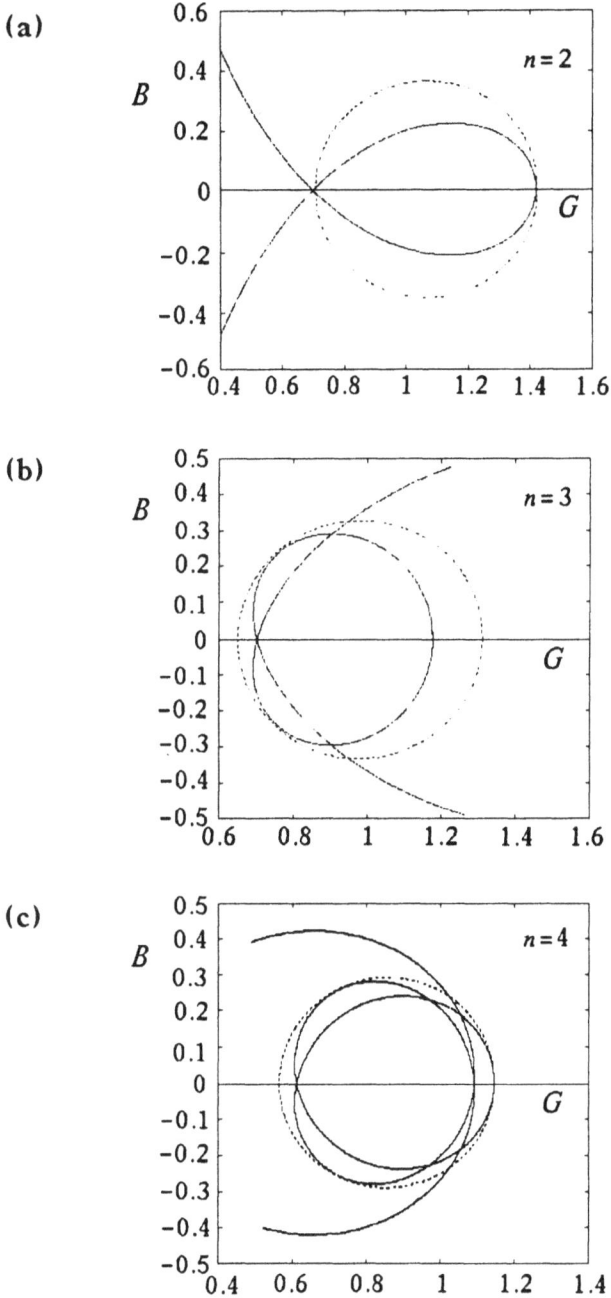

Fig 5.13   Admittance curves for optimised band pass filter networks: (a) $n = 2$; (b) $n = 3$; (c) $n = 4$.

An example.

Suppose that we wish to design a transducer having a centre frequency given by $\omega_S = 3 \times 10^4$ (ie $f_S = 4.775\text{kHz}$), and a bandwidth which is as large as possible within the circle corresponding to a 2:1 range of admittance. As noted above, this is equivalent to choosing $|\rho|_{max} = 0.172$. Considering firstly the case of $n = 3$, we find from *Fig 5.10* that the appropriate value of d_1 is 0.77, and the corresponding values of the other parameters are $k_{12} = 0.82$, $k_{13} = 0.72$, and $\alpha_1{}^2 = 1.30$. Assume that a transducer based on a stack of lead zirconate titanate is to be used, for which a typical value of $W^2 = 0.3$ can be taken (corresponding to a coupling coefficient of 0.48). Substituting these values in equations (5.35) to (5.40) gives values for the network components, normalised to R_1, as below:-

$$L_1 = 7.16 \times 10^{-5} R_1 \qquad C_1 = 1.55 \times 10^{-5}/R_1$$

$$L_2 = 2.15 \times 10^{-5} R_1 \qquad C_2 = 5.18 \times 10^{-5}/R_1$$

$$L_3 = 5.15 \times 10^{-5} R_1 \qquad C_3 = 2.16 \times 10^{-5}/R_1$$

$$R_S = 1.30 R_1$$

To achieve these values, the transducer must have $Q_M = 2.15$ (from (5.41)). The resulting fractional bandwidth is 0.60 (from (5.33)).

Similar calculations can be carried out for networks with other numbers (n) of arms. Results have also been calculated for the cases $n = 4$ and $n = 2$, and admittance curves for all three cases are shown in *Fig 5.13a - c* illustrating how the curves approximate more nearly to the $|\rho| = 0.172$ circle (shown dashed) as n increases. The bandwidths according to the $|\rho|_{max}$ criterion, and the required values of Q_M, are:-

For $n =$	2	3	4
$\omega_B/\omega_S =$	0.38	0.60	0.67
$Q_M =$	2.6	2.15	2.1

The $n = 2$ case represents the simple tuned transducer discussed earlier in this chapter, and is the most commonly used. Most of the improvement in bandwidth compared to this simple case is achieved by using $n = 3$, with reducing benefits as n is increased further.

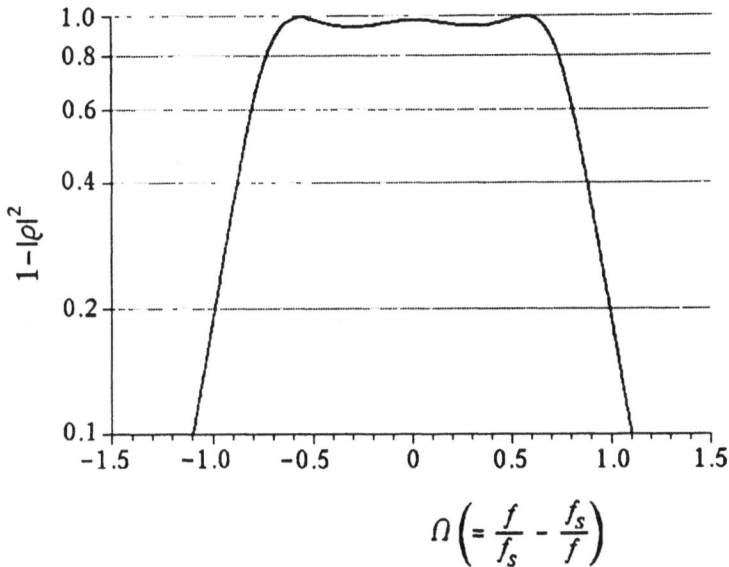

$$\Omega \left(= \frac{f}{f_s} - \frac{f_s}{f} \right)$$

Fig 5.14   Power delivered into load, $(1 - |\rho|^2)$ versus frequency parameter Ω for $n = 3$ example.

Fig 5.14 shows the variation with frequency of the relative power into the load $(1 - |\rho|^2)$ for this example, with $n = 3$, illustrating the ripple within the pass band which is characteristic of the Tchebychev filter design. The allowed variation of 2:1 in admittance magnitude corresponds to a ripple of 0.13dB in the pass band (from (5.48)).

Coupled resonator networks.

The networks discussed in the previous section can be realised by adding electrical elements to the piezoelectric transducer. Difficulties can sometimes be experienced, however, in meeting the high power requirements for these electrical components in the space available, and it is therefore worth enquiring whether these networks could be realised instead by mechanical elements. We saw in section 4.9 that there is no mechanical equivalent to an inductor with one terminal at ground potential (in the impedance analogue system), and consequently many filter networks cannot be realised mechanically. In

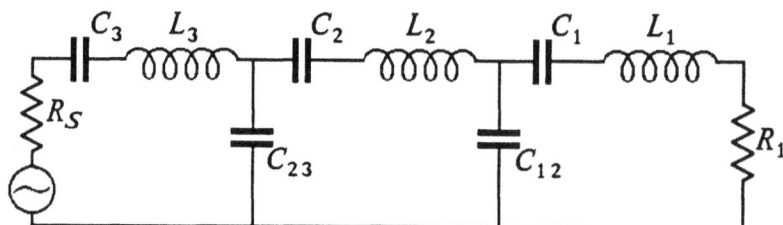

Fig 5.15   Coupled resonator network.

particular, the basic (or "canonical") type of filter in the previous section is not suitable, since there is no mechanical equivalent to inductors such as L_2. However, a "coupled resonator" network of the form shown in *Fig 5.15* can be a good approximation to the full canonical circuit over a narrow band, and does not involve the use of inductors connected to ground. Rodrigo [5.4], in a study of broad band transducer methods, concluded that such a coupled resonator network, with a Tchebychev response, was the most suitable filter type for a primarily mechanical realisation. In the basic network shown in *Fig 5.15*, the piezoelectric ceramic is represented by the capacitors C_2 and C_{23}, with all elements to the right of C_{23} corresponding to mechanical elements and those to the left to electrical elements. The ratio C_2/C_{23} thus cannot exceed the value of W^2 for the ceramic.

In order to obtain the desired response for this type of network, the resonance frequencies of the meshes must be made equal when the other meshes are open–circuited, and the design considered is also symmetrical, with $L_1 = L_3$ and $C_{12} = C_{23}$. Rodrigo's analysis, using filter methods such as those in [5.5], showed that the widest bandwidth is achieved when C_1 and C_3 become infinitely large, so that they are effectively short–circuited. Thus, for the right-hand and left-hand meshes, $L_1 C_{12} = L_3 C_{23} = 1/\omega_1^2$, and for the middle mesh

$$\omega_1{}^2 = \frac{1}{L_2}\left\{\frac{1}{C_2} + \frac{1}{C_{23}} + \frac{1}{C_{12}}\right\}$$

$$= \frac{1}{L_2 C_2}\left\{1 + 2W^2\right\} \qquad\qquad (5.49)$$

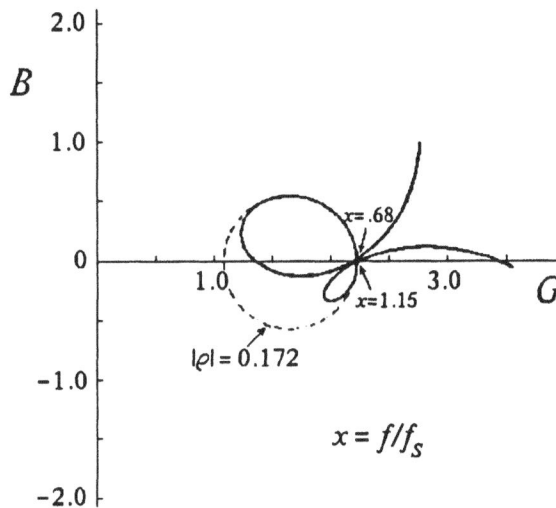

Fig 5.16 Admittance curve for coupled resonator network, for $W^2 = C_2/C_{23} = 0.3$, $Q_M = 2.22$.

since $C_2/C_{23} = C_2/C_{12} = W^2$. Hence, $L_2C_2 = (1 + 2W^2)/\omega_1^2$. If the components are scaled in terms of ω_1C_{23}, the terminating resistor is expressed as $\omega_1C_{23}R_1 = d_1 = R_1/\omega_1L_1$. The transducer is then designed to obtain the best combination of d_1 and W^2.

This is best done empirically, an example being illustrated in *Fig 5.16* which shows the admittance curve for a network for which $W^2 = 0.3$ and $d_1 = 0.45$ (corresponding to $Q_M = 1/d_1 = 2.22$). This curve just fits inside a circle (shown dashed) for $|\rho| = 0.172$. The bandwidth, defined in terms of the frequencies where the curve intersects the circle, is from $0.68\omega_1$ to $1.15\omega_1$. The mid-band frequency is given by $(0.68 \times 1.15)^{1/2}\omega_S$, ie $0.88\omega_S$, and the fractional bandwidth is therefore $0.47/0.88 = 0.53$. This is an improvement over the value of 0.38 for a simple parallel tuned transducer having $W^2 = 0.3$, but is not as good as the Fano circuit described in the example above. Since the coupled resonator circuit represents a narrow band approximation to the more complete Fano network, this degradation from the performance of the full circuit is not unexpected, and its extension to wider band applications leads to the deviations from an ideal admittance curve which are evident in *Fig 5.16*. The potential advantage of this circuit, however, is that it is possible

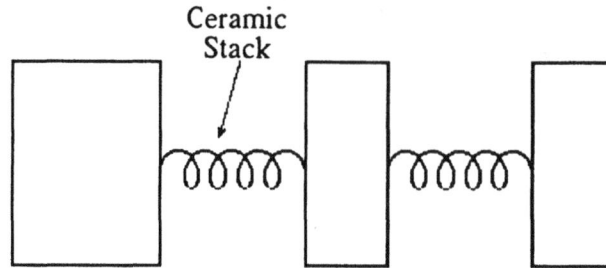

Fig 5.17   Mechanical realisation of coupled resonator design.

to realise the part of the circuit to the right of C_{23} mechanically, with C_2 representing the compliance of the piezoelectric ceramic. This leaves L_3 as the only network component to be realised electrically. Using the equivalences of *Fig 4.14*, we can see that L_2, C_{12}, and L_1, correspond to a series mass, spring, mass combination, giving an arrangement as illustrated in *Fig 5.17*; C_{23} is the clamped capacitance of the stack, and L_3 is the added electrical inductance. The practical realisation of such a design may be tackled by the methods described in Chapter 7.

These broad band designs have been based on methods of filter synthesis derived for electrical networks. In those networks, the usual assumption is made that the components are themselves independent of frequency. We should remember that this assumption is not entirely justified when we are dealing with transducers, since for example the loading on the radiating face is a function of frequency, and some of the mechanical parameters may also depend to some extent on frequency. The effect of such variations on the performance of these circuits is generally fairly small, and their evaluation is left to the designer if considered necessary.

5-5   SUMMARY

If a piezoelectric ceramic transducer is electrically tuned by means of a parallel inductor, its equivalent circuit resembles that of a half-section band-pass filter, and its electrical behaviour may be analysed by standard filter methods. The shape of the admittance loop is determined by the parameter $\beta = 1/WQ_M$. For maximum power output from the amplifier driving the transducer, it is commonly required that the nominal load should lie within limits such as a modulus range not greater than 2:1, and a power factor not less than 0.8. The maximum bandwidth satisfying these limits is achieved if $WQ_M \simeq 1.2$ (ie $\beta \simeq 0.8$); the fractional bandwidth is then approximately equal to k. This condition corresponds to a filter with a ripple of about 0.13dB in the same pass band.

The effects of deviations from the nominal component values in the equivalent circuit (arising from manufacturing variations, environmental changes, etc) need to be taken into account. Typically, these may increase the range of admittance modulus of the load to 6:1, and increase phase angles up to 60°. Design of a transducer according to these filter methods, and use of a driving amplifier whose output impedance matches the filter network, improves the reliability of the system and prevents large deviations of the acoustic power output of the transducer from its design value.

The methods of filter analysis may be extended to 3 or 4-section networks to achieve wider bandwidth designs. These may be realised either by adding electrical elements, or by mechanical modifications to the transducer. These techniques offer the possibility of increasing the bandwidth of the system by some 50%.

REFS Chapter 5

References

5.1 Green,E., *Amplitude-Frequency Characteristics of Ladder Networks*, Marconi's Wireless Telegraph Co., 1954.
5.2 Bode,H.W., *Network Analysis and Feedback Amplifier Design*, Van Nostrand, 1945.
5.3 Fano,R.M., "Theoretical limitations on the broadband matching of arbitrary impedances", *J Franklin Inst*, <u>249</u>, 57–83 and 139–154, (1950).
5.4 Redwood,M., and Rodrigo,G.C., "Analysis and design of piezoelectric sonar transducers", Ph.D. Thesis, Queen Mary College, London, 1970.
5.5 Zverev,A.I., *Handbook of Filter Synthesis*, Wiley and Sons, 1967.

Additional Reading

1 Stansfield,D., British Acoustical Society, Spring Meeting 1971, "Transducer bandwidth".
2 Morris,J.C., British Acoustical Society, Spring Meeting 1971, "Transducer design based on filter theory".

Chapter 6
ACOUSTIC RADIATION

Although the analysis of Chapter 5 is in terms of electrical networks, their realisation in a transducer system implies that the terminating resistor at the mechanical end of the network represents the motional resistance of the transducer. A starting point in the design of a transducer is therefore to evaluate the motional impedance predicted for an array geometry which has been derived from the considerations of Chapter 2. This impedance will be in mechanical terms, and can subsequently be converted to electrical terms by the relationships of section 4.7. If the overall electro-acoustic efficiency of the transducer is high, the main contribution to the motional resistance is due to the acoustic radiation resistance of the transducer. The calculation of radiation impedance is thus the effective starting point in the design, and is the subject of the first part of this chapter. When transducers are assembled into arrays, the calculation must allow for the interactions between elements, and these are considered in sections 6.2 and 6.3. Section 6.4 deals with some aspects of beam forming, already referred to briefly in Chapter 2. The final section gives a brief discussion of the characteristics of parametric arrays based on the non-linear interactions in an acoustic field.

6-1 RADIATION IMPEDANCE OF SINGLE PISTON

The radiation impedance of a vibrating surface is the ratio of the force exerted by the surface on the medium to the normal velocity of the surface. It is analogous to the electrical impedance, which is the ratio of voltage to current, and is thus consistent with the (force–voltage) analogy between acoustic and electrical quantities. Because the force (F) and velocity (u) (normal to the radiating surface) are phasors, their ratio is in general complex. The radiation impedance Z_r, defined by

$$Z_r = F/u \qquad (6.1)$$

is therefore complex, and may be separated into its real and imaginary components, the radiation resistance (R_r) and radiation reactance (X_r), by writing

$$Z_r = R_r + jX_r \qquad (6.2)$$

(Note that these denote quantities in mechanical terms in this chapter.)

The value of the radiation impedance depends on the size and shape of the radiating surface, and the nature of its surroundings. The classical calculation is for a circular piston in an infinitely large and rigid baffle, and this is considered first. The effects of finite size baffles and different shapes of pistons and baffles are summarised in the later part of the section.

Piston in infinite rigid baffle.

The calculation of Z_r for a circular piston in an infinite rigid baffle is given in standard texts such as Ref [6.1]. The treatment derives the pressure across the face due to motion of an infinitesimal area of the piston, and then integrates the total contributions generated by motion of the whole piston, assuming the velocity to be uniform over the whole piston area. The result can be normalised by extracting a factor $\rho c A_p$, the remaining functions being dependent on ($2ka$), where a is the piston radius (hence $A_p = \pi a^2$). ($2ka$) is proportional to the diameter of the piston in terms of wavelengths, since

$$k = \omega/c$$

$$= 2\pi/\lambda \quad (= 2\pi f/c) \qquad (6.3)$$

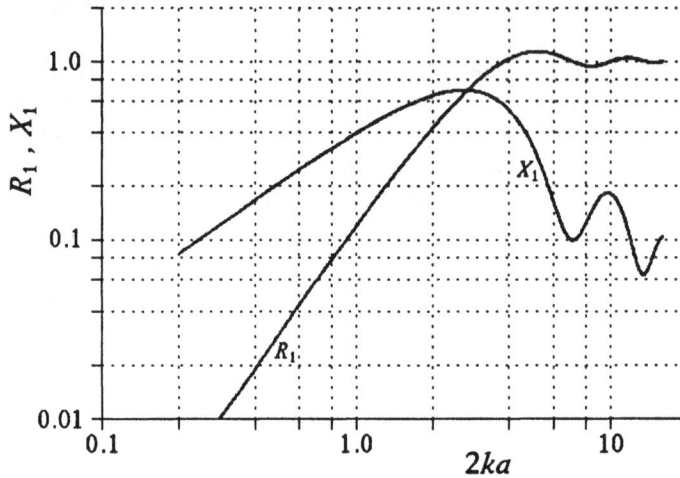

Fig 6.1 **Normalised radiation impedance functions for circular piston in infinite rigid baffle.**

The radiation resistance and radiation reactance functions can thus be written as,

$$R_r = \rho c A_p R_1(2ka) \qquad\qquad (6.4)$$

and

$$X_r = \rho c A_p X_1(2ka) \qquad\qquad (6.5)$$

Values of these normalised radiation impedance functions R_1, X_1, are shown in *Fig 6.1*, plotted against the parameter *2ka*. This parameter may be regarded as representing the variation of either frequency or piston size; it is equal to $4\pi a/\lambda$, the ratio of twice the piston circumference to the acoustic wavelength, or 2π(piston diameter/λ). For example, a value for *2ka* of π corresponds to a piston diameter of $\lambda/2$. Ref [6.1] tabulates values of the functions R_1 and X_1 for *2ka* up to 16.

Useful approximations can be derived for two extremes [6.1]. When *2ka* is large, $R_1 \simeq 1$, and hence $R_r \simeq \rho c A_p$, independent of frequency. Also for large *2ka*, $X_1 \simeq 2/\pi ka$, and thus X_1 gradually decreases towards zero as frequency or piston size increase. For very small pistons, ie when *2ka* < 1, $R_1 \simeq k^2 a^2/2$, and hence

$$R_r = \rho c A_p k^2 a^2 / 2$$

$$= \frac{\rho c k^2}{2\pi} A_P^2 \qquad (6.6)$$

Thus, for pistons which are very small compared with a wavelength, R_r is proportional to the square of the piston area; alternatively, for a given piston size, R_r is proportional to the square of the frequency, and this strong dependence on ka for small pistons is the basic cause of the difficulty of generating acoustic power at low frequencies. The radiated acoustic power is given by $u^2 R_r$, by analogy with the electrical case. As R_r becomes very small for low ka, the radiation of appreciable acoustic power from small pistons at low frequencies can therefore involve large piston amplitudes. These are generally difficult to generate efficiently, particularly if using a relatively stiff material such as a piezoelectric ceramic, and for this reason it is difficult to produce high power output from pistons which are small compared with a wavelength. For the same condition of $ka < 1$, $X_1 \simeq 8ka/3\pi$, and thus

$$X_r = \rho c A_p (8ka/3\pi) \qquad (6.7)$$

The radiation reactance X_r may be regarded as an additional vibrating mass (M_r) given by

$$M_r = X_r/\omega$$

$$= \rho c A_p \frac{8ka}{3\pi\omega}$$

$$= \frac{8}{3} \rho a^3 \qquad (6.8)$$

This mass is equivalent to that of a cylinder of water having the same cross sectional area as the piston and a length of $8a/3\pi$ (ie a mass some 25% more than that of a hemisphere equal in diameter to the piston.) This "associated mass" can be quite significant in reducing the resonance frequency, and increasing the mechanical Q-factor, of low frequency transducers when the piston is small and lightweight.

If the drive to a transducer is such as to generate a constant piston velocity, independent of frequency, – ie a constant motional current in the equivalent circuit, – then the

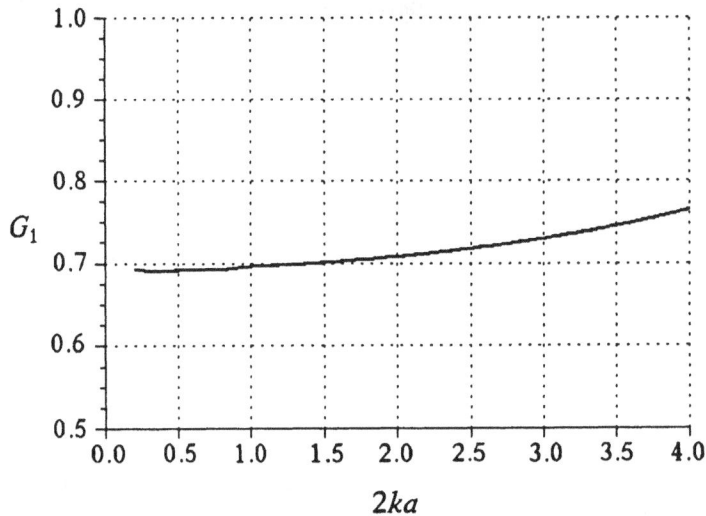

Fig 6.2 Normalised radiation conductance of circular piston in infinite rigid baffle.

output power which is given by $u^2 R_r$, will be proportional to R_r, which varies markedly with frequency for small pistons at low frequency. However, it is worth noting that if the drive generates a force (F) on the medium, then the power output is given by $F^2 G_r$, where G_r is the motional equivalent of the conductance, ie

$$G_r = \frac{R_r}{R_r^2 + X_r^2} \qquad (6.9)$$

as in equation (1.7a). This may be expressed in normalised terms as $G_r = G_1/\varrho c A_p$, and G_1 may then be evaluated by using

$$G_1 = \frac{R_1}{R_1^2 + X_1^2} \qquad (6.10)$$

The variation of G_1 with frequency is illustrated in *Fig 6.2*, which shows that if the force at the front face of the piston could be kept constant, the output would vary only slowly with frequency. The necessity for large piston velocities to achieve reasonable output at low frequencies would not be avoided, however; as the frequency is reduced, the radiation impedance decreases, and a constant force drive at the front face of the piston would generate the greater piston displacement needed to maintain the output power.

Ref [6.1] shows that the power radiated from a piston which is small compared with a wavelength is identical to that for a simple source having the same volume displacement $A_p u$. This is to be expected, since a small piston acts as a volume source, with characteristics independent of its exact shape. The radiation from a rectangular piston in an infinite rigid baffle has been treated in [6.2]. The treatment is essentially the same as for circular pistons, except that the limits in the integrals need to be changed, and the results show that the radiation impedance of a square piston is almost the same as for a circular piston of equal area.

The radiation impedance of a piston in a cylindrical baffle depends on the diameter of the cylinder in terms of wavelengths as well as the size of the piston itself [6.3]. In the usual case where the cylindrical baffle has a diameter several wavelengths across, there are only small differences between the impedance values for a piston in a cylinder and those for a piston in a plane baffle, and it is therefore usually acceptable to use the values for a piston in an infinite plane baffle. A similar conclusion applies to the case of a circular piston in a large spherical baffle [6.3]. The radiation impedance of a piston in a small spherical baffle has been calculated by Morse [6.4].

Piston in finite baffle.

Real baffles deviate to some extent from the assumptions of infinite size and infinite rigidity, and it is sometimes necessary to introduce appropriate corrections to allow for the resulting effects. The volume displacement generated by vibration of a piston in an infinite rigid baffle gives rise to an associated pressure field and acoustic radiation into the half-space in front of the baffle. If the piston dimensions are large compared with a wavelength, the acoustic energy is concentrated into a beam, and provided this is aimed well away from the baffle there is only a small interaction of the acoustic field with the baffle. In that case, the presence or absence of the baffle has little effect. For a small piston, however, the pressure field along the baffle is very significant; if the piston is small enough to be regarded as a point source, the acoustic pressure field is omni-directional over the front half-space and zero for the rear half-space. If the baffle is removed, the pressure field can be dissipated into the rear half-space, and the forward acoustic pressures caused by a given volume displacement are thus reduced. In addition to the changes in the directionality of the field, there is also a reduction of the back-pressure on the radiating piston, and hence of its radiation impedance.

Calculations of the radiation impedance of a circular piston in a finite size baffle have been carried out by various authors, including especially Crane [6.5]. The basic approach is to represent a piston radiating only from one side as a combination of a pulsating disc, with both sides vibrating in antiphase, and an oscillating disc, with both sides vibrating in phase. Each component has half the amplitude of the original source. They therefore combine to produce the full amplitude at the front surface, and zero at the rear – ie to represent a piston radiating from one side only. The pulsating disc, which has the opposite faces moving in opposition, generates a pressure field which is symmetrical about the plane of the disc, and is hence not affected by the presence of a baffle in that plane. The oscillating disc generates a pressure field with a dipole characteristic, for which there is an acoustic particle velocity normal to the plane of the baffle, and which would therefore be disturbed by the introduction of a baffle. The oscillating disc component in the model is surrounded by a baffle of the required size, and the problem is transformed to that of calculating the effect of a continuous virtual source in the plane of the baffle extending from the baffle edge to infinity, and with a source level distribution which would just cancel the normal velocity components of the oscillating source. The combination would be equivalent to a piston in an infinite rigid baffle, for which the solution is known, and the difference between the field for an infinite baffle and the planar virtual source distribution gives the components needed to derive the field for the partially baffled source.

Ref [6.5] tabulates values of radiation impedance as a function of piston and baffle radii, including the case where the baffle shrinks to the size of the piston, so that it is equivalent to an unbaffled piston radiating from one side only. *Fig 6.3* shows values of the normalised radiation resistance and reactance for the unbaffled circular piston. Comparison with the curves of *Fig 6.1* shows that the value of R_1 for a small unbaffled piston is about half that of a baffled piston of the same size, as might be expected from the reduction in field at the front face. The results of [6.5] show that baffles larger than about 2λ in diameter are approximately equivalent to an

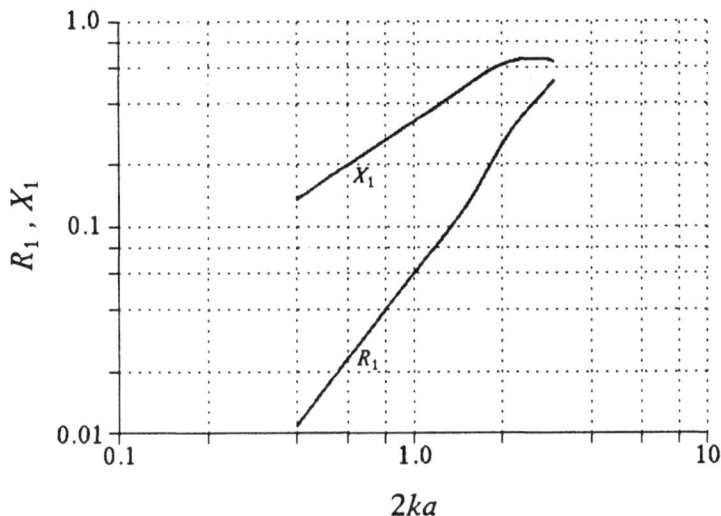

Fig 6.3 Normalised radiation impedance functions for unbaffled circular piston.

infinitely large baffle. Although the model of a piston radiating from one face only does not appear to correspond to any practical transducer design, it is probably the nearest simple approximation to the common arrangement of a piston transducer fitted into a rigid casing.

The effects of finite rigidity of the baffle are difficult to evaluate, and involve a knowledge of the baffle details [6.6]. In practice, the aim must generally be to design a baffle which is a reasonable approximation to a rigid baffle – ie to have an impedance which is high compared with that of water, – and if that is achieved the assumption of infinite rigidity is usually satisfactory. The major problems arise if a very low acoustic compliance exists near the radiating piston. Such a situation can arise, for example, because of the presence of air-filled cellular rubber, or of air behind compliant joints around the piston. These situations should be avoided if possible, especially for radiating faces which are small compared with a wavelength. In those cases where such arrangements are unavoidable, evaluation of the effects is best carried out experimentally.

Transducer in air.

The radiation impedance of a transducer in air is very small, and is generally negligible compared with the internal losses within the transducer. Derivation of the equivalent circuit for a transducer in air therefore depends on estimating these internal losses rather than the calculation of radiation impedance. In practice, experimental measurements usually provide a better basis for these estimates than do theoretical calculations.

Cylinders and Spheres.

The radiation impedance of a cylindrical radiator which is long compared with a wavelength is given [6.7] by

$$Z_r = j2\pi ah\rho c \, \frac{H_0^{(2)}(ka)}{H_1^{(2)}(ka)} \qquad (6.11)$$

where a and h are the radius and height respectively of the cylinder, (whence the surface area of the cylinder is $2\pi ah$,) and

$H_0^{(2)}(ka)$ is a Hankel function = $J_0(ka) - jN_0(ka)$

$H_1^{(2)}(ka)$ is a Hankel function = $J_1(ka) - jN_1(ka)$

$J_0(ka)$ and $J_1(ka)$ are Bessel functions

$N_0(ka)$ and $N_1(ka)$ are Neumann functions.

The radiation impedance of a spherical source (of radius a) is given [6.8, 6.9] by

$$Z_r = 4\pi ka^3 \rho c \, \frac{(ka + j)}{1 + (ka)^2} \qquad (6.12)$$

6-2 RADIATION IMPEDANCE OF ELEMENTS IN ARRAYS

When numbers of transducers are assembled together to form an array, the acoustic pressure generated by any one of the transducer elements exerts a force on all the other elements. The overall result of these complex interactions between the elements is to modify the radiation impedances of the individual transducers in the array. The calculation of these effects in detail can involve extensive computations (eg [6.10 - 6.16]), and a number of these have been undertaken in efforts to predict the performance of arrays of closely spaced projector elements in sufficient detail to avoid catastrophic failure. However, the dangers of failure can be reduced by engineering approaches ([6.17], see also Sections 5.3 and 6.3), and for most applications it is the average value of the radiation impedance which needs to be known, rather than the detailed individual values. This allows considerable simplification of the calculations, and clearer presentation of the results. In this section, the basic principles of the analysis will be described, together with some of the results which are of greatest use in array design.

Mutual Interaction Impedances.

A piston vibrating with a normal velocity U_1 in a fluid medium gives rise to a reactive force on its own surface which we shall denote by F_{11}. Its radiation impedance is then given by $Z_{11} = F_{11}/U_1$ (as in (6.1)). Suppose that a second piston, near the first, vibrates with a velocity U_2 and produces a force on the first piston of F_{12}. Then, using a similar notation, the mutual impedance can be defined as

$$Z_{12} = F_{12}/U_2$$

and hence
$$F_{12} = Z_{12} \cdot U_2$$

If the array is formed of n elements, the total force on the ith piston is thus given by the general expression

$$F_i = \sum_{j=1}^{n} Z_{ij} \cdot U_j$$

and the total impedance of the ith piston by

$$Z_i = \frac{1}{U_i} \sum_{j=1}^{n} Z_{ij} \cdot U_j \tag{6.13}$$

This is composed of the "self radiation impedance" Z_{ii} and the sum of the mutual impedances multiplied by the velocity ratios, ie

$$\sum Z_{ij} U_j / U_i \quad (i \neq j)$$

The full calculation of the radiation impedance would thus require a knowledge of the values of Z_{ij} for all pairs of transducers in the array, and a knowledge of all the piston velocities. Unfortunately, the velocities are themselves dependent on the mutual interactions, and there is no simple method for calculating them. Instead, successive approximation techniques have been employed to give solutions of adequate accuracy, although the computations involved in calculating all the velocities are even then formidable. Calculations carried out by Freedman [6.13] allowed him to derive approximations for the mutual radiation impedances for pistons in a plane baffle. The self and mutual radiation impedances may be separated into their real and imaginary parts by writing

$$Z_{11} = R_{11} + jX_{11}$$

and
$$Z_{12} = R_{12} + jX_{12}$$

Approximate expressions for the mutual impedance of circular pistons in an infinite rigid baffle are, from [6.13],

$$R_{12} = R_{11} \frac{1}{N_1} \cdot \frac{\sin kd}{kd} \qquad (6.14)$$

$$X_{12} = R_{11} \frac{1}{N_2} \cdot \frac{\cos kd}{kd} \qquad (6.15)$$

where d is the centre-to-centre spacing of the elements,

$$N_1 = 1 - 0.0411ka + 0.138(ka)^2$$

$$N_2 = 1 + 0.009ka + 0.118(ka)^2$$

and a is the radius of the piston.

Values of N_1 and N_2 are plotted in *Fig 6.4*. It is evident from equations (6.14) and (6.15), and the curves in *Fig 6.4*, that the interaction effects become relatively small for separations beyond about $\lambda/2$.

The exact mutual impedance of square or rectangular pistons depends not only on the size of the pistons but also on their shape and orientation relative to each other [6.12]. In practice, an adequate approximation is usually obtained by treating square

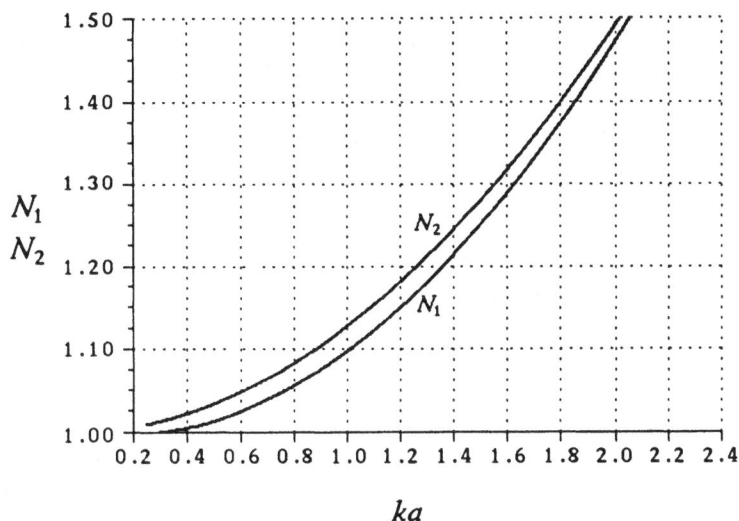

Fig 6.4   **Mutual interaction factors N_1, N_2, as functions of ka.**

pistons as though they were circular pistons having the same area and same centre-to-centre spacing. Non-rigidity of the baffle greatly increases the complexity of the calculation of mutual interaction effects [6.14], although it may have some advantages in causing the mutual impedance to fall off more rapidly with separation between pistons. Calculations of mutual impedance for elements in cylindrical and spherical baffles have been carried out (See [6.11]), and show that curvature of the baffle also causes the interaction effects to fall off more rapidly with separation than for a plane baffle.

The expressions above may be used to evaluate the mutual impedances for all pairs of elements in an array, and hence in principle to evaluate all the element impedances and the piston velocities. This is practical only for relatively small numbers of elements, because of the rapidly escalating computational demands, and for larger arrays it is common to adopt the simplifying assumption that all the pistons have the same velocity. This permits the element impedances to be calculated as the sum of the self and all the mutual impedances – ie equation (6.13) with all the values of U_j equal to U_i. The impedances may then be used to calculate approximate values of piston velocities, and if necessary a further round of calculations based on these velocities may be carried out to obtain a better approximation to the real values.

Evaluation of the piston velocities is necessary for two main purposes; firstly to allow accurate calculations of the radiated acoustic field, and secondly to calculate the internal stresses in each transducer, in order to identify highly stressed elements and thus take steps to prevent their failure. Results of such computations may be presented in terms of the equivalent electrical admittance curves, and results for closely spaced elements may show very marked deviations from the well-behaved loops representative of a single transducer. The curves are typically very frequency dependent, and in some cases negative conductance values have been observed, indicating that a particular transducer is actually absorbing power instead of transmitting it. Elements having especially high or low values of admittance magnitude are particularly susceptible to mechanical or electrical breakdown, and failure of individual elements may even lead to runaway failure of a complete array.

These calculations should be carried out for all elements and all frequencies of interest. They are however very demanding on computer time, and in practice such computations of the individual impedances and velocities are worth while only in very rare cases, when a major development is planned. It is generally preferable to adopt one of the engineering approaches described in the next section to reduce the effects of the interactions. Although the effect of the interactions in producing variations across the array has received most attention in the literature, it is generally their effect on the average radiation impedance which has the greatest influence on the characteristics of the array. Indeed, this effect is usually beneficial, by increasing the mean radiation resistance and hence helping to achieve a wide bandwidth, despite the uncertainties and problems which may well be introduced by the variations across the array.

Toulis [6.15] considered the radiation from small circular pistons in a plane rigid baffle by treating each piston as though it were radiating into an infinitely long tube. He thus obtained a simple equation for the **average radiation impedance** $\langle Z_r \rangle$ of an element in an array, viz.,

$$\langle Z_r \rangle = \rho c A_p \chi + Z_{11}' (1 - \sqrt{\chi}) \qquad (6.16)$$

where Z_{11}' is the self radiation impedance of the piston without a baffle, and χ is the "packing factor" of the array, defined as the ratio of the piston area A_p to the area of array per piston (ie total array area divided by the number of pistons). This treatment is valid only for small pistons, eg for piston diameter less than 0.3λ, but this is sufficiently accurate for most cases where interaction effects are serious. The same result was obtained for pistons on a spherical baffle. A simple approximation which is often adequate is given by $\langle R_r \rangle = \rho c A_p \chi$, which is considerably greater than the value for a single piston in a baffle (for such small pistons).

Freedman [6.13] and Morris [6.16] carried out detailed calculations of the radiation impedances of transducers in a range of typical array geometries, and hence derived approximate expressions for their average radiation impedance. The important parameters are the piston diameter and the centre-to-centre spacing between pistons, both in terms of wavelengths. The average radiation resistance is then given by

$$\langle R_r \rangle = \rho c A_p [R_{11}(1 + D_1/N_1)] \qquad (6.17a)$$

and the average radiation reactance by

$$\langle X_r \rangle = \rho c A_p [X_{11} + R_{11}D_2/N_2] \qquad (6.17b)$$

where R_{11} and X_{11} are the normalised self radiation resistance and reactance of the piston in a baffle, N_1 and N_2 are factors depending on piston diameter, as in (6.14) and (6.15), and D_1 and D_2 are factors which depend on the element spacing. R_{11} and X_{11} are given for pistons in an infinite plane rigid baffle in *Fig 6.1*, and N_1, N_2 in *Fig 6.4*. Values of D_1 and D_2 depend on the geometry of the array, but can be expressed with sufficient accuracy as functions of the area of array per element in terms of (wavelength)^2, as in *Fig 6.5*. The same curves can then be applied to give approximate values for circular pistons in either a square or a triangular arrangement. These curves extend the design curves to piston diameters and separations up to 0.6λ. For small pistons there is good agreement with values derived from the simple equation (6.16) given by Toulis.

6-3   VELOCITY CONTROL

We have seen in the previous section that difficulties can arise from the variations in piston velocity across the array, and that the predictability of the array performance would be greatly improved if the velocities were known. Methods of achieving this "velocity control" have been investigated by several workers [6.17,6.11], the earliest reported work being that by Carson. One possibility would be to measure the velocities of

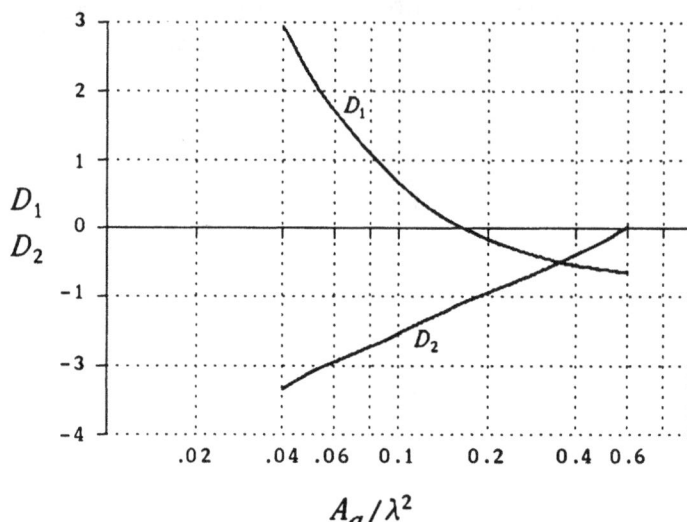

$$A_a / \lambda^2$$

Fig 6.5   Mutual interaction factors D_1, D_2, as functions of array area per element.

each piston and use a feedback system to adjust the drive to each to the desired value, but the practical complexities and cost of such a system make it an unlikely solution. A more practical solution was proposed by Carson, who used an electrical network to control the piston velocities. The basis of the method can be understood in terms of the equivalent circuit shown in *Fig 5.1*, in which the right hand (C_1, L_1, R_1) combination represents the motional arm, and the current (i_1) through it represents the motional velocity. If the countermass is much heavier than the front mass, the motional current i_1 is equivalent to the piston velocity. R_1 and L_1 contain contributions from the mutual impedance effects, and the aim is thus to make the current i_1 known and independent of variations in R_1 and L_1. At the resonance frequency ω_S, the impedance of the motional arm is small compared with the parallel combination of L_0 and C_0. If the source resistance R_S is made large, the source behaves as a current generator, and virtually all of this current must pass through the motional arm. The motional current, and hence the piston velocity, is thus determined by the output current from the drive amplifier, and is almost independent of variations in the radiation loading.

   This simplified explanation of the method is applicable at the resonance frequency of a parallel tuned transducer. At other

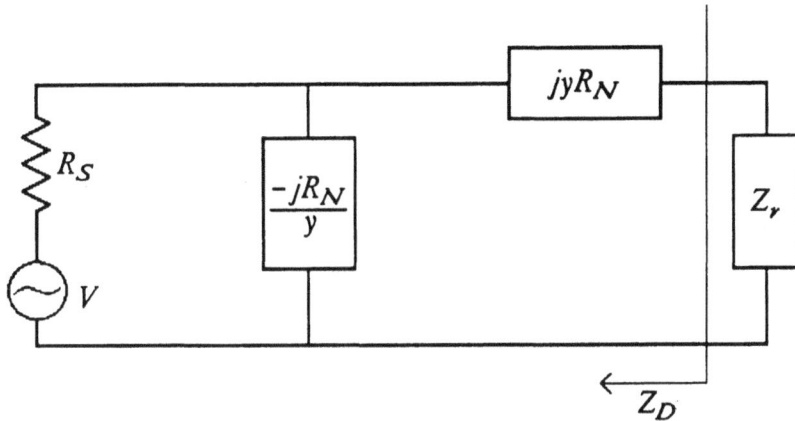

Fig 6.6 Band pass filter network, showing impedance of drive network (Z_D) seen from output terminals.

frequencies, calculation of the impedances of the arms would be needed, and in general it may be necessary to add a series reactance in the feed circuit. The more general treatment is given in the references. The basic principle is to make the impedance of the network high when looking back into it from the radiation load, at the right hand end in *Fig 5.1*. If this is achieved, – ie the effective source impedance is high compared with the load, – the current through the load will be near to its specified value. The method can be applied with good expectations of success at a single frequency or over a narrow band of frequencies, but the frequency dependence of the network impedance and the mutual interaction effects make its performance over wider bands open to more doubts, and extensive calculations may again be necessary.

The filter methods of Chapter 5 may be used to analyse the problem, and suggest a simple practical approach. In *Fig 6.6*, which is similar to the filter network of *Fig 5.3*, we are interested in the impedance (Z_D) of the drive circuit, as seen by the radiation load. This can readily be shown to be given by

$$Z_D = jyR_N \; \frac{\dfrac{jR_N}{\beta_S y}}{\dfrac{1}{\beta_S} - \dfrac{j}{y}}$$

$$= jR_N\{y - 1/(y - j\beta_S)\}$$

$$= \frac{R_N}{\beta_S^2 + y^2}\{\beta_S + jy(\beta_S^2 + y^2 - 1)\} \qquad (6.18)$$

where y and R_N are defined as before, in (5.3)–(5.6), and β_S is in this case the parameter representing the normalised value of the internal resistance of the voltage source, ie $\beta_S = R_N/R_S$. The modulus of the drive impedance, normalised by reference to R_N, is then given by

$$\frac{|Z_D|^2}{R_N^2} = \frac{1}{(\beta_S^2 + y^2)^2}\{\beta_S^2 + y^2(\beta_S^2 + y^2 - 1)^2\}$$

$$= \frac{1}{\beta_S^2 + y^2}\{1 + y^2(\beta_S^2 + y^2) - 2y^2\} \qquad (6.19)$$

Fig 6.7 illustrates results for $|Z_D|/R_N$ plotted against the frequency parameter y, for various values of β_S. A small value of β_S corresponds to a high source impedance (ie a current

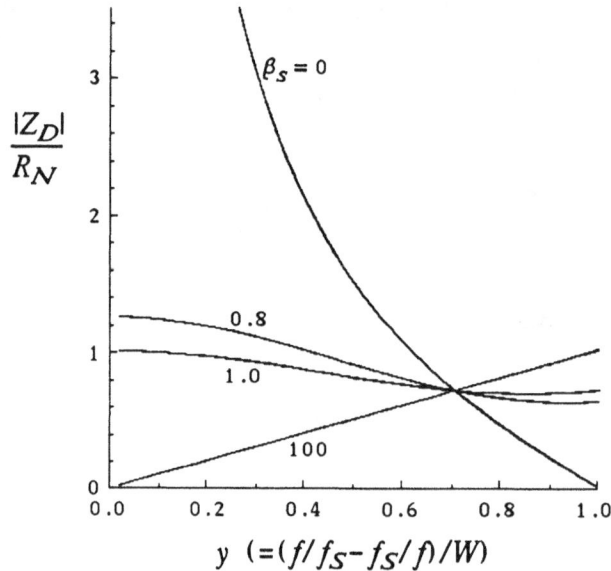

Fig 6.7 Normalised drive impedance versus frequency parameter y, for values of β_S.

source), whilst a voltage source is represented by a high β_S. The curves thus show that a high generator impedance produces a high drive impedance (Z_D) for small values of y, ie for frequencies close to resonance, but that Z_D falls to low values as y approaches unity. At the other extreme, a voltage source produces a small Z_D at resonance, increasing as the frequency deviates from resonance. Intermediate values of β_S cause less variation in $|Z_D|$ over the band up to $y = 1$. The condition for velocity control, that the drive impedance should be high compared with the radiation load, is thus met for a narrow band around resonance by using a current source, as in Carson's technique. However, if we wish to maximise the bandwidth by using the methods in Chapter 5, the operating frequency band extends to $y \simeq 1/Q_M W$, which should be about 0.8 for the optimum admittance loop of *Fig 5.5*. At the edges of the band, $|Z_D|$ falls to much lower values, and the velocity control is much less effective. If a bandwidth out to $y = 1$ were wanted, the velocity control at the edges of the band would be quite ineffective. An approximately matched generator, having $\beta_S \simeq 1$, avoids extreme velocity deviations for wide bandwidths, although giving less velocity control at any particular frequency. This is an example of the tolerance of a matched system to variations of the load, as noted in section 5.3. A choice of β_S equal to 0.8, to match the optimised filter network in section 5.3, would appear a good choice if the transmission bandwidth is to be maximised.

6-4   DIRECTIVITY PATTERNS

The influence of the specified directivity pattern of a transducer or array on its overall size and geometry was discussed briefly in Chapter 2. The main purpose of array directivity for a transmitter is to concentrate the acoustic power into the direction of interest, and hence increase the source level in that direction. The direction of maximum sensitivity is known as the acoustic axis. For a receiver, the aim of directivity is generally to allow the resolution of signals from sources in different directions, or to reduce the noise received from angles other than the desired direction. Directivity also permits the angle of an incoming signal to be measured, but the accuracy of such measurements can often be appreciably better than the beam width itself, depending on the available processing, and the critical factors are therefore the angular resolution or noise rejection which is required. In this section, the beam patterns of transducers and arrays will be considered in more detail, to clarify their influence on the design of the transducers themselves.

Derivations of the theoretical beam patterns of transducers of various shapes are given in several texts, some of the most useful being listed in the bibliography at the end of this chapter, and this section will summarise some of the results. These beam patterns are commonly expressed in terms of far-field directivity functions, which relate the acoustic sensitivity to the angle (θ) relative to a line normal to the array or transducer face, and assuming the field to be measured far enough away from the transducer to be free from near-field effects. The range of these near-field effects will be discussed later in this section.

Circular piston in infinite baffle.

The directivity function $F(\theta)$ of a circular piston in an infinite rigid baffle is given [6.18] by

$$F(\theta) = \frac{2J_1(ka\sin\theta)}{ka\sin\theta} \qquad (6.20)$$

where J_1 is a Bessel function, a is the piston radius, and $k = 2\pi/\lambda$. Values of the function $2J_1(x)/(x)$ are tabulated in [6.18], and are illustrated in *Fig 6.8*, using $x = ka\sin\theta$ as the variable. The curve crosses the axis when $x = 3.83$, 7.02, 10.15, etc. The directivity function should thus have its first zero at the angle θ_1 corresponding to $ka\sin\theta = 3.83$, ie

$$\sin\theta_1 = 3.83/ka = 0.61\lambda/a \qquad (6.21)$$

These zeros (or nulls) exist symmetrically on either side of the maximum, and the beam between them is called the major lobe. At larger angles, the first side lobe exists between θ_1 and the angle given by $ka\sin\theta = 7.02$, and so on for the remaining minor lobes. The values of $2J_1(x)/x$ within this first side lobe region are negative, but this has no significance for the beam shape, since it is the square of this quantity which is evaluated in calculating the radiated intensity as a function of angle. Beam patterns are usually presented in logarithmic terms, in dB relative to the intensity on the acoustic axis, and may be plotted on rectangular axes or as a polar plot.

Although the angular width between the nulls defining the major lobe is sometimes of importance, it is more common to characterise the beam width in terms of the angle between the

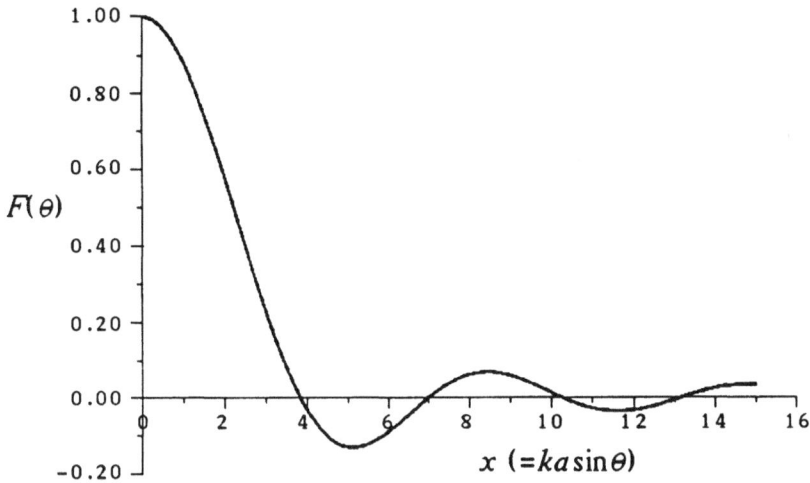

Fig 6.8   Directivity function $F(\theta)$ of circular piston of radius a in infinite rigid baffle.

two directions where the intensity first falls to some specific fraction of its maximum. The angle between the "half-power" directions (the −3dB points) is usually taken as the definition of **beam width** − and will generally be understood in this book unless otherwise stated, − but the −6dB or −10dB points have sometimes been used by other authors. The total beam width between the −3dB points may be derived from the directivity function (6.20), giving the result for a circular piston in an infinite baffle as

$$\theta(-3\text{dB}) = 2\sin^{-1}(0.512\lambda/2a) \qquad (6.22a)$$

$$\simeq 60°\,(\lambda/2a) \qquad (6.22b)$$

Equation (6.22b) is a good approximation for piston diameters greater than about one wavelength. Total beam widths to other levels are given approximately by

$$\theta(-6\text{dB}) \simeq 1.4\,\theta(-3\text{dB}) \qquad (6.23a)$$

$$\theta(-10\text{dB}) \simeq 1.8\,\theta(-3\text{dB}) \qquad (6.23b)$$

and $$\theta(\text{between nulls}) \simeq 2.3\,\theta(-3\text{dB}) \qquad (6.23c)$$

The calculated value for the peak of the first side lobe for a circular piston in an infinite baffle is −18dB relative to the level on the acoustic axis. The **directivity index** may be calculated by integrating the acoustic field over all directions. Provided that the piston diameter exceeds $\lambda/2$, the directivity index is given to within 1dB by

$$\text{DI} = 10\log(4\pi A_p/\lambda^2) \qquad\qquad (6.24)$$

where A_p is the area of the piston. The relationship between DI and beam width for rectangular and elliptical transducers was shown in Fig 2.5.

The expressions in the paragraphs above were derived on the assumption that the piston vibrated with uniform velocity and phase over its whole surface. If the velocity is not uniform over the face, the prediction of radiation pattern becomes much more complex, and will not be considered any further here. As a general rule, if the amplitude of vibration near the edges is smaller than at the centre, the beam width is wider than that calculated above, since the effective piston diameter is reduced. Removal of the baffle reduces the beam width for small pistons, though it has little effect on the beam width for pistons larger than 2λ in diameter. Replacement of the rigid baffle by a "pressure release" surface narrows the beam still further, to form a null along the baffle. *Fig 6.9*, from [6.19], shows total beam widths for the three different baffle conditions.

Effect of finite baffle.

The radiation patterns of a source in a finite size baffle have been studied experimentally by Delaney [6.24,6.25]. The edges of a baffle may be regarded as giving rise to virtual sources around the periphery, which are in anti-phase with the source because of the "pressure release" nature of the freedom to radiate in all directions from the baffle edge. Significant effects thus occur if the baffle is symmetrical around the source, and especially for a baffle radius of a few integral multiples of $\lambda/2$. In general, the beam pattern of a small source in a modest size baffle (ie a few wavelengths across) tends to be narrower than would be expected for the source in an infinite baffle. Beam patterns of a rectangular source in a cylindrical baffle have been treated theoretically by Laird and Cohen [6.26].

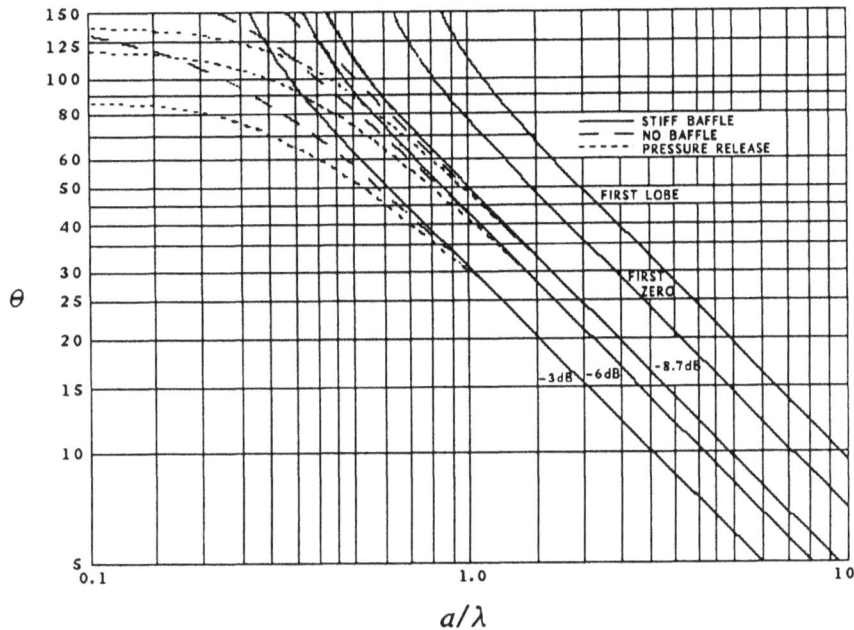

Fig 6.9 Total beam width (θ) for circular piston of radius (a) in a plane baffle, for an acoustic wavelength λ; from NRDC Div 6, Summary Tech Report Vol 13, *Design and Construction of Magnetostrictive Transducers*, 1946, p 128.

Continuous line source.

The directivity function for a continuous line source of length L, in a plane containing the line [6.19], is

$$F(\theta) = \frac{\sin\{(\pi L/\lambda)\sin\theta\}}{(\pi L/\lambda)\sin\theta} \qquad (6.25a)$$

which may also be written [6.20] as

$$F(\theta) = \frac{\sin(KL/2)}{KL/2} \qquad (6.25b)$$

where $K=(2\pi/\lambda)\sin\theta$. K thus represents the phase shift per unit length along the array due to the angle of an incident plane wave front.

In this case, the total beam width between the -3dB points is given by

$$\theta(-3\text{dB}) \simeq 50°\lambda/L \qquad (6.26)$$

The height of the first side lobe is -13dB relative to the main lobe. The directivity index of a continuous line source is given by

$$\text{DI} \simeq 10\log(2L/\lambda) \qquad (6.27)$$

provided that the length is at least one wavelength.

Rectangular source in infinite baffle.

The directivity function for a rectangular source, in a plane parallel to an edge (of length L_1) and normal to the face of the source, is the same as for a continuous line source of length L_1. If the length of the edge normal to the first is L_2, a similar expression for the directivity function in the plane parallel to L_2 is applicable, but with L_2 substituted for L_1. Note that the narrower beam is formed in the plane containing the longer side. The effect on the beam width of removing the baffle, or operating the source in a pressure release baffle, is similar to that for a circular piston [6.19]. The directivity index for a rectangular source in a baffle is also given by (6.24), provided neither side is less than about $\lambda/2$.

The directivity function of a rectangular source in a plane parallel to a diagonal and normal to the source is typical of an array with tapering, and shows reduced side lobe levels [6.19, 6.20].

Beam steering of a linear array of point sources.

For a linear array of N point sources, equally spaced with separation d, and of equal amplitude and phase, the directivity function is given by

$$F(\theta) = \frac{\sin(NKd/2)}{\sin(Kd/2)} \qquad (6.28a)$$

where $K = (2\pi/\lambda)\sin\theta$ [6.20]. When the phases of the sources are all equal, the peak response is in the broadside direction, perpendicular to the line of the array. The angle of the peak response may be steered away from broadside by applying a progressive phase shift φ between adjacent elements along the array. Equation (6.28a) then becomes

$$F(\theta) = \frac{\sin\{N(Kd + \varphi)/2\}}{\sin\{(Kd + \varphi)/2\}}$$ (6.28b)

The applied phase shift is related to the **steering angle** (θ_0) by

$$\varphi = -(2\pi/\lambda)d\sin\theta_0$$

and the general directivity function for a linear array of point sources steered to an angle θ_0 may thus be written as

$$F(\theta) = \frac{\sin\{(N\pi d/\lambda)(\sin\theta - \sin\theta_0)\}}{\sin\{(\pi d/\lambda)(\sin\theta - \sin\theta_0)\}}$$ (6.29)

Modifying K to $K' = (2\pi/\lambda)(\sin\theta - \sin\theta_0)$, this may be expressed as

$$F(\theta) = \frac{\sin(NK'd/2)}{\sin(K'd/2)}$$ (6.30)

which resembles (6.28a) except that $\sin\theta$ is replaced by $(\sin\theta - \sin\theta_0)$; ie the angles (in their sine form) are referred to the steer angle θ_0 instead of to the broadside direction. The length of the array is $(N - 1)d$.

Equation (6.30) is illustrated in *Fig 6.10* for the particular example of a linear array of nine point sources, $F(\theta)$ being plotted against $K'd/2$. The width of the main beam is determined by the value of Nd, which is approximately the total length of the array. This figure shows how the side lobes decrease as the angle moves away from the steer direction out to $K'd/2 = \pi/2$, and then rise again to form a diffraction peak at $K'd/2 = \pi$. This behaviour is then repeated cyclically over further intervals of π. Although it is convenient to use the variable $K'd/2$ for the abscissa in plotting this function, in order to obtain curves with general applicability, the actual application of the curves to any particular array involves some consideration of this variable. From the expression above,

$$K'd/2 = (\pi d/\lambda)(\sin\theta - \sin\theta_0)$$ (6.31)

For the broadside case, with $\sin\theta_0 = 0$, the maximum real value of $K'd/2$ is $\pi d/\lambda$, since $\sin\theta$ cannot exceed unity for real

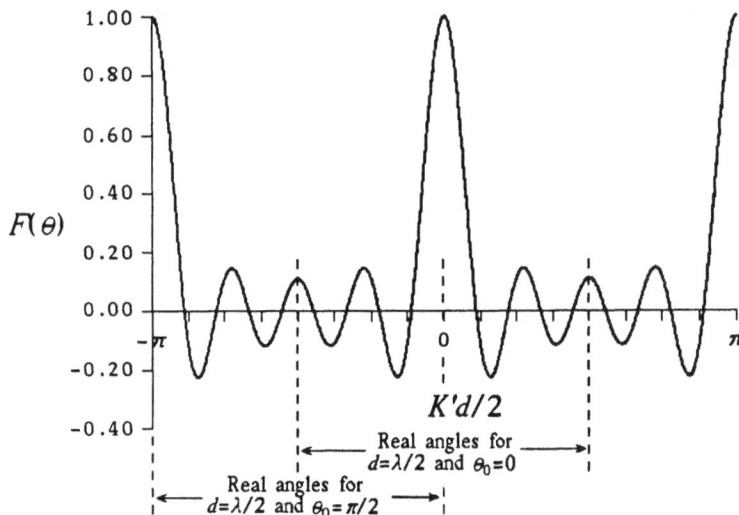

Fig 6.10   Directivity function $F(\theta)$ of linear array of point sources, $(n=9)$.

angles. Thus, if d is such that $\pi d/\lambda$ is less than π, then $K'd/2$ can never reach the value (π) at which the second diffraction peak occurs, and in this case only one major lobe occurs (in real space). This corresponds to the condition that the spacing should be less than λ, in order to avoid the existence of any secondary diffraction peaks, as derived in Section 2.1. For larger inter-element spacings, diffraction lobes (sometimes called "aliassing lobes") may appear, and must be taken into account. If the array is steered to $\theta_0 = -\pi/2$, the first diffraction lobe will occur when $(d/\lambda)(\sin\theta + 1) = 1$, ie $\sin\theta = (\lambda/d) - 1$. This will be a real angle if $\lambda/d - 1 \leq 1$, ie if $d \geq \lambda/2$, again as in Section 2.1. Thus, for an array steered to end-fire, the inter-element spacing should be less than $\lambda/2$ to avoid any diffraction lobes.

The zeros of $F(\theta)$ are equally spaced along the $K'd/2$ axis, and are therefore equally spaced in terms of $\sin\theta$. If an array is steered towards end-fire, the width of the main beam thus increases in terms of the actual angles (θ) in the plane of the array. The beamwidth between -3dB points for a uniformly spaced linear array (of length L) steered to end-fire is given approximately by

$$\theta(-3\text{dB}) \simeq 108°\sqrt{(\lambda/L)} \qquad\qquad (6.32)$$

In contrast to equation (6.26) for the broadside case, the beamwidth for the end-fire array is proportional to the square root of λ/L, and a narrow end-fire beam could therefore only be achieved by using a very long array. The same generalised curve (*Fig 6.10*) can be used to derive beam patterns, with the steered beam centred on $K'd/2 = 0$, and with corresponding adjustments to the limits of real angles, as indicated in *Fig 6.10*.

Because of the linear nature of the array, the beam patterns are symmetrical about the array, and in general form beams which are conical about the line of the array. In the case of the broadside beam, the main beam forms a toroid normal to the array; as the beam is steered towards end-fire, it widens in the plane of the array but becomes narrower in a plane through the main beam and normal to the array, and at end-fire it forms a single beam (for $d < \lambda/2$) with equal widths in both planes. As a result, the gain of the array remains approximately constant as the beam is steered, despite the broadening of the main beam in the plane of the array. For an array in which the element spacing is about $\lambda/2$ or greater, the directivity index is approximately equal to $10\log N$ (where N is the number of elements in the array) for all steering angles.

The above expressions apply to the case of a line array of point sources in which all the sources have the same amplitude, although their phases may vary linearly according to the applied steering angle. However, it is sometimes desirable to apply "shading" along the array, by varying the effective sensitivities of the elements, in order to achieve better control of the beam pattern. The commonest form of shading is to reduce the amplitudes towards the edges, since this has the effect of reducing the side lobes of the array [6.20], at the cost of some broadening of the width of the main beam, and the effects of various shading schemes have been analysed by several authors (see eg [6.20, 6.23, 6.27]). Varying both amplitude and phase of the elements is also used on occasions, for example to achieve narrower beams by means of "superdirective" techniques [eg 6.21]. For further details of such applications, the reader is referred to the list of suggested reading at the end of this chapter. The main point of importance concerning the transducer design itself is the greater sensitivity of superdirective arrays to variations in the characteristics of the elements of the array. Such arrays therefore require much care in controlling the tolerances of the individual elements. This is easier to achieve for non-resonant receiving arrays than for transmitting arrays operating at resonance.

Although the discussion has been in terms of an array of point sources, the linearity of transducer behaviour means that exactly the same results apply to a line array of omni-directional receivers. In that case, the beam pattern is that of the hydrophone response.

Woollett [6.22], in a useful review of various types of transducers, gave a classification of radiators according to their directional patterns. There are often several ways of achieving a particular beam pattern, and the choice to be adopted may depend on a balance of practical factors such as size, power requirements, and cost. Some particular examples are worth mentioning here, especially for receiving systems. It has already been noted that a vertical line array of co-phasal elements would have a beam pattern which is toroidal in the horizontal plane. If the array is reduced to only two elements with $\lambda/2$ spacing between them, the beam pattern will again be a toroid in the horizontal plane, with nulls along the vertical axis and a wide beam width in the vertical plane. If the spacing is much less than $\lambda/2$, the pattern will become more nearly omni-directional in the vertical. If the elements are closer than $\lambda/2$ and are connected in anti-phase, they will cancel signals arriving in the horizontal plane, but will give an output for signals arriving along the line joining the pair of elements. This is a version of an end-fire array, although in this case the design condition is to produce a null in the horizontal plane, and a small signal is produced for signals in either end-fire direction. Such a beam pattern is called a dipole pattern. It may be combined with an omni-directional sensor to produce a "cardioid" pattern with a peak response in one direction only, thus giving a small though directional sensor. The price to be paid for this super-directivity is a lower overall hydrophone sensitivity and more stringent requirements for close tolerances (See Section 10.10). Such arrangements are generally less suitable for projector systems than for hydrophones, because of the effects of array interactions and the need for transducer surfaces which are large enough to radiate the required power.

Linear array of elements each of finite length.

Equations (6.28)–(6.31) apply to a linear array of point sources. If the sources are of finite size, so that they have some directivity in themselves, they modify the overall beam pattern. If the elements are all of the same length L, with centre-to-centre separation d, the resulting overall directivity function is given [6.20] by

$$F(\theta) = \left\{ \frac{\sin(KL/2)}{KL/2} \right\} \cdot \left\{ \frac{\sin(NK'd/2)}{\sin(K'd/2)} \right\} \tag{6.33}$$

The directivity function is thus the product of (1) the beam pattern of any individual element by itself, and (2) the pattern for a corresponding linear point array. Although equation (6.33) is for a linear array of elements of finite length, the statement is an expression of the rather more general **product theorem**, which applies to linear arrays of elements of identical directivity patterns however they are produced [6.23]. A similar result applies to rectangular arrays of identical sources.

Equation (6.33) is directly applicable to an array which can radiate freely in all directions, as for the line array of point sources. It is more commonly applied to an array of elements mounted in a baffle. If the baffle is ideal, – ie infinitely large and rigid, – it prevents radiation from the rear of the baffle, but the directivity pattern in front of the baffle is correctly described by (6.33). The effect of finite impedance of the baffle is to multiply the directivity function by an extra factor of $\cos\theta/(\Delta + \cos\theta)$, where $\Delta = \rho c/Z(\omega)$, and $Z(\omega)$ is the acoustic impedance of the baffle. For a rigid baffle, $\Delta \ll 1$, and the effect is small except very near the plane of the baffle. For a soft baffle ($\Delta > 1$), the factor is approximately proportional to $\cos\theta$. In practice, the effect only becomes serious when the main lobe is steered far enough to have a significant contribution in the plane of the baffle.

The product theorem has considerable significance for practical planar arrays. For projector arrays, it is usual to make the elements fill as much as possible of the available radiating surface. If the element length becomes equal to the separation, the resulting pattern should be the same as for a continuous line, and this is indeed the result from (6.33) for an unsteered array with $d = L$. *Fig 6.11* illustrates the array pattern for a nine element array as before, with the dotted curve indicating the pattern for an element with $d = L$. The nulls of the element pattern coincide with the diffraction lobes of the array pattern, so that the diffraction lobes are cancelled for a completely filled array, and the pattern is that of a continuous line source. If the elements only occupy 0.9 of the separation distance, however, the element pattern is somewhat wider, as shown by the dot–dashed curve, and the diffraction lobe is not entirely cancelled. The effect of gaps between elements is thus to increase the potential problems from the major diffraction lobes, if the inter–element separation exceeds λ. The danger is eliminated

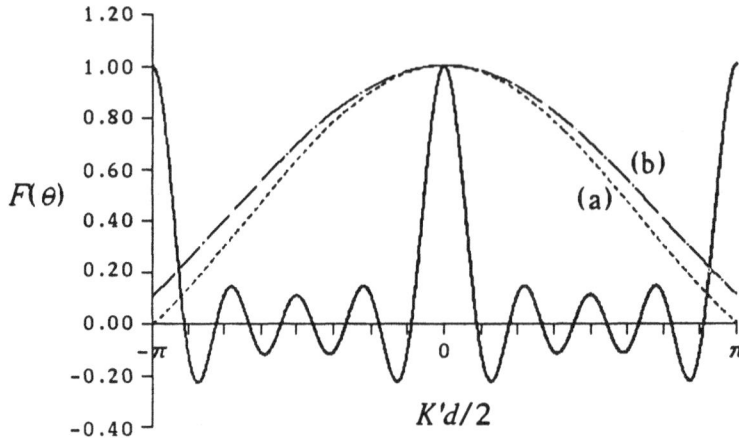

Fig 6.11 Directivity functions for linear array of finite length sources, $(n = 9)$: (a) $L = d$; (b) $L = 0.9d$.

if the spacing is less than λ, but this sometimes involves increased costs, and the chosen separation is often a balance between cost and technical problems. For arrays using circular pistons in a baffle, it is clear that the surface cannot be completely filled, and the possible existence of diffraction lobes must be taken into account.

The potential problems are more severe for arrays which are to be steered. In this case, the pattern due to the array is steered away from broadside, whilst the element pattern remains as before, since it is not affected by the beam steering. The amplitude of the main beam is therefore reduced by the element pattern, but the amplitude of the side lobe nearest to the broadside position increases. The beam pattern thus becomes unsymmetrical, and some of the side lobes may become appreciably larger, the effects being larger as the steering angle is increased. Again, the problems are reduced most effectively if the spacing can be made less than $\lambda/2$. This solution may be adopted fairly readily for low frequencies, where wavelengths are large; it is for higher frequency applications that element sizes to meet this condition become so small that greater difficulties arise in terms of practical arrangements for mounting and connecting the elements, and the associated costs.

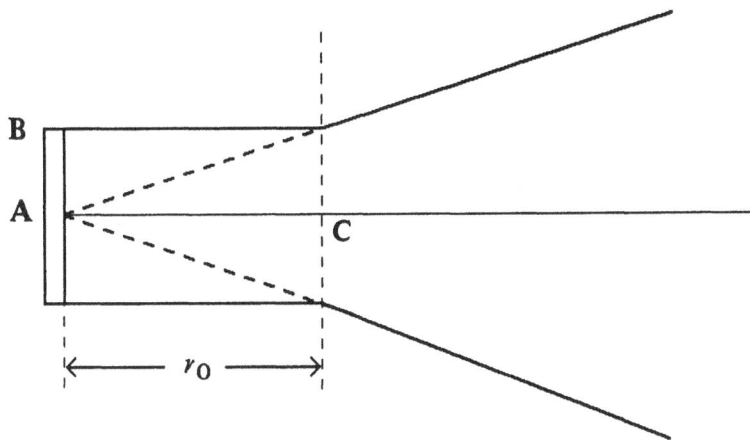

Fig 6.12 Acoustic field regions in front of a plane radiating surface.

Near field effects.

The acoustic field produced by an array is the summation of the contributions from all the elements, allowing for the differing acoustic path lengths from each element to the point of interest in the field. At short ranges from the array, the distances to each element vary rapidly with position, and the acoustic field then exhibits a very complex structure. This region, within which the acoustic field shows rapid variations with position, is known as the near field, or Fresnel region. As the range increases, the path length differences settle down to a more gradual variation, and the field becomes that of the far field, or Fraunhofer region. The beam patterns quoted in the earlier parts of this section are all for this far field. The elements of the array may also be interpreted as elemental areas of a single piston, and the same considerations therefore apply to the field in front of a single radiating source.

The usual illustration of the regions is shown in *Fig 6.12*. For a circular radiating face, the acoustic field near the transducer is confined mostly within a cylinder extending outwards from the face to a transition range (r_0), where it starts to diverge and form its far field pattern. Although the acoustic intensity within the near field region shows rapid fluctuations with position, it remains generally confined within the cylinder, with an average intensity which is nearly constant with range, and for high power transmissions it is possible for cavitation to occur throughout this region. The transition range at which the far field may be considered to start is determined by the range where the path difference between the shortest and longest paths (AC and BC in *Fig 6.12*) becomes a small fraction of a wavelength. This path difference is given approximately by $L^2/8r$ (for $r \gg L$), where L is the length of the transducer face and r is the range. How small a fraction of λ is needed to achieve far field conditions is a matter of judgement, and various authors quote different values, but a choice of $\lambda/8$ is fairly typical, and gives the simple result that the transition range is approximately L^2/λ [6.29]. This range (often called the Fresnel range) depends on the dimensions of the radiating face rather than its shape, and the same expression may be used for other shapes of transducer faces if L is taken as the maximum dimension. For a narrow rectangular array, there is a short transition range corresponding to the array width, and beyond that range the beam diverges in the plane containing the short array dimension; there is then a further transition range, corresponding to the longer array dimension, which indicates the end of the near field region .

Zemanek [6.30] showed that the actual acoustic field near to a radiating piston differs appreciably from the idealised diagram of *Fig 6.12*. By calculating the acoustic field in front of a flat circular piston mounted in an infinite baffle, he showed that the field is concentrated to pass through an area of diameter typically one-quarter of the piston diameter, at a range of about $0.2L^2/\lambda$, as illustrated in *Fig 6.13* (from [6.30]), which shows the −3dB and the −6dB contours for a piston of diameter equal to 5λ. Zemanek's calculations suggest that the acoustic field settles down to its far-field behaviour well within the generally accepted Fresnel range of L^2/λ. However, this value of L^2/λ is usually taken as the range of the boundary between near and far field, even though it may be unnecessarily conservative.

6-5   PARAMETRIC ARRAYS

The relationship between pressure and density for water is not exactly linear, and this produces some distortion of the acoustic waveform for signals of large amplitude. This non-linear behaviour is the basis for the "parametric arrays", which have particular uses for generating narrow beams at low

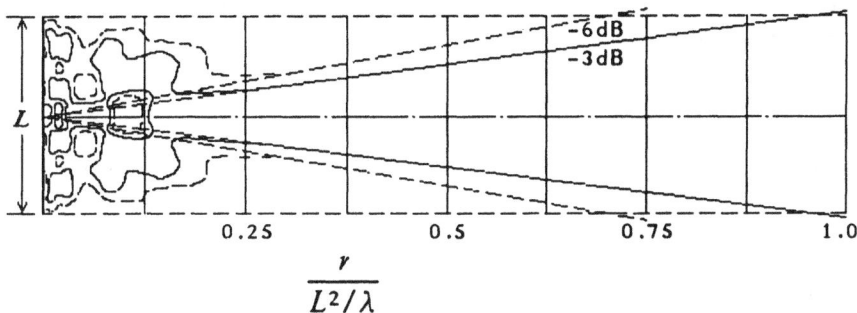

Fig 6.13 Acoustic field contours in front of radiating piston of diameter 5λ; from *J Acoust Soc Am*, in Zemanek,J., *JASA* 49, 181-191, (1971).

frequencies from relatively small arrays. If two sound waves at frequencies f_1 and f_2 propagate in the same direction through water, the non-linearity of the medium results in the generation of sound at the sum $(f_1 + f_2)$ and difference $(f_1 - f_2)$ frequencies. The primary frequencies f_1 and f_2 are generally chosen to be close together, so that $f_1 - f_2$ is much less than either f_1 or f_2. The sum frequency component is rapidly absorbed, whilst the difference frequency propagates with less attenuation than either of the primary frequencies, so that at long range the field is dominated by the difference frequency.

This basic mechanism is realised in practice by transmitting the two primary frequencies from a single array. Much of the interaction takes place in the near field, where the amplitude is highest, and the effect is to produce a virtual array extending from the face of the transmitter into the medium. The propagation of the primary waves through this region causes the virtual array to behave as an end-fire source, and radiation of the difference frequency therefore occurs, with its acoustic axis normal to the face of the transmitter. Since the length of the virtual array can be quite long, the beamwidth of the difference frequency component can be very narrow, and much narrower than would be achieved for a conventional array of the same radiating area operating directly at the difference frequency. In addition to this potential for generating narrow beams at low frequencies, parametric arrays possess several other advantages:-

a)   Because the virtual array behaves as an end-fire array with exponential shading, it has no significant side lobes outside the main lobe.

b)   The beam width is approximately constant over a wide band of frequencies.

c)   Radiation of broad band signals is possible, since small fractional modulations of the primary frequencies become much larger fractional modulations at the difference frequency. For example, suppose that the nominal primary frequencies are 100kHz and 110kHz; a change of 5kHz in either primary frequency can thus produce a variation in the difference frequency from 5 to 10kHz. This large proportional bandwidth at the difference frequency is easily achieved, since it requires only a 5% swing in the driving frequency, well within the bandwidth of a practical transducer.

d)   Cavitation is rarely a problem, since the cavitation level increases rapidly with frequency, and is hence much higher at the primary frequencies than at the difference frequency.

These desirable features are of course obtained at the cost of some disadvantages, the most serious being the very low conversion efficiency from electrical power to acoustic power at the difference frequency. This is to some extent counteracted by the directivity achievable by the array, so that quite useful source levels can be produced. The source level is very sensitive to the power output at the primary frequencies, because the non-linearity is very dependent on the acoustic pressure amplitudes. And if steering of the beam is wanted, it is usually necessary to effect it by mechanical rotation of the transmitter.

The design of parametric arrays is described by Berktay [6.31], and a useful review has also been given by Bjørnø [6.32]. The arrays may be classified into two types according to the value of the parameter $\alpha_T R_0$, in which R_0 is the Fresnel range, (L^2/λ or (array area)$/\lambda$), and α_T is an absorption coefficient given by $\alpha_T \simeq \alpha_1 + \alpha_2 + \alpha_d$, where α_1, α_2, α_d are the absorption coefficients at the two primary frequencies and at the difference frequency. (R_0 is referred to as the Rayleigh range in [6.31] and [6.32].) If the absorption is strong, or the Fresnel range long, so that $\alpha_T R_0 \gg 1$ (Np), then most of the non-linear interaction takes place within the near field region, and the array is said to be "absorption limited." If $\alpha_T R_0$ is small, much of the interaction takes place beyond R_0, and the array is said to be "spreading-loss limited."

For absorption limited arrays, the radiated beamwidth at the difference frequency is given approximately by

$$\theta_d \simeq (\alpha_T \lambda_d / \pi)^{1/2} \qquad (6.34)$$

where λ_d is the wavelength at the difference frequency $(f_1 - f_2)$. The beamwidth thus increases as the primary frequencies are increased (corresponding to higher absorption), or as the difference frequency is reduced (ie for larger λ_d). This is as expected, since the first effect reduces the length of the virtual array, whilst the second reduces the ratio of array length to wavelength. For spreading–loss limited arrays, the beamwidth increases with range, and becomes asymptotic to the half–power beamwidth of the product of the primary beam directivity patterns. The source level at the difference frequency $((SL)_d)$ is given [6.31] by an expression of the form

$$(SL)_d = (SL)_1 + (SL)_2 + 20\log F_d + Q$$

$$- 286.5 \text{ dB re } 1\mu\text{Pa at } 1\text{m} \qquad (6.35)$$

where F_d is the difference frequency in kHz, and Q is a parameter varying from about +10 to −10dB as $\alpha_T R_0$ increases. (Its values are plotted in [6.31].) The source level thus increases as the difference frequency is increased – although it is necessary to take into account the influence of F_d on the value of Q in order to evaluate the effect fully.

Berktay [6.31] commented that there was no unique solution for the design of a parametric array to meet a particular specification, and discussed some of the factors involved. The ratio between primary and difference frequencies is typically between 10 and 20. Many of the parametric arrays have been designed for high frequencies, but Bjørnø quoted an example of a large array, using a projector with a radiating face 0.5 × 2m, operating at primary frequencies around 24kHz, to generate signals between 250Hz and 5kHz. The primary source level was 259dB re 1μPa at 1m, and the beamwidths of the primary frequencies were 2° × 8°. The difference frequency source levels varied from 186dB at 250Hz to 230dB (re 1μPa at 1m) at 5kHz. Quite useful source levels can thus be produced, even at low frequencies, despite the low overall efficiency of the method.

The discussion above is a very brief and simplified treatment of the subject of parametric acoustic sources, fuller information being given in references such as [6.31] and [6.32]. Parametric receivers have also been described [6.32], but have not been widely used.

Once the requirements of the basic transmitter for a parametric array have been determined, the aim of the transducer design is to satisfy these requirements at the primary frequency. The fact that the array is to be used as a parametric source has little, if any, influence on the design, which can thus be tackled using methods as for other arrays at similar frequencies.

REFS Chapter 6

References

6.1 Kinsler,L.E., and Frey,A.R., *Fundamentals of Acoustics*, (2nd Ed), Wiley & Sons, 1962, p180.

6.2 Morse,P.M., and Ingard,K.U., *Theoretical Acoustics*, McGraw-Hill, 1968, Sect 7.4.

6.3 Porter,D.T., "Two Fortran programs (0577 and 0717) for computing electroacoustic behaviour of transmitting sonar arrays", Underwater Sound Laboratory Report No 791, (1967).

6.4 Morse,P.M., *Vibration and Sound*, McGraw-Hill (2nd Ed), 1948, p326 (Reprint published by Acoust Soc Am, 1981).

6.5 Crane,P.H.G., "Method for the calculation of the acoustic radiation impedance of unbaffled and partially baffled piston sources", *J. Sound Vib*, $\underline{5}$, 257-277, (1967).

6.6 Mangulis,V., "On the radiation of sound from a piston in a non-rigid baffle", *J Acoust Soc Am*, $\underline{35}$, 115-116, (1963).

6.7 Beranek,L.L., *Acoustic Measurements*, Wiley and Sons, 1949, p60 (Revised edition published by Acoust Soc Am, 1988).

6.8 Beranek,L.L., op. cit., p62.

6.9 Hueter,T.F., and Bolt,R.H., *Sonics*, Wiley & Sons, 1955, p56.

6.10 Pritchard,R.L., "Mutual acoustic impedance between radiators in an infinite rigid baffle", *J Acoust Soc Am*, $\underline{32}$, 730-737, (1960).

6.11 Sherman,C.H., "Analysis of Acoustic Interactions in Transducer Arrays", *IEEE Transactions on Sonics and Ultrasonics*, $\underline{SU-13}$, 9-15, (1966).

6.12 Arase,E.M., and Hahn,P.D., "Mutual radiation impedance of square and rectangular pistons in a rigid infinite baffle", *J Acoust Soc Am*, $\underline{36}$, 1521-1525, (1964).

6.13 Freedman,A.F., "Approximations for the mutual radiation impedance coefficients between baffled and unbaffled circular pistons", Admiralty Underwater Weapons Establishment, AUWE Tech Note 320/68, (1968).

6.14 Sherman,C.H. "Theoretical model for mutual radiation resistance of small transducers at an air-water surface", *J Acoust Soc Am*, $\underline{37}$, 532-533, (1965).

6.15   Toulis,W.J., "Radiation load on arrays of small pistons", *J Acoust Soc Am*, <u>29</u>, 346–348, (1957).

6.16   Morris,J.C., "Average radiation impedance of circular pistons in a broadside array", Admiralty Underwater Weapons Establishment, AUWE Tech Note 326/68, (1968).

6.17   Carson,D.L. , "Diagnosis and cure of erratic velocity distributions in sonar projector arrays", *J Acoust Soc Am*, <u>34</u>, 1191–1196, (1962).

6.18   Kinsler,L.E., and Frey,A.R., op. cit. Ch 8.

6.19   Nat. Def. Res. Comm. Div. 6, Summary Technical Report, Vol 13, *Design and Construction of Magnetostriction Transducers*, 1946.

6.20   Tucker,D.G., and Gazey,B.K., *Applied Underwater Acoustics*, Pergamon Press, 1966, Ch 6.

6.21   Pritchard,R.L., "Optimum directivity patterns for linear point arrays", *J Acoust Soc Am*, <u>25</u>, 879–891, (1953).

6.22   Woollett,R.S., "Ultrasonic transducers: 2. Underwater sound transducers", *Ultrasonics*, <u>8</u>, 243–253, (1970).

6.23   Urick,R.J., *Principles of Underwater Sound*, McGraw–Hill (3rd Ed), 1983, Ch 3.

6.24   Delaney,M.E., Burton,A.J., and Rennie,A.J., "Radiation from a point source of sound on the surface of rigid spheres and discs", National Physical Laboratory, Report No AP 21, (1965).

6.25   Delaney,M.E., and Rennie,A.J., "Radiation from a point sound source on the surface of rigid cylindrical baffles", National Physical Laboratory, Report No AP 26, (1967).

6.26   Laird,D.T., and Cohen,H., "Directivity patterns for acoustic radiation from a source on a rigid cylinder",

6.27   *J Acoust Soc Am*, <u>24</u>, 46–49, (1952).
         Drane,C.J., "Derivation of excitation coefficients for Chebychev arrays", *Proc IEE*, <u>110</u>, 1755–1758, (1963).

6.28   Morse,P.M., and Ingard,K.U. , op. cit., Section 7.4.

6.29   Bobber,R.J., *Underwater Electroacoustic Measurements*, US Government Printing Office, 1970, Section 3.4.

6.30   Zemanek,J., "Beam behaviour within the nearfield of a vibrating piston", *J Acoust Soc Am*, <u>49</u>, 181–191, (1971).

6.31   Berktay,H.O., "Parametric Sources – Design Considerations in the Generation of Low Frequency Signals", in Akal,T., and Berkson,J.M. (Eds), *Ocean Seismo-Acoustics; Low Frequency Underwater Acoustics*, Plenum Press, 1986.

6.32   Bjørnø,L., "Parametric Acoustic Arrays", in Tacconi,G. (Ed), *Aspects of Signal Processing (Proc NATO Advanced Study Institute, Portovenere, La Spezia)*, Reidel, 1977.

6.33   Urick,R.J., op. cit., Section 4.3.

Additional Reading.

1   Skudrzyk,E., *Simple and Complex Vibratory Systems*, Pennsylvania State Press, 1968, Chap 12. (Radiation Fields).

2   Hanish,S., *A Treatise on Acoustic Radiation*, Vol I, 2nd Ed, Naval Research Laboratory, 1981.

3   Camp,L. (Ed), *Underwater Acoustics*, Wiley Interscience, 1970, Chap 7. (Beam patterns and shading).

4   Albers,V.M. (Ed), *Underwater Acoustics Handbook*, Pennsylvania University Press, 1965, Chap 13. (Beam patterns and shading).

5   Vanderborck,G.,   Steichem,W.,   and   Fromont,B., "Computation of the self and mutual radiation impedance between transducers in an array in presence of thick elastic skin using a mixed finite element-plane wave method", *Proc IoA*, 6, Pt 3, 86(and A1), (1984).

Chapter 7
ELEMENT DESIGN

7-1   INTRODUCTION

Chapters 4 and 5 considered the transducer as a component in the overall system for transferring power from the electrical source to a load representing the acoustic radiation. The value of this impedance in acoustic terms was treated in Chapter 6. In this chapter we discuss how these approaches are brought together to derive the design of the transducer element itself. As a general rule, transducer design does not need to be treated as a precise science. Although it is based on established physical principles, and exact equations could be derived for the design, the parameters obtained rarely need to be realised precisely, and various approximations can be used to make the analysis simpler and easier to interpret. We shall therefore start by assuming a rather simplified model of a transducer in order to establish the basic design, and improve the approximations later, when necessary, to obtain more exact values. In practice it is usual to find that this simplified approach is quite adequate, since for many applications the only parameter whose exact value is critical is the resonance frequency, and this is often achieved partly by empirical means.

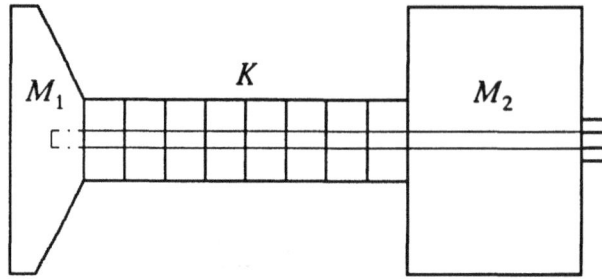

Fig 7.1   Basic design of piezoelectric transducer element.

The treatment will assume the element to have a basic design as indicated in *Fig 7.1*, in which a piezoelectric stack is assembled between a front piston (M_1) which radiates into water, and a countermass (M_2) at the rear, which provides inertial backing. Such a design is typical of low frequency piston-type transducers, operating between about 2kHz and 30kHz, and is often referred to as a "Tonpilz" design. (This term (literally a "vibrating mushroom") was used by Fischer [7.1] in one of the earliest treatments of transducer designs which incorporated end-masses to lower the resonance frequency, but he attributes the expression to Hahnemann and Hecht in 1920 [7.2].) Particularly towards the low end of this frequency range, the components of the element have dimensions which are small compared to a wavelength of sound in the appropriate material, and it is possible to adopt the "lumped mass" approximation. The end masses can then be represented simply by their static masses, without any effects due to their elasticity, and the piezoelectric stack as a massless spring (of stiffness K). The stack is assembled from a number of piezoelectric rings, cemented together, with electrodes to allow the electric field to be applied. When an alternating field is applied to the stack, it causes the end masses to be displaced in opposing directions, and in the absence of any external force or reaction the displacements of the two masses are inversely proportional to their masses, thus keeping the centre of gravity stationary. If the two masses are equal, the design is said to be balanced and the displacements of the front and rear have the same amplitude. If however the rear mass is much heavier than the front piston, most of the displacement will occur at the front piston, which radiates into the water, and such unbalanced designs have some advantages, which will become evident later.

During part of the vibration cycle, the piezoelectric stack is under tension as it is accelerating the masses inwards towards each other. Unfortunately, ceramic materials are notoriously unreliable in tension, and although a tensile strength may be quoted for them, it is safer if possible to avoid any tensile stress being applied. This may be effected by inserting through the middle of the stack a "centre bolt", which applies a compressive force greater than the peak alternating stress, so that it prevents the stress in the ceramic ever becoming tensile rather than compressive. This simple model of a transducer element will be used to derive the basic design equations, leaving some of the more practical aspects to be discussed in the next chapter.

7-2   BASIC ELEMENT DESIGN

Consideration of the overall acoustic characteristics required for the array leads to a choice of array geometry. In particular, the size and separation of the transducer pistons are chosen from the beam steering requirements, together with some practical aspects such as size, weight, and cost of the array, as indicated in Chapter 2. Once these parameters have been selected, the equations in Chapter 6 are used to determine the radiation impedance of the transducer. If the element forms part of an array, it is the mean radiation impedance which needs to be evaluated. The required system performance determines what tranducer bandwidth is necessary, and we shall assume that the requirement is fairly demanding, so that a good match to the source impedance is wanted over as wide a range of frequency as possible. This implies that we should aim for the optimum relationship $Q_M W \simeq 1.2$, as in Chapter 5. The required bandwidth thus indicates the necessary coupling coefficient, and hence the value of Q_M. The system characteristics also specify the resonant frequency of the transducer in water. These specified parameters can now be used to determine the main features of the transducer, assuming the lumped mass approximation.

The resonance frequency for a mechanical system as in *Fig 7.1* is given by

$$\omega_S^2 = (2\pi f_S)^2 = K \frac{(M_1 + M_2)}{M_1 M_2} \qquad (7.1)$$

where K is the stiffness of the stack between the end masses. K is related to the dimensions of the stack by

$$K = \frac{E_e A_c}{\ell_c} \qquad\qquad (7.2)$$

where E_e is the effective Young's modulus of the stack, and A_c and ℓ_c represent its area and length. To a first approximation E_e could be taken as the Young's modulus of the piezoelectric ceramic alone, although it will be shown later that better accuracy can be obtained by including the effects of the other components of the stack, such as the centre bolt and the joints between the ceramic rings.

Fig 7.1 represents a transducer which radiates acoustic energy from a piston (M_1), whilst the "countermass" (M_2) is supported in air within the outer casing. We thus assume that all of the energy dissipation is associated with the motion of M_1, and may be represented by a (mechanical) resistor (r). An "effective vibrating mass" (M_e) can be defined by $M_e = M_1(1 + M_1/M_2)$, and the mechanical Q-factor of the system is then given by

$$Q_M = \frac{\omega_S M_e}{r} = \frac{\omega_S M_1}{r}(1 + M_1/M_2) \qquad\qquad (7.3)$$

For a transducer with low internal losses, the value of r is approximately equal to the radiation resistance calculated for the particular piston size and array geometry by the methods described in Chapter 6. This value can be substituted, together with the required values of Q_M and ω_S, in equation (7.3) to calculate what value of M_e is wanted. We now have to make a choice of the ratio of M_1/M_2, the "head-to-tail" mass ratio.

It is quite common to use a balanced design, in which the head and tail masses are equal, ie $M_1/M_2 = 1$. In that case, $M_e = 2M_1$. However, in those applications when a low Q_M is needed, it is clear from equation (7.3) that a low effective mass is desirable. It is often difficult to make a piston of the required size with a sufficiently low piston mass, and in such cases there is some advantage in choosing an unbalanced design with a much heavier countermass, ie $M_1/M_2 \ll 1$. For example, if the tail mass is ten times the piston mass, so that $M_1/M_2 = 0.1$, the effective mass M_e is only $1.1M_1$, and the problems of achieving a satisfactory piston design are much reduced. We therefore select a value of M_1/M_2 on these practical grounds, the chosen value usually lying between 1 and about 0.1. The value of M_e derived from equation (7.3), and the chosen M_1/M_2, then determine M_1 and M_2.

Equation (7.1) is then used to calculate the stiffness of the stack. Assuming a value for the effective Young's modulus of the stack, this gives the ratio of area to length of the stack, A_c/ℓ_c.

ie:-
$$\frac{A_c}{\ell_c} = \frac{K}{E_e}$$

$$= \frac{M_1 \omega_S^2}{E_e(1 + M_1/M_2)} \qquad (7.4)$$

In order to determine the actual dimensions of the stack we need another relationship, and this can be obtained by considering what volume of ceramic is needed to handle the input power.

Power limits

The role of cavitation in limiting the output power of a transducer has already been discussed (Sect 2.3). In this section, however, the limitations which arise from the internal features of the design are considered. The most important of these factors in limiting the power output of a piezoelectric transducer are:-

(a) Mechanical tensile stress in the ceramic stack. This can usually be counteracted by the use of a centre bolt, the design of which will be treated in more detail later.

(b) Mechanical hysteresis losses in the stack, either in the ceramic itself or in the joints.

(c) Electrical stress in the ceramic stack. This may limit the power for the following reasons:-

(1) the dielectric loss in the ceramic may cause heating, in addition to the heating due to mechanical losses as in (b). If the temperature rises too far, depoling of the ceramic may occur.

(2) depoling of the ceramic may occur if an excessively high field is applied, even at room temperature.

(3) Electrical breakdown may occur either through the ceramic or along its surface.

Of these factors, (2) requires fields which are appreciably larger than are applied in practice. Heating of the ceramic due to the combination of dielectric and mechanical losses is likely to be important only when long pulses or continuous transmissions are needed. In the more common cases when the transmitted pulses are short compared with the thermal time constant of the stack, the time–averaged power is usually low enough for heating not to cause serious limitations. The danger of over–heating is minimised by choosing a piezoelectric ceramic with low internal losses, and by careful assembly of the stack. In practice, for most piezoelectric transducers, the critical limiting factor is (3), the electrical breakdown which can occur, usually along the surface of the ceramic.

For most applications it is therefore reasonable to assume that the power is limited by the maximum electrical field which can safely be applied. The value of this limiting field cannot be specified accurately, as it depends on the ceramic and construction details, but a fairly conservative figure would be about 2kV/cm rms, and this value will be used where appropriate in the examples. Some caution may be needed in particular cases to ensure that the assumption that the power is limited by electric field rather than mechanical heating is justified.

Ceramic dimensions.

The power dissipated in the motional arm for an applied voltage V is given by $W_m = V^2 G_m$. The acoustic power radiated at resonance is then given by

$$W_{aS} = \eta_{ma} V^2 G_{mS}$$

$$= \eta_{ma} V^2 Q_M \omega_S k^2 C_{LF} \qquad (7.5)$$

(using equation (4.27)). If the ceramic stack is composed of a number (n) of rings of thickness t connected electrically in parallel, the length and capacitance of the stack are given by

$$\ell_c = nt \qquad (7.6)$$

$$C_{LF} = \frac{\varepsilon A_c n^2}{\ell_c} \qquad (7.7)$$

Thus

$$W_{aS} = \eta_{ma}\, V^2 Q_M \omega_S k^2 \cdot \frac{\varepsilon A_c n^2}{\ell_c}$$

$$= \eta_{ma}\, Q_M \omega_S\, k^2 \varepsilon \cdot \frac{V^2}{t^2} \cdot (A_c \ell_c)$$

$A_c \ell_c$ is the volume of the ceramic, and the acoustic power handling capacity of the stack is therefore given in terms of power/unit volume of ceramic by

$$\frac{W_{aS}}{A_c \ell_c} = \eta_{ma}\, \omega_S Q_M k^2 \varepsilon E^2 \qquad (7.8)$$

where $E = V/t$ is the applied rms electrical field.

Multiplication of equations (7.4) and (7.8) allows the ceramic length to be derived. Thus

$$\frac{W_{aS}}{\ell_c^2} = \frac{M_1 \omega_S^2}{E_e(1 + M_1/M_2)}\, (\eta_{ma}\, \omega_S Q_M k^2 \varepsilon E^2)$$

$$\ell_c^2 = \frac{W_{aS} E_e(1 + M_1/M_2)}{M_1 \omega_S^3 \eta_{ma} k^2 Q_M \varepsilon E^2} \qquad (7.9)$$

Similarly, the ceramic cross sectional area is given by

$$A_c^2 = \frac{W_{aS} M_1 \omega_S}{E_e(1 + M_1/M_2)\eta_{ma} k^2 Q_M \varepsilon E^2} \qquad (7.10)$$

These two equations allow us to calculate the dimensions of the ceramic stack, using the previously determined value of M_1. It is of interest, however, to evaluate these dimensions in terms of the basic transducer parameters. To do this, equation (7.3) is used to substitute for M_1, and r is expressed in normalised form as $r = \rho c A_p \tilde{R}$. The power output (W_{aS}) can also be written as $W_{aS} = W_0 A_p$, where W_0 is the power density, ie the power per unit area of the piston. Then

$$\ell_c = \frac{(1 + M_1/M_2)}{\omega_S k Q_M E} \left\{ \frac{W_0 E_e}{\eta_{ma} \varepsilon \rho c \tilde{R}} \right\}^{1/2} \qquad (7.11)$$

$$\frac{A_p}{A_c} = (1 + M_1/M_2) k E \left\{ \frac{E_e \eta_{ma} \varepsilon}{W_0 \rho c \tilde{R}} \right\}^{1/2} \qquad (7.12)$$

Note that in equation (7.12) it is the ratio of piston to ceramic area which is derived, and that this ratio is independent of both frequency and Q_M. Predicted values of A_p/A_c may thus be calculated by assuming typical values of the appropriate constants, and should apply over a wide range of resonance frequencies and Q_M values.

For example, assume $\rho c = 1.5 \times 10^6 \text{kg.m}^{-2}.\text{sec}^{-1}$ for sea water, an idealised case of $M_1/M_2 = 0$, and a typical value of the efficiency $\eta_{ma} = 0.9$. From (6.16), $\tilde{R}$ for an element in an array is given approximately by the value of χ, which is defined as A_p/A_a (where A_a is the area of array per piston). Thus $W_0\chi$ is equal to W_a/A_a, and is thus equal to the power output per unit area of array. This will be denoted by W_0' and it is convenient to use this instead of W_0 as a parameter relating to the power output of the array. For a single piston, $\tilde{R}$ should be replaced by the value of R_1 from Section 6.1.

For a transducer made of a typical barium titanate projector material, it is reasonable to assume its relative dielectric constant to be $\varepsilon_r = 1200$, so that $\varepsilon = \varepsilon_r\varepsilon_0 = 1.062 \times 10^{-8}\text{F/m}$ (using $\varepsilon_0 = 8.85 \times 10^{-12}\text{F/m}$), and that the effective Young's modulus of the stack, making a small allowance for the joints, may be taken as $E_e = 11 \times 10^{10}\text{Pa}$. Equation (7.12) then becomes

$$\frac{A_p}{A_c} = 2.65 \times 10^{-2} \cdot \frac{kE}{(W_0')^{1/2}} \qquad (7.13)$$

Some typical operating values for the transducer can now be assumed; eg a maximum applied field of 2 kV/cm, an effective coupling coefficient $k = 0.28$, and a power output W_0' of 1 watt/cm^2 of array area. Then the ratio of piston to ceramic area has the value

$$\frac{A_p}{A_c} = 14.8 \qquad (7.13\text{a})$$

Similar expressions can be evaluated for a transducer using a typical lead zirconate titanate projector material (eg Type 1 material). In this case, we assume for the stack,

$$\varepsilon_r = 1300, \text{ giving } \varepsilon = 1.15 \times 10^{-8} \text{ F/m}$$
$$E_e = 6.0 \times 10^{10} \text{ Pa}$$

Then
$$\frac{A_p}{A_c} = 2.03 \times 10^{-2} \cdot \frac{kE}{(W_0')^{1/2}} \qquad (7.14)$$

And for the assumed operating values of $E = 2\text{kV/cm}$,

$W_0' = 1$ watt/cm^2, and a typical coupling coefficient of $k = 0.5$, this becomes, for a LZT transducer,

$$\frac{A_p}{A_c} = 20.3 \qquad (7.14a)$$

These expressions thus give typical values for the ratio of piston to ceramic area for transducers meeting the rather idealised assumptions quoted.

Assuming the matched condition of $kQ_M = 1.2$, and the typical values above, the power handling capacity of the ceramic can be calculated from (7.8), giving:–

for barium titanate,

$$W_a/\text{ceramic volume} = 0.8F_S \text{ watts/cm}^3 \qquad (7.13b)$$

for lead zirconate titanate,

$$W_a/\text{ceramic volume} = 1.6F_S \text{ watts/cm}^3 \qquad (7.14b)$$

where F_S is the resonance frequency in kHz.

Centre Bolt.

The function of the centre bolt through the piezoelectric ceramic stack is to prevent the ceramic being subjected to a tensile stress during part of the alternating stress cycle, as explained in the introduction to this chapter. The bolt is used to apply to the stack a steady compressive force which is greater than the peak alternating force in the stack. As a general rule it is advisable to allow a factor of safety by making this compressive force twice the peak value of the alternating force in the ceramic at maximum working power. The bolt has the undesirable effect of tending to clamp the end masses by restricting their motion, but this can be tolerated if the stiffness of the bolt is appreciably less than that of the stack. The design requirement is therefore for a bolt which is strong enough to apply the required force, yet is not so stiff as to restrain motion of the end masses too firmly. In this section the basic design of the centre bolt is considered, and the feasibility of the technique is confirmed.

The peak alternating force in the stack may be deduced from the amplitude of vibration of the piston. If the power output is $W_a = W_0 A_p$, then

$$W_0 A_p = u^2 \rho c A_p \tilde{R}$$

where u^2 is the mean square velocity of the radiating piston and $\tilde{R}$ is the normalised radiation resistance, ie $R_r = \rho c A_p \tilde{R}$. Thus

$$u = \left\{ \frac{W_0}{\rho c \tilde{R}} \right\}^{1/2} \qquad (7.15a)$$

If the motion of the piston is sinusoidal and of angular frequency ω, the rms displacement (x) is simply related to the rms velocity by

$$x = u/\omega \qquad (1.4a)$$

$$= \frac{1}{\omega} \left\{ \frac{W_0}{\rho c \tilde{R}} \right\}^{1/2}$$

The peak value of the displacement is therefore

$$\hat{x} = \frac{1}{\omega} \left\{ \frac{2W_0}{\rho c \tilde{R}} \right\}^{1/2} \qquad (7.15b)$$

This represents the displacement of the piston itself; the displacement of the rear mass is M_1/M_2 times that of the piston and is in the opposite phase, so that the extension of the whole stack is calculated by multiplying the piston displacement by $(1 + M_1/M_2)$.

The force in a spring is given by the product of stiffness and extension. Thus the peak dynamic force in the ceramic stack is

$$\hat{F}_c = \frac{K}{\omega} \left\{ \frac{2W_0}{\rho c \tilde{R}} \right\}^{1/2} (1 + M_1/M_2)$$

At resonance, when $\omega = \omega_S$, this becomes

$$\hat{F}_{cS} = M_1 \omega_S \left\{ \frac{2W_0}{\rho c \tilde{R}} \right\}^{1/2} \qquad (7.16)$$

The force applied by the centre bolt should be significantly greater than this to allow an adequate safety margin, and a factor of two is often used. The force to be applied by the centre bolt (F_b) is then given by

$$F_b = 2M_1\omega_S \left\{ \frac{2W_0}{\rho c \tilde{R}} \right\}^{1/2} \tag{7.17}$$

In applying this steady force to the stack, the centre bolt is stretched by an amount $y_b = F_b/K_b$, where K_b is the stiffness constant of the bolt, and the stress in the bolt is given by

$$\frac{F_b}{A_b} = \frac{2M_1\omega_S}{A_b} \left\{ \frac{2W_0}{\rho c \tilde{R}} \right\}^{1/2} = \frac{2K}{A_b\omega_S} \left\{ \frac{2W_0}{\rho c \tilde{R}} \right\}^{1/2} \tag{7.18}$$

where A_b is the effective cross-section of the centre bolt.

The centre bolt restrains the vibration of the stack to some degree, and to avoid this restraint having too large an effect the stiffness of the bolt must be small compared with that of the stack. Let the ratio of the stack stiffness (K) to that of the bolt (K_b) be denoted by b, and assume that the length of the centre bolt is equal to the length of the ceramic stack (ℓ_c). Then,

$$E_e A_c = b E_b A_b$$

where E_b is the Young's modulus of the centre bolt material. The stress in the centre bolt given by (7.18) may then be expressed as

$$\frac{F_b}{A_b} = \frac{2K}{\omega_S A_b} \left\{ \frac{2W_0}{\rho c \tilde{R}} \right\}^{1/2}$$

$$= \frac{2E_e A_c(1+M_1/M_2)}{\omega_S \ell_c} \cdot \frac{b E_b}{E_e A_c} \cdot \left\{ \frac{2W_0}{\rho c \tilde{R}} \right\}^{1/2}$$

$$= \frac{2(1+M_1/M_2)b E_b}{\omega_S \ell_c} \cdot \left\{ \frac{2W_0}{\rho c \tilde{R}} \right\}^{1/2}$$

Using equation (7.11) for ℓ_c, this becomes

$$\frac{F_b}{A_b} = 2b E_b k Q_M E \cdot \left\{ \frac{2\eta_{ma}\varepsilon}{E_e} \right\}^{1/2} \tag{7.19}$$

We can use this expression to evaluate typical values of the stress in the centre bolt. Thus, taking the values for barium titanate (as before) of $\varepsilon_r = 1200$ and $E_e = 11 \times 10^{10}\,\text{Pa}$, and assuming $\eta_{ma} = 0.9$, the stress in the bolt is given by

$$\frac{F_b}{A_b} = 8.34 \times 10^{-10} \, E_b b k Q_M E \qquad (7.20a)$$

For a beryllium copper bolt the value of E_b is approximately 12.7×10^{10}Pa, and the stress is then

$$\frac{F_b}{A_b} = 106 \, b k Q_M E \quad \text{(Pa)}$$

The value of b ($= K/K_b$) is typically between 5 and 15 in practical designs. Thus, assuming $b = 10$, a matched design in which $k Q_M = 1.2$, and an applied field of $E = 2 \times 10^5$ V/m, the stress in the bolt is

$$\frac{F_b}{A_b} = 2.5 \times 10^8 \text{ Pa} = 250 \text{ MPa}$$

Note that this value is independent of frequency and of the ratio M_1/M_2. A typical value for the yield stress of beryllium copper is 10^3MPa, so that an applied stress of about 250MPa does not appear unreasonable.

For a typical lead zirconate titanate ceramic (as before), the corresponding relationships are

$$\frac{F_b}{A_b} = 11.8 \times 10^{-10} \, E_b b k Q_M E \qquad (7.20b)$$

$$= 149 \, b k Q_M E \quad \text{for beryllium copper,}$$

$$= 3.6 \times 10^8 \text{ Pa} = 360 \text{ MPa, for a matched design} \\ \text{as above.}$$

These expressions thus confirm that it is possible to use a centre bolt to apply sufficient stress to maintain the ceramic under compression without clamping the motion of the stack too severely. In practice, the alternating stresses in the centre bolt lead to some risk of fatigue failure of the bolt, and this possibility must be taken into account during the design. Practical aspects of the design of the centre bolt itself are discussed in Section 8.4.

Transformation ratio.

The transformation ratio introduced in Section 4.8 can be derived from the above expressions. Equation (4.30) gives one relationship between electrical and mechanical quantities; ie :—

$$Q_M = \frac{\omega_S M_e}{r_m} = \frac{\omega_S L_1}{R_1}$$

$$= \frac{1}{\omega_S C_1 R_1} \qquad (4.11)$$

$$= \frac{1}{\omega_S k^2 C_{LF} R_1} \qquad (4.13)$$

The ratio between the resistive components expressed in mechanical and electrical terms is equal to φ^2 (4.36). Thus,

$$\varphi^2 = r_m / R_1$$

$$= \omega_S^2 k^2 C_{LF} M_e \qquad (7.21)$$

This factor may be evaluated by calculating C_{LF} from (7.7), and M_e from (7.3).

7-3   CONDITIONS

In these calculations of the basic design we have simplified the treatment by making use of the "lumped constants" approximation, in which the stress is assumed to be constant along the stack. This is effectively equivalent to assuming the length of the stack to be small compared with the wavelength of longitudinal vibrations in the stack. This assumption can now be tested by using the expressions derived above for the transducer element dimensions. Thus, equation (7.11) gives the stack length ℓ_c as

$$\ell_c = \frac{(1 + M_1/M_2)}{\omega_S k Q_M E} \left\{ \frac{W_0 E_e}{\eta_{ma} \varepsilon \rho c \tilde{R}} \right\}^{1/2} \qquad (7.11)$$

The wavelength of longitudinal vibrations in the stack is given by

$$\lambda_c = \frac{2\pi c_e}{\omega_S}$$

where c_e is the speed of longitudinal vibrations along the stack. Since c_e is given approximately by $c_e^2 = E_e/\rho_c$ this may be expressed as

$$\lambda_c \simeq \frac{2\pi}{\omega_S} \left\{ \frac{E_e}{\rho_c} \right\}^{1/2}$$

The ratio of stack length to wavelength is thus

$$\frac{\ell_c}{\lambda_c} \simeq \frac{\omega_S}{2\pi} \left\{ \frac{\rho_c}{E_e} \right\}^{1/2} \frac{(1 + M_1/M_2)}{\omega_S k Q_M E} \left\{ \frac{W_0 E_e}{\eta_{ma} \varepsilon \rho c \tilde{R}} \right\}^{1/2}$$

$$= \frac{(1 + M_1/M_2)}{2\pi k Q_M E} \left\{ \frac{\rho_c W_0}{\eta_{ma} \varepsilon \rho c \tilde{R}} \right\}^{1/2} \tag{7.22}$$

Assuming $\eta_{ma} = 0.9$, $M_1/M_2 = 0.1$, and $\rho c = 1.5 \times 10^6 \text{kg.m}^{-2}.\text{s}^{-1}$, and taking typical values for barium titanate of $\varepsilon = \varepsilon_r \varepsilon_0 = 1.062 \times 10^{-8} \text{F/m}$ and $\rho_c = 5.6 \times 10^3 \text{kg/m}^3$, this becomes

$$\frac{\ell_c}{\lambda_c} \simeq \frac{109}{k Q_M E} \left\{ \frac{W_0}{\tilde{R}} \right\}^{1/2} \tag{7.22a}$$

For example, if $k Q_M = 1.2$, $E = 2 \times 10^5 \text{V/m}$, $W_0 = 10^4 \text{W/m}^2$, and $\tilde{R} = 0.5$,

$$\ell_c/\lambda_c \simeq 0.064$$

thus confirming that the stack length is sufficiently short compared with λ_c for the assumption of constant stress to be valid. For a lead zirconate titanate stack, taking values of $\varepsilon = 1.15 \times 10^{-8} \text{F/m}$ and $\rho_c = 7.6 \times 10^3 \text{kg/m}^3$, equation (7.22) becomes

$$\frac{\ell_c}{\lambda_c} \simeq \frac{122}{k Q_M E} \left\{ \frac{W_0}{\tilde{R}} \right\}^{1/2} \tag{7.22b}$$

And for $k Q_M = 1.2$, $E = 2 \times 10^5 \text{V/m}$, $W_0 = 10^4 \text{W/m}^2$, and $\tilde{R} = 0.5$, as before, this becomes

$$\ell_c/\lambda_c \simeq 0.072$$

confirming that the assumption is also valid for an LZT stack in a lumped mass design.

7-4   SOME TYPICAL VALUES

Stack length, ℓ_c

For a barium titanate stack, assuming values as above for ε, E_e, ρc, η_{ma}, E, and M_1/M_2, equation (7.11) reduces to

$$\ell_c = \frac{15.2}{\omega_s k Q_M} \left\{ \frac{W_0}{\tilde{R}} \right\}^{1/2} \quad \text{(m)} \qquad (7.23a)$$

If $kQ_M = 1.2$, and ω_s is expressed in terms of the resonance frequency F_s in kHz, then

$$\ell_c = \frac{2.02 \times 10^{-3}}{F_s} \left\{ \frac{W_0}{\tilde{R}} \right\}^{1/2} \quad \text{(m)}$$

$$= \frac{2.02}{F_s} \left\{ \frac{W_0}{\tilde{R}} \right\}^{1/2} \quad \text{where } \ell_c \text{ is in mm.} \qquad (7.24a)$$

The corresponding relationships for a lead zirconate titanate stack, assuming again values as in Section 7.3, are

$$\ell_c = \frac{10.8}{\omega_s k Q_M} \left\{ \frac{W_0}{\tilde{R}} \right\}^{1/2} \quad \text{(m)} \qquad (7.23b)$$

and for $kQ_M = 1.2$,

$$\ell_c = \frac{1.43}{F_s} \left\{ \frac{W_0}{\tilde{R}} \right\}^{1/2} \qquad (7.24a)$$

where ℓ_c is in mm, and F_s is in kHz.

Transducer volume

An estimate of the volume of a single element in its casing may be derived by taking the cross-sectional area of the casing as a little greater than the area (A_p) of the piston, and by noting that the length of the casing is usually roughly twice the length (ℓ_c) of the stack. The spacing between elements is commonly about half the acoustic wavelength in water, and a typical piston diameter may therefore be taken as a half wavelength, except at very low frequencies when this becomes too large to be practical. Thus the piston area may be assumed to be given approximately by

$$A_p = \frac{\pi}{4} \left\{ \frac{\lambda}{2} \right\}^2$$

$$= \frac{\pi \lambda^2}{16}$$

$$= \frac{\pi^3 c^2}{4 \omega_S^2}$$

The volume of the transducer casing for a barium titanate design can then be estimated as

$$V_t \simeq 2 \ell_c A_p$$

$$\simeq \frac{5 \times 10^8}{k Q_M \omega_S^3} \left\{ \frac{W_0}{\tilde{R}} \right\}^{1/2} \quad \text{using equation (7.23a),}$$

$$\simeq \frac{2 \times 10^{-3}}{k Q_M F_S^3} \left\{ \frac{W_0}{\tilde{R}} \right\}^{1/2} \qquad\qquad (7.25)$$

where F_S is in kHz and the volume V_t is in m^3. A design using lead zirconate titanate would be slightly smaller, but since these expressions must be regarded as giving only a very rough approximation for the volume, the difference in the calculated value would not be significant.

A rough estimate of the mass of the transducer can then be derived by multiplying the volume by an average density for such devices. For a transducer using a bronze casing, experience suggests a typical value for the overall relative density of about 4, and for an aluminium housing about 2. Thus, using equation (7.25), the mass of the transducer element in a bronze housing should be about

$$\text{Mass} \simeq \frac{8}{k Q_M F_S^3} \left\{ \frac{W_0}{\tilde{R}} \right\}^{1/2} \qquad \text{(kg)}$$

As an example, a transducer resonant at 5kHz, with $\tilde{R} = 0.5$, $k Q_M = 1.2$, and with a power output of 180watts, (equivalent to 1W/cm^2 of piston area for a piston diameter of $\lambda/2$,) would have an estimated volume of 1980cm^3. This would correspond to a power output of just under 0.1watt/cm^3, or about 25watts/kg for a barium titanate design.

7-5 SUMMARY

a) Using the value of Q_M from Chap 5, and the value of R_r from Chap 6, calculate the required effective mass (Eqn 7.3). Choose a value of M_1/M_2.

b) Calculate the stack stiffness to give the required resonance frequency (7.1).

c) Calculate the volume of ceramic (7.8)

d) Hence calculate the dimensions of the stack (7.11, 7.12).

e) Design a centre bolt to apply a force given by (7.17), but having a stiffness less than 20% of the stack.

f) Application of these methods also permits simple though crude estimates of the volume and weight of the transducer.

REFS Chapter 7

References

7.1 Fischer,F.A., *Gründzuge der Elektro-akustik*, 1st Ed, 1949, (2nd Ed, 1959), p29.
7.2 Hahnemann,W., and Hecht,H., "Die Grundform des mechanisch-akustischen Schwingungskörpers. (Der Tonpilz.)," Physik. Zeitschr., <u>21</u>, 187-192, (1920).

Chapter 8
PRACTICAL TRANSDUCER DESIGN

The basic design of a piston transducer element was considered in Chapter 7. In this chapter we shall discuss how this basic design can be converted into a practical device, and it is perhaps worth repeating that attention will again be focussed on low frequency piston type projectors. The first part of the chapter will deal with the corrections which it is advisable to introduce into the simple treatment of the previous chapter. The following sections discuss various engineering aspects, which are aimed mainly at achieving predictable and reliable performance. They are often based on experience, and are hence largely matters of judgement rather than being deducible from established scientific laws. They are nevertheless important, and the reasons for some of the choices will be explained in this chapter.

A piston transducer has a typical construction as shown in *Fig 7.1*, and before the various design equations can be used it is necessary to make several choices of parameters. One of the major options is the ratio of tail to head mass. If the masses are equal, and the radiation loads are the same at both ends, the symmetry of the design will cause the masses to move with equal amplitude in opposite phase to each other, and the centre point of the stack will remain stationary. The plane through the centre point, perpendicular to the axis of the stack, thus behaves as a nodal plane, and the element could be fixed to a housing at that point without affecting the motion of the element itself. If the element radiates from only one end, energy needs to be transmitted along the stack and there is no true node. It is however common to describe an element which is mounted at its point of minimum displacement as having a nodal mount. If the tail mass is heavier than the effective piston mass, the centre of gravity, and hence the "nodal" plane, is displaced towards the tail mass. The position of this nodal plane is approximately $M_2/(M_1 + M_2)$ of the distance along the stack towards M_2, and in this case a nodally-mounted design would be supported at this point. Whether this is done or not, it is often a useful simplification to consider the motion of the element in two parts, one in front of the node and the other behind it.

8-1 EFFECTIVE VIBRATING MASS OF STACK

In the equations so far, the stack has been assumed to be an ideal compliance of zero mass. In practice, of course, the stack does make a contribution to the total vibrating mass, and this should be taken into account, particularly for designs which have lightweight pistons to achieve a wide transmitting bandwidth. This contribution is only a proportion of the static mass of the stack because the velocity varies with position along the stack. It is easy to show that the effective vibrating mass of the stack is approximately one-third of its static mass.

Consider only the part of the stack in front of the node, and assume that the mass of the stack is small (though not negligible) compared with that of the end mass. This implies that the resonance frequency is low, and that the length of the stack is small compared with a wavelength of sound in the ceramic at that frequency. The displacement of the stack is then approximately proportional to its distance from the nodal plane. Thus, considering an element of length δx at a distance x from the node, as in *Fig 8.1*, the displacement velocity is approximately vx/L, where L is the total length of the stack and v is the velocity of the end mass M. If the static mass of the stack is M_S, the mass of the element is given by $M_S \delta x/L$, and the kinetic energy of the spring by the integral

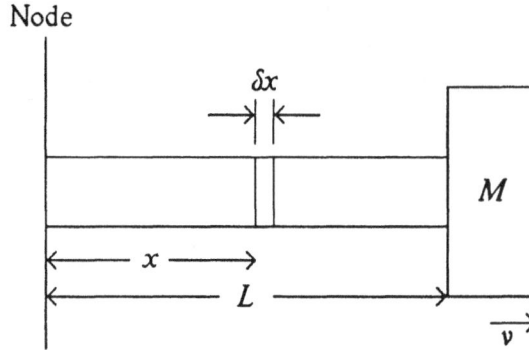

Fig 8.1   Basic stack and end mass.

$$\frac{1}{2} \int_0^L \frac{M_S}{L} \, dx \, \frac{(vx)^2}{L^2}$$

$$= \frac{1}{2} \frac{M_S}{L} \frac{v^2}{L^2} \frac{L^3}{3}$$

$$= (1/6)M_S v^2$$

The kinetic energy of the end mass is $Mv^2/2$, and the total kinetic energy of the system is thus $(1/2)(M + M_S/3)v^2$. The effective vibrating mass of the stack is therefore $M_S/3$, and this should be added to the end masses in the appropriate equations. For example, the resonance frequency of the system in *Fig 8.1* is given by

$$\omega_S^2 = \frac{K}{M + M_S/3} \tag{8.1}$$

A similar contribution of one-third of the stack mass behind the nodal plane should be added to the tail mass. When the tail mass is much greater than the head mass, however, this contribution has little effect.

8-2   EFFECT OF JOINTS AND CENTRE BOLT

The basic design of a transducer element is illustrated in *Fig 7.1*, showing a centre bolt through the stack. The stack

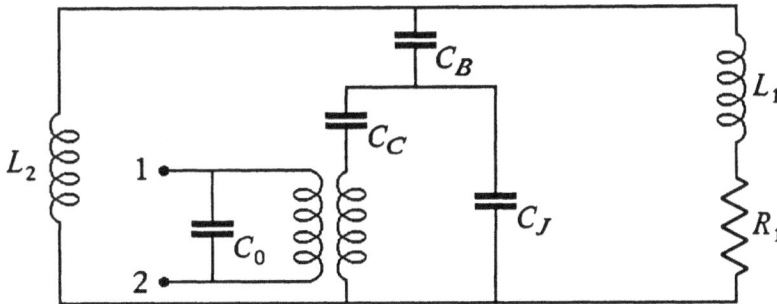

Fig 8.2 Equivalent circuit including joints and centre bolt.

itself is usually assembled from a number of rings, using an adhesive resin to form joints between the rings. Details of this type of construction will be described later. The purpose of this section is to investigate how the bolt and the joints modify the characteristics of the transducer from the idealised design previously assumed.

It is convenient to tackle this question by deriving the equivalent circuit of the element, using the results obtained in Section 4.9. It was shown there that mechanical compliances in series can be represented by electrical capacitors in parallel; it is therefore reasonable to represent all the ceramic rings by one capacitor, which is in parallel with a second capacitor representing the total compliance of the joints (together with any insulators or spacers which are inserted into the stack in series with the ceramic). The centre bolt is mechanically in parallel with the ceramic stack, so its electrical equivalent capacitor should be in series with those for the ceramic plus joints. The equivalent circuit for an element including joints and centre bolt is thus as shown in *Fig 8.2*, in which

L_1 represents the effective radiating mass (including any contributions from the stack and radiation load).

R_1 represents the resistive loading for the radiating piston.

L_2 represents the tail mass (assumed to have no significant energy dissipation associated with it).

C_C represents the compliance of the ceramic rings (all in series).

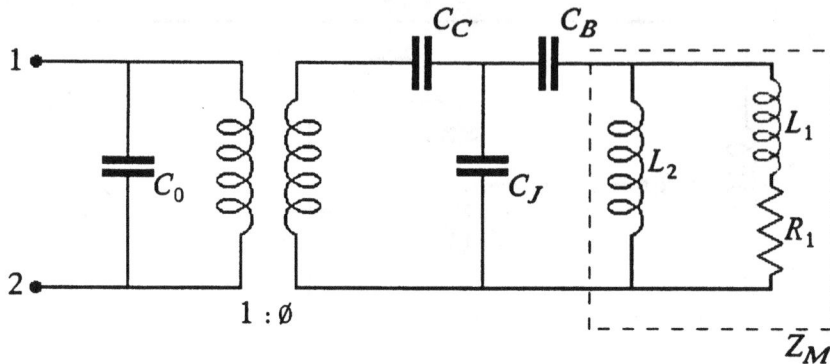

Fig 8.3   Re-arrangement of Fig 8.2.

C_J represents the total compliance of the joints plus insulators, spacers, etc.

C_B represents the compliance of the centre bolt.

C_0 is the clamped capacitance across the electric terminals 1,2.

and the transformer represents an ideal electro-mechanical transformer of ratio φ.

(Note that the relationship with the equivalent circuits is clearer if the mechanical springs are characterised in terms of their compliances rather than their stiffness coefficients.)

This circuit may be rearranged into the form shown in *Fig 8.3*, and can then be transformed (by a combination of network transforms) into the circuit of *Fig 8.4*, in which:-

$$C_1 = \frac{C_C C_J}{C_C + C_J} \tag{8.2a}$$

$$C_2 = \frac{1}{n^2} \frac{(C_C + C_J)C_B}{(C_C + C_J + C_B)}$$

$$= \frac{C_C^2 C_B}{(C_C + C_J)(C_C + C_J + C_B)} \tag{8.2b}$$

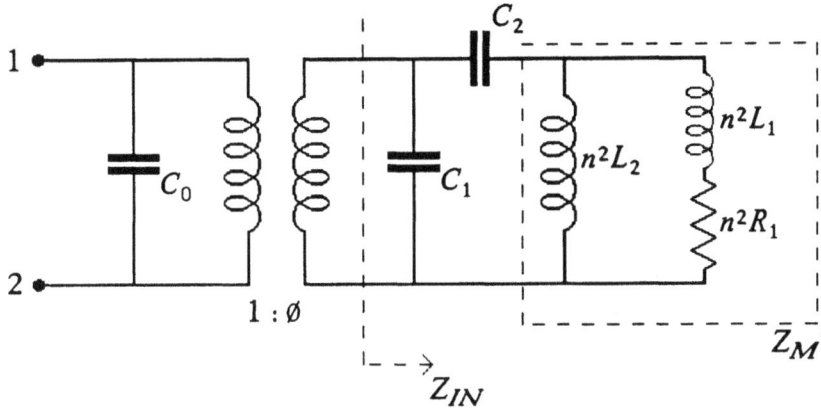

Fig 8.4   Transformation of Fig 8.3.

and $$n^2 = \frac{Z_M'}{Z_M} = \frac{(C_C + C_J)^2}{C_C^2} \qquad (8.2c)$$

Although this transformation may not be obvious, it is fairly straightforward (though tedious) to confirm that it is correct by evaluating the input impedance for the two circuits of *Figs 8.3* and *8.4*, as follows. Let the combination of L_1, R_1, and L_2 be denoted by Z_M, as indicated in *Fig 8.3*, and the corresponding combination in *Fig 8.4* by Z_M'. Also, let the input impedance of the circuit seen from the right hand side of the transformer be Z_{IN}. Then, for the circuit of *Fig 8.4*, the input admittance is given by

$$1/Z_{IN} = j\omega C_1 + \frac{j\omega C_2}{1 + j\omega C_2 Z_M'}$$

Therefore,

$$Z_{IN} = \frac{1 + j\omega C_2 Z_M'}{j\omega C_1(1 + j\omega C_2 Z_M') + j\omega C_2}$$

$$= \frac{1 + j\omega C_2 Z_M'}{j\omega(C_1 + C_2) - \omega^2 C_1 C_2 Z_M'} \qquad (8.3)$$

For the circuit of *Fig 8.3*, the derivation is a little longer and is obtained by moving progressively to the left from Z_M, adding one component at a time. This gives

$$Z_{IN} = \frac{1}{j\omega C_C} + \frac{1 + j\omega C_B Z_M}{j\omega C_J(1 + j\omega C_B Z_M) + j\omega C_B}$$

$$= \frac{j\omega(C_J + C_B) - \omega^2 C_J C_B Z_M + j\omega C_C(1 + j\omega C_B Z_M)}{j\omega C_C\{j\omega(C_J + C_B) - \omega^2 C_J C_B Z_M\}}$$

$$= \frac{j\omega(C_C + C_J + C_B) - \omega^2 Z_M C_B(C_C + C_J)}{j\omega C_C\{j\omega(C_J + C_B) - \omega^2 C_J C_B Z_M\}}$$

$$= \frac{1 + j\omega Z_M \cdot \dfrac{C_B(C_C + C_J)}{C_C + C_J + C_B}}{\dfrac{C_C}{C_C + C_J + C_B}\{j\omega(C_J + C_B) - \omega^2 C_J C_B Z_M\}} \qquad (8.4)$$

Using the expressions of Equation (8.2) to substitute in (8.3) gives

$$Z_{IN} = \frac{1 + j\omega Z_M \cdot \dfrac{C_B(C_C + C_J)}{C_C + C_J + C_B}}{\dfrac{C_C}{C_C+C_J+C_B}\left\{j\omega \dfrac{C_J(C_C+C_J+C_B)+C_B C_C}{C_C + C_J} - \omega^2 C_J C_B Z_M\right\}}$$

$$= \frac{1 + j\omega Z_M \cdot \dfrac{C_B(C_C + C_J)}{C_C + C_J + C_B}}{\dfrac{C_C}{C_C + C_J + C_B}\{j\omega(C_J + C_B) - \omega^2 C_J C_B Z_M\}}$$

which is identical to (8.4), thus confirming the equivalence of the circuits in *Figs 8.3* and *8.4*, with the transformations of Equation (8.2).

In *Fig 8.4*, C_1 is effectively in parallel with C_0, and the mechanical resonance frequency occurs at the resonance of the series components, ie when C_2 resonates with $n^2 L$, where L is the effective inductance of the parallel combination of L_1 and L_2. Thus, the resonance frequency of the system including the effects of the joints and centre bolt is

$$\omega_S^2 = \frac{1}{C_2(n^2 L)}$$

$$= \frac{C_C + C_J + C_B}{C_B(C_C + C_J) L}$$

$$= \frac{(C_C + C_J + C_B)(L_1 + L_2)}{C_B(C_C + C_J)(L_1 L_2)} \tag{8.5}$$

$$= \frac{\{(C_J/C_C) + (C_B/C_C) + 1\}}{(C_B/C_C)\{(C_J/C_C) + 1\}} \, \omega_{S0}^2 \tag{8.5a}$$

where $\omega_{S0}^2 = (L_1 + L_2)/C_C L_1 L_2$ is the resonance frequency (squared) of the ideal system without joints and bolt. The effect of the bolt is to raise the resonance frequency, whilst the joints cause a reduction, the overall effect being given by Equation (8.5a).

If we denote by C_0' the clamped capacitance C_0 when it is transferred to the motional side of the transformer in *Fig 8.3*, then the coupling coefficient of the ideal element (without joints and bolt) is given by

$$k^2 = C_C/(C_0' + C_C) \tag{4.12}$$

Using the circuit of *Fig 8.4* to allow for the joints and bolt, and noting that C_1 is in parallel with C_0', the effective coupling coefficient is given by

$$k_e^2 = \frac{C_2}{C_0' + C_1 + C_2}$$

$$= \frac{\dfrac{C_C^2 C_B}{(C_C + C_J)(C_C + C_J + C_B)}}{\dfrac{C_C(1 - k^2)}{k^2} + \dfrac{C_C C_J}{C_C + C_J} + \dfrac{C_C^2 C_B}{(C_C + C_J)(C_C + C_J + C_B)}}$$

in which (8.2) has been used to substitute for C_1 and C_2, and (4.12) to substitute $C_0' = C_C(1 - k^2)/k^2$. This expression can then be reduced to

$$k_e^2 = \frac{k^2 C_C C_B}{(C_C + C_J)\{(1 - k^2)C_C + C_B + C_J\}} \tag{8.6}$$

This can be expressed in a simpler form by writing the compliance of the ceramic plus joints as $C_S = C_C + C_J$, and the ratio of the bolt compliance C_B to C_S as b $(= C_B/C_S)$ (as in Sect 7.2). Then

$$C_C + C_J + C_B = C_S(1 + b)$$

and (8.6) becomes (after some simplification)

$$\frac{k_e^2}{k^2} = \frac{b}{(C_S/C_C)(1 + b) - k^2} \qquad (8.7)$$

This equation thus allows the calculation of how much the coupling coefficient of the ceramic itself (k) is reduced by the effects of the joints and the centre bolt. It was noted above that the joints and bolt had opposite effects on the resonance frequency, but both factors cause a reduction in the coupling coefficient, since both reduce the useful mechanical stored energy.

It is sometimes desirable to achieve a particular value of k_e, in order to optimise the admittance loop, and equation (8.7) can be inverted to derive the value of C_S/C_C needed to reduce k_e to its required value. Thus, re-arranging (8.7),

$$1 + \frac{C_J}{C_C} = \frac{C_S}{C_C}$$

$$= \frac{k^2}{(1 + b)}(1 + b/k_e^2) \qquad (8.8)$$

Examples of curves to show required values of C_J/C_C are illustrated in *Fig 8.5*. If the required value of C_J is greater than that due to the joints themselves, a spacer may be used in the stack, its compliance being included with the joints in C_J. This provides a simple and controllable way of reducing the coupling coefficient when necessary, although its effect on the resonance frequency must of course also be taken into account.

8-3   STACK CONSTRUCTION

The equations of Chapter 7 have shown how the length and cross-sectional area of the ceramic stack can be determined for any particular design. The need to apply a static compressive force in order to prevent dynamic tensile forces in the ceramic

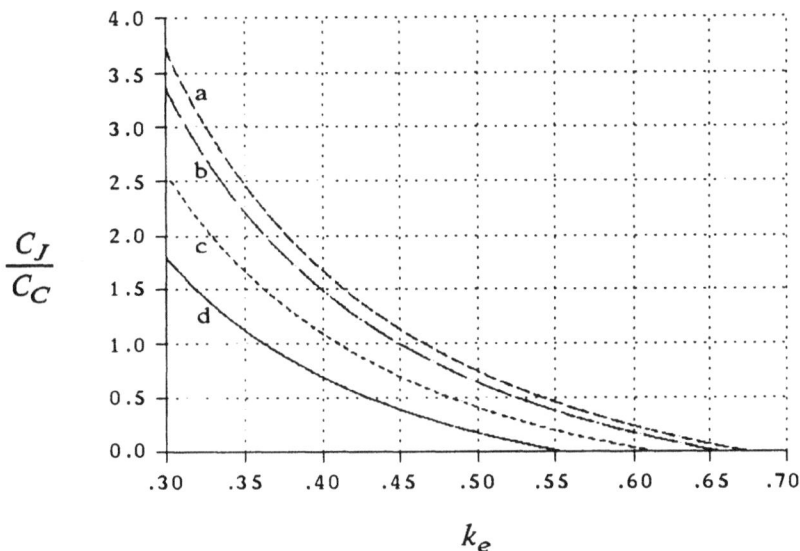

Fig 8.5   Relationship   between C_J/C_C and   effective
coupling   coefficient k_e,   for   various   values   of   the
ratio $b = C_B/C_S$:   (a) $b = 10$;   (b) $b = 5$;   (c) $b = 2$;
(d) $b = 1$.

has also been described. A bolt through the centre of the stack
is the most usual and convenient technique for achieving this, a
typical stack thus consisting of a column of ceramic rings with
a centre bolt through the middle. The question that now arises is
how many rings the stack should be divided into.

Two opposing factors need to be taken into account. If the
stack is composed of only a few rings the number of joints will
be low, and this will minimise the mechanical dissipation in the
joints and the degradation in coupling coefficient caused by
them. There is therefore some incentive to make the stack of
only a few thick rings, to minimise the effects of the joints.
However, these rings need to be poled during manufacture by
having a very high electric field applied – of at least 1kV/mm
– and a maximum ring thickness may therefore be set by the
maximum voltage which can safely be applied by the ceramic
manufacturer. Also, the capacitance of a stack composed of n
rings in parallel is given by $C_{LF} = \varepsilon A_c n^2/\ell_c$ (Eqn 7.7). If the
number of rings is low, the transducer capacitance will be low

and its impedance high, and a high driving voltage will then be needed to produce a large acoustic output. An upper limit to this voltage may arise from various causes such as flash-over between conductors.

The choice of ring thickness is thus a compromise between these factors. In practice, rings are usually between about 5 and 15mm in thickness. The lower values tend to be limited by the finite size of the connections to the electrodes between the rings, which increases the danger of flash-over to the neighbouring electrodes. The number of rings is therefore selected to give a suitable thickness within this range. The rings are normally connected electrically in parallel to minimise the impedance, although other (eg series-parallel) connections are occasionally used for special purposes such as to increase the transducer's voltage sensitivity.

The joints are crucial in determining the characteristics of the transducer, and much care is needed during manufacture to ensure satisfactory and reproducible properties. The equations of Section 8.2 show the effects of the joints in reducing the resonance frequency and coupling coefficient. The effect on resonance frequency could in principle be allowed for, if the joints could be well enough controlled, by making the ceramic components slightly shorter to compensate for the joint compliance. The effect on the coupling coefficient, however, is always to reduce it below that of the ceramic, and this can be serious when a maximum value of k_e is sought. The joints also cause energy dissipation because of their mechanical hysteresis losses, the resulting temperature rise depending on the drive levels and pulse lengths. For high powers and long pulses, the heating of the joints tends to increase still further the energy losses and compliance in the joints and can thus lead to runaway failure. The associated heating of the ceramic changes its characteristics and may also lead to depoling and catastrophic failure of the transducer. It is therefore clear that the joints need to be as stiff as possible, and to have minimum mechanical hysteresis losses, to reduce both the degradation of transducer performance and also its sensitivity to variations in the quality of the joints.

It is worth pausing for a moment to consider what might be the result of assembling the stack without any material in the joints - a technique which has considerable appeal for quick experiments. If the ceramic rings have been ground on their flat faces, the surface roughness is likely to be of the order of 10^{-3}mm. If two such rings are placed together, the high spots on the surfaces make contact over only a very small proportion of the mating faces – typically 10^{-3} – 10^{-4} of the nominal area [8.1] – and for the remaining area there is effectively an air gap between the two rings. The stiffness of the joint is then approximately $10^{-3}E_cA_c/10^{-6}$, in which a joint thickness of 10^{-6}m (the surface roughness scale) has been assumed, and A_c, E_c represent the nominal area and Young's modulus of the ceramic. By comparison, the stiffness of a ceramic ring 5mm thick would be $E_cA_c/5 \times 10^{-3}$, so that the joint has only 5 times the stiffness of the ceramic ring. Each joint thus contributes an extra compliance which is equivalent to a slice of ceramic about 1mm thick; this can have a substantial effect on the characteristics, and is very difficult to control. In a practical element, the joints need also to include an electrode, so that each joint has two pairs of mating surfaces, which makes the effects even more significant. Dry joints, although tempting to use for convenience, are thus very likely to lead to serious deviations from the predicted characteristics, and should be avoided. If the interstices of the joints are filled with grease, their stiffness is much higher than for the air–filled joints, and this technique can be used in experimental stacks to give a rough approximation for the resonance frequency. Such joints are usually rather lossy, however, and should not be used for any high stress measurements, or where stability of the properties is important.

The materials most commonly found to be suitable for the joints are the epoxy resins, usually with an added powdered slate filler to increase the stiffness. Considerable experimentation may be needed to derive the optimum mixture for the more demanding applications where high stiffness and low losses are important. Curing at an elevated temperature is usually desirable for best results, but the temperature must not be so high as to cause depoling of the piezoelectric ceramic. The epoxy resin manufacturers generally suggest curing conditions and optimum joint thickness to achieve maximum tensile strength, and although in principle the joints in a pre–biassed stack should never be in tension, greatest reliability in the transducer seems to be obtained when the quality of the joints is confirmed by achieving high bond strengths. For most applications, a tensile strength of the bond of at least 10MPa is desirable.

Each joint needs to incorporate some form of electrode as well as providing the mechanical bond between rings. Some designers have used a thin metal ring with wires laid across, to ensure electrical contact to the silvering on the ceramic rings themselves and to give the specified joint thickness. Others have used a conducting resin made by mixing graphite into the resin,

Fig 8.6 Typical transducer element assembly. Courtesy of Cray Sonics Ltd.

although it is difficult to obtain strong joints when this technique is employed. A design which has been found to produce very good results uses a thin gauze of Monel metal, 0.08mm thick, which gives good electrical contact and a well-controlled joint thickness. An extension of the gauze outside the rings can serve as a means for attaching the electrical leads. However, although this type of connection is acceptable for some hydrophones, it is not very robust and may be subject to fatigue failure in high power applications. A more reliable technique is to use reinforcing metal (eg brass) rings which have an inside diameter just larger than the ceramic diameter, and spot weld pairs of these rings together with the gauze sandwiched between them. A slot is made in the rings to facilitate connection to the leads, which may then be taken in a spiral around the stack. Braided conductors may be used to reduce the chances of fatigue failure of the leads or electrodes. A typical assembly is shown in *Fig 8.6*.

Before assembly of the stack, all the components must be very carefully degreased. The gauze electrodes must be well impregnated with resin, preferably under vacuum, and the stack should then be assembled and compressed to apply the necessary bias, and to extrude the resin from the joints. It is important to use sufficient resin to form a small shoulder to support the ring electrodes, but not so much as to bridge across the ceramic rings themselves. This support for the electrodes is necessary to prevent excessive vibration, which may cause fatigue failure of the gauze.

High quality joints are essential to the satisfactory operation of a transducer, and much care needs to be taken to obtain good and consistent results. Establishing an acceptable technique requires a long period of testing, and changes should be introduced only when adequately proved. Achieving consistently good joints is probably the most critical feature of transducer manufacture, and insufficient attention to this aspect is likely to cause many defective elements, especially for high power transmitting transducers.

Insulators.

It is usually desirable to insulate the stack electrically from earth, partly to reduce the danger of pick-up caused by earth loops, and partly to avoid electrolytic corrosion resulting from the connection of the radiating piston to other metals in the mounting. Insulators are therefore included in the stack between the ceramic rings and the end masses. These insulators contribute to the stack stiffness, and their compliance should be included with that of the joints in calculating the stack compliance. They need to be as hard as possible, and preferably similar in properties to the piezoelectric ceramic, to avoid thermal stresses arising from temperature changes. Selected grades of steatite or aluminium oxide have been found to be satisfactory, with the surfaces ground and treated to obtain good joints. If the applied voltage is balanced about earth, the insulator can be about half the thickness of the piezoelectric rings, with an allowance for the size of the connector, but in general it is safer to use insulators which are about as thick as the rings. The arrangement in which the connections are balanced about earth can be advantageous when the transducer is also used for reception, since it tends to reduce the susceptibility to electrical pick-up.

An insulator may usefully be fitted around the centre bolt to prevent flash-over to the bolt, and also to serve as a jig to locate the rings and electrodes whilst the stack is being assembled. This insulator needs to be stiff enough only to form a suitable location device, and some form of plastic which can be machined to good tolerances is satisfactory.

Applied field.

The maximum electric field which can safely be applied to piezoelectric ceramics such as barium titanate and lead zirconate titanate without causing depoling is about 10^6V/m (10kV/cm) at room temperature. Alternating fields of this magnitude also cause heating of the ceramic, and the depoling field decreases as the temperature rises. The other factor which limits the applied field is the possibility of flash-over between the electrodes in the stack, which may occur through the air or, more probably, across the surface of the ceramic. A working field of 2×10^5V/m (2kV/cm) may generally be taken as a safe design guide if the surfaces of the ceramic are clean, and sharp points are avoided on the electrodes. This will permit a field of twice the working field to be applied for a short test period, as described in Chapter 9, without deleterious effects on the ceramic. Although higher values are occasionally used, great care is needed to avoid electrical breakdown or gradual degradation.

Provided that heating of the stack is not excessive, it is the danger of flash-over which usually limits the applied field, and it is then reasonable to consider whether the field could be raised by applying an insulating or anti-tracking coating to the ceramic. Some increase can be achieved by such coatings, but the increase is not dramatic and the life of these coatings in the vibrating environment is difficult to guarantee. The main disadvantage, however, is that the simplest method of leak-testing depends on measuring the effect of water on the insulation resistance of the stack (See Chapter 9), and this is complicated by the use of an insulating coating. An alternative method is to fill the transducer case with a gas having a high dielectric strength, such as sulphur hexafluoride, and this technique may be used where the application demands the highest possible field.

8-4   CENTRE BOLT

The basic characteristics of the centre bolt were described in Section 7.2. It needs to be strong enough to apply the required force to the stack, yet not so stiff that it causes too large a degradation of coupling coefficient. In the example quoted in Section 7.2 of a matched lead zirconate titanate transducer with a beryllium copper centre bolt, the static stress in the bolt was shown to be about 1/3rd of its yield stress. Suppose that steel had been used for the bolt instead of beryllium copper. Substituting E_b = 20.7 × 10^{10}Pa in Equation (7.20b) then gives

$$F_b/A_b = 243 \, bkQ_mE$$

For kQ_m = 1.2, b = 10, and E = 2 × 10^5V/m, the stress in the bolt is then 580MPa. The 0.1% proof stress for a good quality carbon steel may typically be about 800MPa, so that the stress in the bolt appears tolerable at about 72% of the proof stress. However, this equation is for the static stress in the bolt, and there is superimposed on this an alternating stress with a peak value of $1/2b$ times the static stress (ie about 26MPa), which means that we should consider more carefully what are the safe conditions when the possibility of fatigue failure is taken into account.

Data on fatigue limits for various materials are given by Frost et al [8.2]. A useful first guide to the fatigue limit for a range of steels is to take it as 45–50% of the tensile strength; for other materials, this factor is more commonly in the range 35–40%. For carbon steel with a tensile strength of 1200MPa, the fatigue limit may thus be estimated at about 600MPa, which appears satisfactorily above the alternating stresses. However, the limit quoted for alternating stresses is for loads which are symmetrical about zero – ie equally positive and negative. In the case of the centre bolt, the dynamic loads are added to the static stress, the stress during a cycle varying by ±26MPa about 580MPa. Reliable guidance for combined stresses such as this is not readily available, but Frost ([8.2], Sect 3.5) summarises results showing that the safe alternating stress is significantly reduced by the addition of the mean static bias. A simple approximation for the reduction in fatigue limit is the modified Goodman relationship, given by

$$\pm T_f = \pm T_{f0}\{1 - (T_m/T_t)\} \qquad (8.9)$$

in which T_f is the fatigue limit when a mean static stress of T_m is applied, T_{f0} is the fatigue limit for zero mean stress, and T_t is the tensile strength. Thus, if the mean stress is 50% of the tensile strength, the fatigue limit is reduced by 50%.

Fig 8.7   Typical design of centre bolt.

Allowance for fatigue effects in the centre bolt design is complicated by two further considerations. The guidance which is available on design against fatigue is generally based on tests carried out up to 10^8 stress cycles. This number of cycles could easily be exceeded by a transducer in regular use. For example, a transducer operating at 5kHz would execute 10^8 cycles in about 330 minutes, ie $5^1/_2$ hours of continuous use. With a duty cycle of 1 in 10 this corresponds to 55 hours of operation. Compare this with the operating hours of a transducer which might be in use for 12 hours per day for 200 days per year – a total of 2400 hours per year. A transducer which is required to operate at this level for several years may thus exceed the 10^8 cycles by two orders of magnitude. Extrapolating curves for fatigue limits to these greater numbers of cycles indicates that the limits should be reduced, although by how much is not well established. In addition to these effects, the likelihood of stress concentrations in the bolt must be remembered. The simplest type of bolt would be a threaded rod screwed into the front piston and with a nut screwed on at the rear mass. A rod of this design would however be subject to appreciable stress concentrations at the thread roots, and these would greatly increase the danger of fatigue failure, particularly for brittle materials such as high tensile steels. It is therefore necessary to increase the diameter of the bolt where it is threaded, and a design as shown in *Fig 8.7* is recommended for reliability, even though it is more expensive to manufacture. The increased diameter in the centre section is to provide support for the insulator along the centre of the stack.

Beryllium copper is less susceptible than high tensile steel to the effects of stress concentrations, and this results in a bolt design with rather better safety factors than for readily available steels. However, the additional cost of beryllium copper, and some potential health hazards in its machining, lead to a general preference for high tensile (about 1200MPa UTS) carbon steel as the bolt material. If necessary, the ratio of bolt to stack compliance may be reduced from the value (b = 10) assumed in the example above, thus reducing the stresses in the bolt, but this will cause a further degradation in the coupling coefficient.

The required force in the centre bolt, and its extension, can be calculated from the expressions in Sect 7.2. The use of a torque wrench on the end nut is not a reliable technique for applying a known force, as it is susceptible to the effects of friction at the mating faces, and it is preferable to determine the force in the bolt by measuring its extension (eg by means of a dial gauge on the end of the bolt). A convenient and reliable method of applying the specified force is to extend the bolt by its predetermined amount by means of a hydraulic ram system operating on the end of the stack, and then tightening the end nut. Another possible technique is to measure the stress applied to the stack directly, by measuring the voltage generated piezoelectrically in the ceramic.

8-5   PISTON

In the design equations, the radiating piston has been assumed to behave as an ideal mass, the front face vibrating with equal velocity over its whole area. For broad band applications, when a transducer with low Q_M is needed, the piston must have a low mass and large radiating area, and it is difficult to meet these requirements without experiencing flexing of the piston. The problems are accentuated by the need to make a seal around the piston which is reliably watertight and also allows the piston to vibrate freely. In this section, the influence of these factors on the design of pistons for broad band projectors is considered.

When the mass of the piston is not unduly restricted, a simple design as in *Fig 8.8a* may be used. This is a good approximation to the ideal rigid piston provided it has adequate thickness and is driven over a reasonably large area by the ceramic. The grooves around the periphery accomodate "piston-type" O-seals to serve as a flexible yet watertight seal. The design of the seal itself is discussed in Section 8.7, but we note here that it is usually desirable to fit two O-rings in series to achieve good reliability. When a low piston mass is important, this design may be modified to that shown in *Fig 8.8b*. The basic shape is a truncated cone, which aims at giving a large area for the minimum mass; however the need for the outer ring to carry the O-seals adds appreciable mass where it is least wanted. The design shown in *Fig 8.8c* represents an attempt to overcome the difficulty associated with the mass

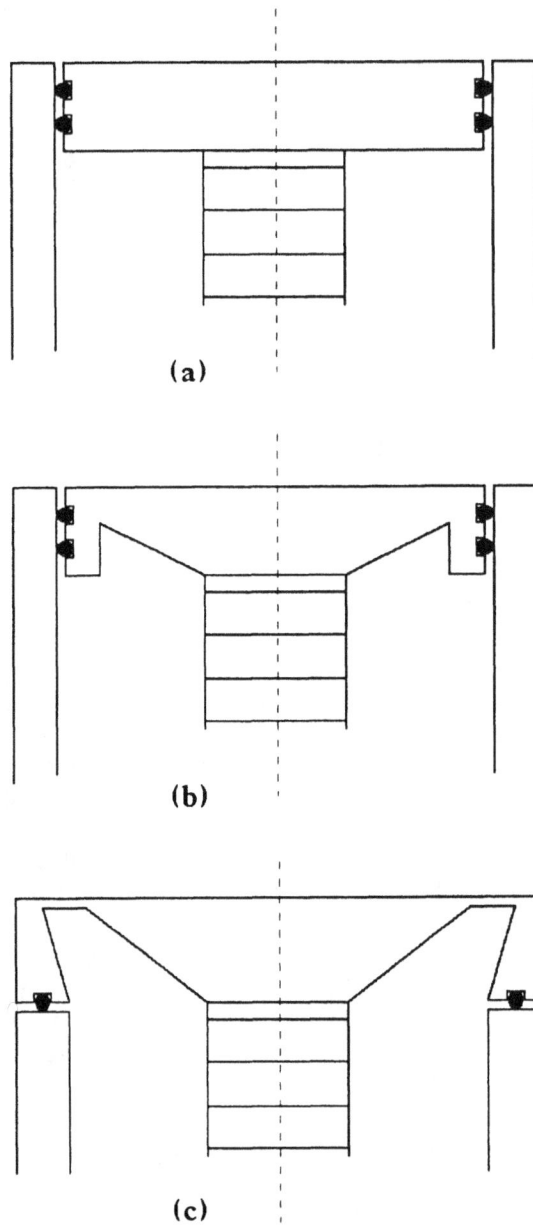

Fig 8.8 Piston designs incorporating flexible watertight seals: (a) O-ring seals around piston; (b) conical piston to reduce mass; (c) flexible diaphragm seal around piston.

around the periphery by using a flexible annular diaphragm with the outer part clamped rigidly to the housing. The effective diameter of the radiating area is somewhere between the diameter of the cone and that of the cone plus diaphragm, so that this arrangement considerably increases the "dead area" between pistons in an array, thus reducing the radiation loading and counteracting the advantages of the low mass. The diaphragm may also be subject to high stresses, either due to static water pressure or to the alternating mechanical loads. Low density materials such as aluminium alloys tend to have rather low tensile strengths, and are susceptible to corrosion in sea water, and they are therefore generally unsuitable for diaphragm seals. This restriction, together with the obvious machining difficulties, limits the usefulness of this diaphragm type of design.

Some of the problems in design of the piston can be reduced by using a rubber seal which is bonded across the front of the piston and its surrounding housing. This can isolate the piston from the sea water, thus eliminating corrosion problems and permitting a freer choice of materials such as the low density aluminium or magnesium alloys. Very reliable bonding over the whole of the piston's radiating face is needed, in order to maintain the acoustic radiation characteristics. For designs in which the piston is in contact with sea water, the dangers of corrosion limit the choice of materials to those which will withstand exposure to this hostile environment. In practice, bronze alloys are very suitable, or titanium when lower density is wanted. Steel is usually avoided because of its susceptibility to corrosion, and corrosion can also be a problem with the aluminium alloys. The choice between the various materials and approaches to the problem of designing a seal around the piston is a matter of engineering judgement, including consideration of the effect on cost and reliability.

The basic assumption that the piston can be represented as a lumped mass implies that it should not flex to any significant extent. It is thus desirable to consider the flexural behaviour of typical pistons. However, since realistic piston shapes do not lend themselves to simple calculations, it is convenient to restrict the analysis to idealised piston shapes, which may be used to give some indication of their design limits. The effects of flexing of the piston show up most severely at its frequency of flexural resonance, and a reasonable design aim to avoid problems is to ensure that this frequency is well above the operating frequencies. The flexural resonance frequency of a flat circular disc driven at its centre is given by

$$f_{fo} = \frac{H}{2\pi a} \left\{ \frac{E_p}{12\rho(1 - \sigma^2)} \right\}^{1/2} \qquad (8.10)$$

where ρ is the density of the material of the piston, σ its Poisson's ratio, E_p its Young's modulus, and H is a (dimensionless) function of the ratio of the thickness (h) of the disc to its radius (a) [8.3]. For a more realistic case where the piston is driven by a finite area of ceramic, the function H needs some modification, and its variation with h/a may then most readily be determined experimentally. Measurements made with typical materials and dimensions have shown that H can be approximated by a linear dependence $H \simeq 3.6h/a$ up to $h/a \simeq 0.5$.

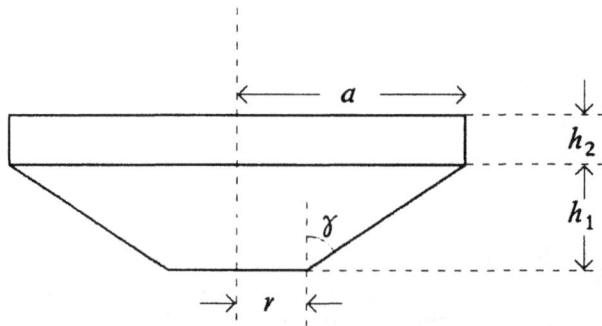

Fig 8.9   Truncated conical piston.

A truncated cone as shown in *Fig 8.9* is more representative of a typical piston, and for this the flexural resonance frequency may be evaluated by first calculating the equivalent thickness (h) of the piston, using

$$h = h_2 + \frac{h_1}{3} \{1 + (r/a) + (r/a)^2\}$$

The flexural resonance ("flapping") frequency is then given [8.4] by

$$f_f = \frac{HG}{ha} \cdot \frac{(h_1 + h_2)}{\sin \gamma} \qquad (8.11a)$$

where
$$G = \frac{1}{2\pi} \left\{ \frac{E_p}{12\rho(1 - \sigma^2)} \right\}^{1/2} \qquad (8.11b)$$

Thus,

$$f_f \simeq 3.6 \frac{G}{a^2} \frac{(h_1 + h_2)}{\sin \gamma} \qquad \text{(For } (h_1 + h_2) < 0.5a) \qquad (8.12)$$

A good design aim for the piston is to make its flexural resonance frequency calculated from this equation at least 1.5 times the highest operating frequency. This may not be easy when a lightweight piston is required, and equation (8.11a) shows the importance of using a material with a high value of G. Typical values of G are 1.9×10^2m/s for Aluminium Bronze, 2.4×10^2m/s for Titanium, and 2.5×10^2m/s for Aluminium alloy (NE6); these show the advantages of using aluminium for the piston, provided corrosion can be avoided. Some of the ceramic materials, such as silicon nitride, also have favourable values of the parameter G and should be free from corrosion effects, but tend to be difficult and expensive to machine in a suitable form. The use of a bonded rubber seal across the front face avoids the need for a ring around the outer edge of the piston to carry the O-ring seals, and permits the use of aluminium alloy. It is therefore very suitable for applications where maximum bandwidth is sought and where this justifies the cost of the bonding process. For less demanding requirements, the simpler arrangements using pistons carrying O-seals are more common.

The resonance frequencies above are calculated for the case where the driving force is applied at the centre of the piston. The simplest mode of vibration of the piston has a nodal ring part way out from the centre, and in principle the resonance frequency can be increased by applying the drive at this nodal ring. In practice, it is usually not easy to calculate the position of the node for realistic piston shapes, or to apply the force precisely at the desired position, so it is preferable to measure the flexing resonance frequency experimentally.

Square pistons have the advantage of increasing the active packing fraction of an array and thus increasing the radiation resistance of the transducers. They do however suffer the disadvantage of being prone to flexing at the corners, and this restricts their usefulness in applications where lightweight pistons are needed. Since O-rings are clearly difficult to fit, the sealing of square pistons is usually effected by a rubber diaphragm extending over the whole array and bonded to the piston faces. This is a fairly simple and inexpensive arrangement for low pressure use, but is liable to stress concentrations in the diaphragm at the corners of the pistons when the pressure is raised, or when the elements are not all driven in phase.

8-6   ELEMENT MOUNTING

In an ideal transducer, the piston would radiate freely by a controlled amplitude, whilst the surrounding case and baffle would remain stationary. This would simplify any calculations of the acoustic field, and also avoid any complications which may arise from coupling between the element and its casing, the effects of such coupling having been neglected in the design equations of the previous chapters. However, it is necessary to support the element in its casing by some technique which ensures that the piston does not touch its housing, and also withstands the hydrostatic pressure of the water. The problem is to devise a suitable technique which allows the piston to vibrate freely, provides the necessary support, and gives good isolation of the alternating forces between the element and the casing.

Two fundamental approaches to this problem are usually considered. One method is to support the element at its nodal plane. In an ideal symmetrical transducer which radiated equally from both ends, the mid-point along the stack would have zero axial velocity, and a thin, though rigid, support at that position would have no effect on the vibration of the element. In practice, as noted earlier, this ideal arrangement is not achievable, either because the radiation conditions are not symmetrical, or because of deviations from the ideal, such as the finite thickness of the nodal plate or variations in the components of the stack. However, although the decoupling may not be perfect, it may be adequate for particular applications, especially if the mounting is itself made compliant. This method is most common for balanced elements, where the front and rear masses are equal and the 'node' is near the mid-point of the stack. For unbalanced designs where the rear mass is much heavier than the front piston, the nodal plane becomes close to the rear of the stack and it is often more difficult to arrange a suitable mount. In such cases a second approach is generally adopted, in which the element is mounted on some form of spring.

A diagram of the arrangement is shown in *Fig 8.10*, in which M_1 and M_2 represent the front and rear masses, F_1 and F_2 the forces in the ceramic stack and supporting spring, K_2 the stiffness of the spring, and x_2 the alternating displacement of the rear mass. The spring is attached to a casing (at the left of the diagram) having a mass which is large enough for its displacement to be neglected. For the vibration of the element to be well decoupled from the casing, the force F_2 on the casing should be small compared with F_1. The ratio F_2/F_1 can be calculated by applying the equations of motion to the system, ie:-

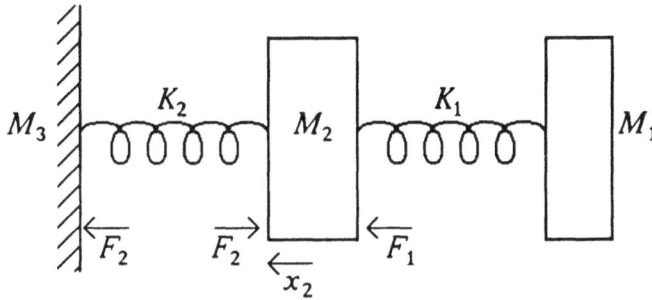

Fig 8.10 Unbalanced element mounting: basic mechanical arrangement.

$$M_2\ddot{x}_2 = F_1 - F_2$$

and
$$F_2 = K_2 x_2$$

where $\ddot{x}_2$ indicates the second differential d^2x_2/dt^2. Thus,

$$\frac{F_1 - F_2}{F_2} = \frac{M_2\ddot{x}_2}{K_2 x_2}$$

For sinusoidal displacements, $\ddot{x}_2 = -\omega^2 x_2$, and therefore

$$F_1/F_2 - 1 = -(M_2/K_2)\omega^2$$

$$= -\omega^2/\omega_2^2 \qquad \text{where } \omega_2^2 = K_2/M_2$$

The transfer function representing the force on the casing is therefore given by

$$F_2/F_1 = (1 - \omega^2/\omega_2^2)^{-1} \qquad\qquad (8.13)$$

The variation of $|F_2/F_1|$ with frequency is illustrated in *Fig 8.11.* For static pressure, the force F_2 is equal to the force in the stack. As the frequency is increased, the transmitted force rises to a peak at the resonance of M_2 with K_2, and then falls steadily, so that when $\omega = 3\omega_2$ the value of $|F_2/F_1|$ has fallen to 1/8. This ratio is usually low enough for the transmitted

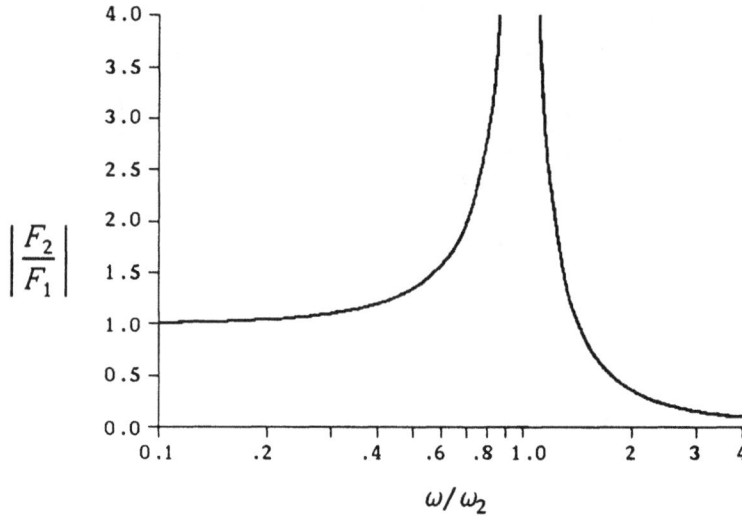

Fig 8.11   Element   decoupling: variation   of   transmitted
force $|F_2/F_1|$ with normalised frequency ω/ω_2.

vibrational force to have only a small effect on the housing, and
the decoupling between element and casing is therefore generally
acceptable if ω_2 is made less than about one third of the lowest
operating frequency. Either of these methods of decoupling the
vibration of the element from the casing also serves as a means
of isolating the element from any vibration of the casing itself,
and hence reducing the acceleration sensitivity of the transducer,
a topic which will be treated in more detail in Chapter 10.

Various forms of springs are used in practical designs. The
basic requirements are for a high compliance along the axis of
the stack, with sufficient strength to withstand the external
pressure, and a low compliance normal to the axis in order to
give good location of the piston in its surround. A metal
diaphragm spring meets these requirements well unless very high
operating pressures are needed. A simple diaphragm consisting
of a thin disc clamped between rings tends to suffer from
unreliable   performance,   because   of   inconsistencies   in   the
clamping, and it is usually preferable for the annular diaphragm
to be machined from a thicker plate so that it has a thicker
inner section and outer ring which are integral with the
diaphragm. The inner may then form part of the stack and the
outer ring be fixed to the housing, as in *Fig 8.18a,b* (p234).

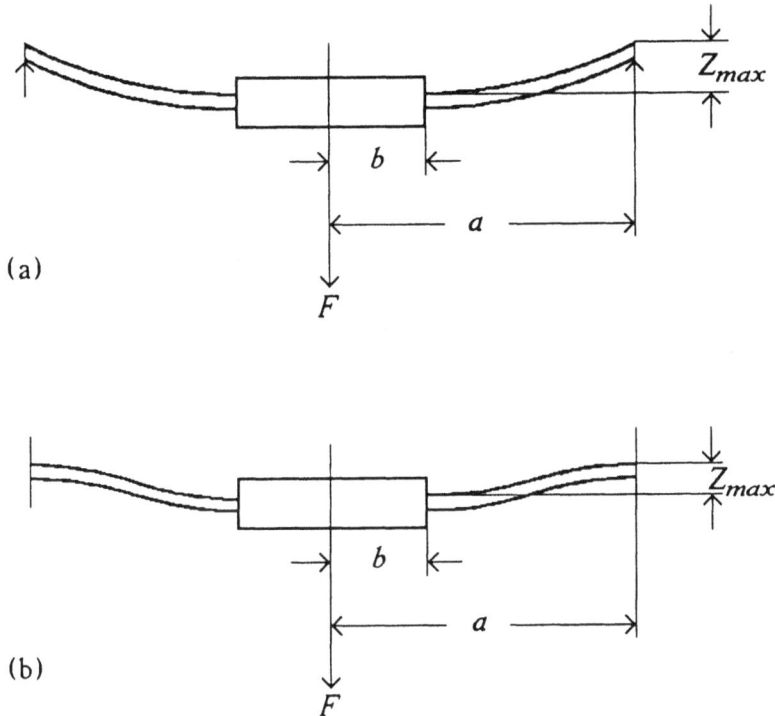

(a)

(b)

**Fig 8.12   Edge clamping conditions for diaphragm springs:
(a) edge supported; (b) edge clamped.**

The characteristics of such springs may be calculated from standard texts such as that by Timoshenko and Woinowsky–Krieger [8.5]. *Fig 8.12* shows diagrams of the relevant cases. A force F is applied axially to the central boss, with the edges of the circular diaphragm either supported (*Fig 8.12a*) or clamped (*Fig 8.12b*). Expressions for the displacement z_{max} of the central boss, and the maximum stress T_{max} in the plate are given by [8.5]

$$z_{max} = k_1 \frac{Fa^2}{Eh^3} \qquad (8.14a)$$

$$T_{max} = k_2 \frac{F}{h^2} \qquad (8.14b)$$

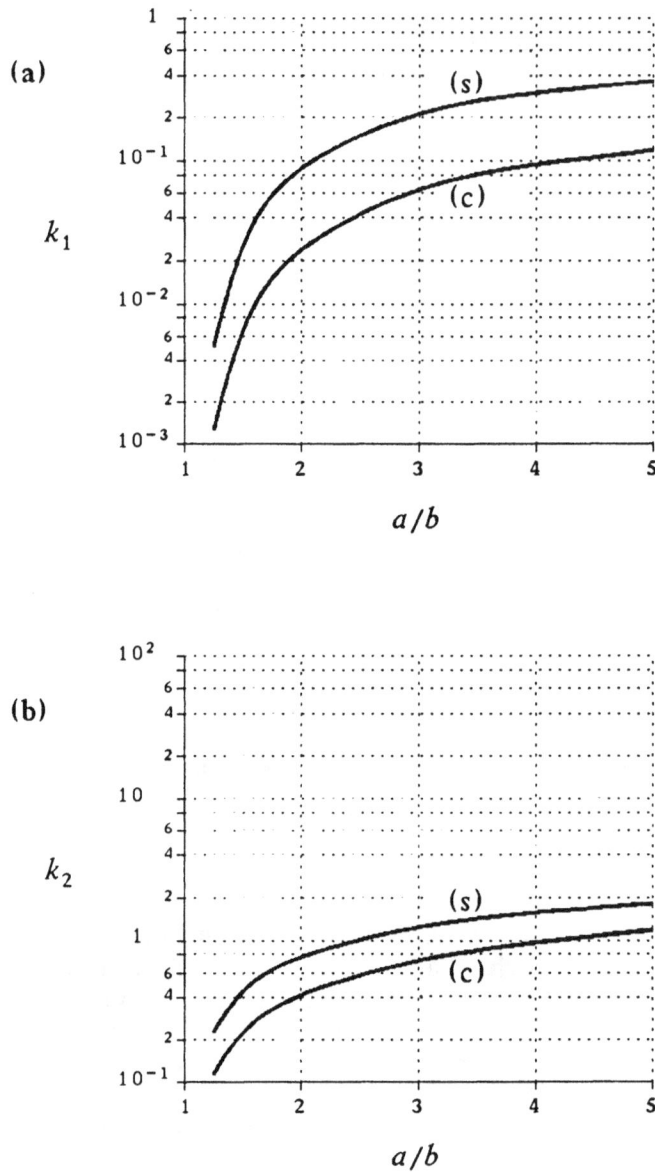

Fig 8.13   Design factors k_1, k_2, for diaphragm springs: (a) k_1; (b) k_2: (s) = supported edge; (c) = clamped edge.

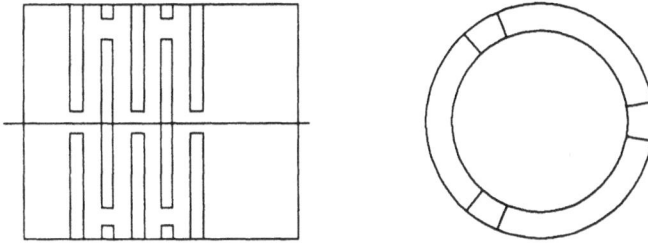

Fig 8.14   Cylindrical slotted spring.

in which a is the radius of the diaphragm, h is its thickness, E is the Youngs modulus of the diaphragm material, and k_1, k_2 are functions of a/b, the ratio of the outer radius of the diaphragm to the radius of the central boss. Values of k_1 and k_2 are tabulated in [8.5], and plotted in *Figs 8.13a and b*. The values of k_1 in *Fig 8.13a* show that a change in edge conditions from being simply supported to being clamped can change the stiffness by a factor of over three. It is therefore important to ensure consistency in the edge conditions, and this is why machining the diaphragm from a thicker plate, so that it is integral with the inner boss and outer ring, is preferred.

The diaphragms are often made from steel to obtain sufficient strength, but the external pressure may generate high stresses in the plate, and this restricts the depths for which they can be used. Diaphragm springs such as this are very suitable as nodal mounts, particularly for balanced elements, combining the benefits of the nodal mount with the decoupling due to the spring compliance.(See Chapter 10 for a fuller treatment.) For an unbalanced element with a heavy rear mass, the node is near the rear of the stack, and a diaphragm spring may be used at the node, if mechanically convenient, or attached immediately to the front or rear of the tail mass (as in *Fig 8.18b*).

If the transducer is to be used at great depths, a more substantial spring is required to withstand the pressure. A conventional coil spring is not very stiff for lateral displacements, and a design which is better suited for such applications is a 'slotted spring', which consists of an annular cylinder with slots cut tangentially around the circumference, as in *Fig 8.14.* The slots in alternate rows are offset so that they form a series of curved beams loaded at their centres, and the stiffness of the spring may be calculated (to a first

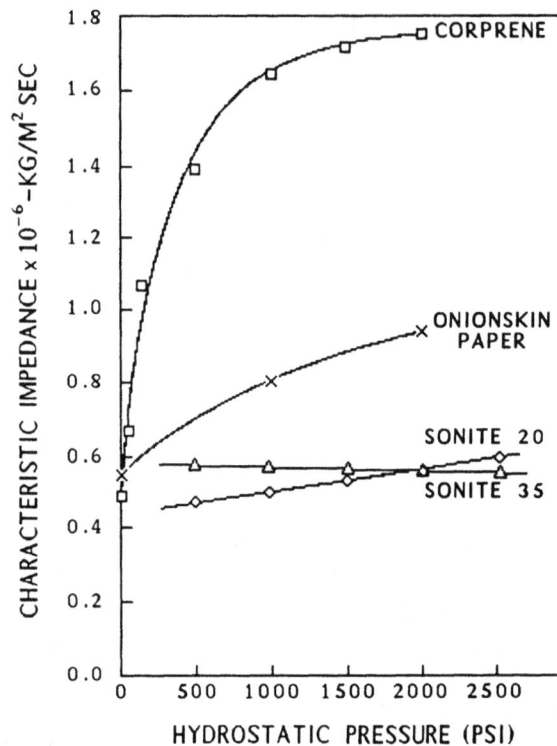

Fig 8.15

Characteristic impedance as a function of pressure
for onionskin paper, Corprene DC 100, Sonite 20,
and Sonite 35 (1000psi $\simeq$ 6.9MPa); from *J Acoust
Soc Am*, in Higgs *et al.*, *JASA* 50, 946-954, (1971).

approximation) by neglecting the curvature and treating it as an
assembly of straight loaded beams (using eg Ref [8.6]).
Obtaining a precise value of the stiffness, and determining the
maximum safe load, is probably best tackled experimentally.
This type of spring is usually attached to the rear mass, and
may even be machined from the same material, if the
requirements for strength and density are compatible.

Both of these designs have the advantages of being very
nearly linear over quite a wide range of loads, and of being
virtually free of mechanical dissipation losses, so that they
conform well to the assumptions used in deriving the equations.
Mounting springs have sometimes been made of materials which

are less linear and more lossy, but may have lower cost than the metal designs. Examples of such materials are Corprene, Sonite, and onionskin paper, which are often referred to as 'pressure release' materials but are more commonly used as decoupling springs. 'Corprene' is a registered trademark of the Armstrong Cork Company for a polychloroprene–cork composite made in various compositions. One type (Corprene DC–100) was originally produced as a gasket material, but has been used for mounting transducer elements for many years, especially in the USA. 'Sonite' is a registered trademark of the Johns Manville Corporation for a glass fibre–asbestos composite which was derived from a thermal insulating material for use in acoustic applications. Onionskin paper is a high quality paper containing about 25% cotton fibre which has found occasional use over many years as a decoupling spring material, almost entirely in the USA. The characteristics of these materials have been measured by various authors ([8.7]–[8.11]). Their results are illustrated in *Fig 8.15*, which shows the marked variation of characteristic impedance as the applied pressure is increased. These variations in properties of the mount cause changes in the resonance frequency and beam patterns of the transducer as the working pressure is raised, and in general these materials are satisfactory only for external pressures up to a few hundred psi (2–3MPa).

In the cases discussed above, the force on the piston due to the external pressure is transmitted through at least part of the stack, and this may apply a high stress to the piezoelectric ceramic because of the magnification ratio between piston and ceramic area. It is then appealing to consider whether the external load could be taken by a support at the piston itself. Such an arrangement could be represented in *Fig 8.10* by taking M_2 as the piston and M_1 as the rear mass, and good decoupling would require the resonance of the spring stiffness with the piston mass to be well below the operating frequencies. Since the mass of the piston is generally appreciably less than that of the rear mass, it implies that the spring would have to be considerably more compliant than if the element were supported at the rear. This would result in greater movement of the front piston as the external pressure was increased, with worse problems in achieving a good watertight seal. The extra restraint on the alternating motion of the piston would also cause a reduction of the coupling coefficient of the transducer. This method is therefore not very suitable for the high pressure applications where it would be most useful. It has occasionally been used for low pressure devices by supporting the piston on a stack of rings of onionskin paper. The increasing stiffness and non-linearity of such an arrangement restricts its use to non-critical low pressure applications [8.10].

Belleville springs are metal washers which are slightly conical, so that they have non-linear stiffness characteristics [8.6]. For example, the washer may be designed to have a high stiffness for relatively small loads, but a low stiffness for moderate loads, increasing again for high loads. The low stiffness region corresponds to the condition where the washer is nearly flat, and it is possible to make the washer have a very low (or even negative) stiffness for loads in that range. These stiffness characteristics have some appeal for use in transducers [8.12], since they should make it possible to withstand the static load on the piston without greatly restricting the motion of the piston for small displacements. Furthermore, because the washers are usually made of metal, they should have low internal losses, in contrast to the onionskin paper or Corprene types of material described above. However, Belleville washers have not in practice been greatly used in transducers, partly because operation over a range of pressures is usually wanted, and partly because considerable stress concentrations can exist at the edges of the washers.

For some high pressure applications, it is possible to use a mechanically balanced element with all the piezoelectric ceramic behind the nodal mount. The front half of the stack has a stiffness and mass to balance that of the rear, but is composed of passive materials; the element is excited into oscillation around resonance by the ceramic behind the node. Such an arrangement allows the external pressure to be taken by the nodal mount (via the front half of the stack), so that the ceramic is not subjected to the stress arising from the hydrostatic pressure. The disadvantage of this design is that the coupling coefficient is markedly reduced by the inactive part of the stack (as in equation (8.7)), so that it is suitable only where a relatively low coupling coefficient is tolerable.

Another approach to the problem of designing for high pressure use is to fill the transducer casing with a fluid such as oil. Whilst this can greatly reduce the static pressure difference across the piston, it also gives rise to alternating pressures within the casing. The stiffness of the fluid is effectively in parallel with that of the stack, and thus reduces its coupling coefficient and increases the resonance frequency. Even small air bubbles within the housing can have a marked – and non-linear – effect on the stiffness of the fluid, and it is therefore not easy to control the characteristics of the transducer. Resonances within the fluid make the effects yet more difficult to control, and should generally be avoided. It is possible to make the compliance of the cavity frequency dependent by introducing orifices in the casing, with the aim of maintaining the pressure balance for static pressures whilst achieving a higher cavity compliance at operating frequencies. Despite these complications, oil-filling has been successfully used by some designers, especially for flexural disc designs (Chapter 11), which are particularly susceptible to external pressure.

8-7 WATERTIGHT SEALS

The action of the piezo-ceramic stack is to cause the piston and its countermass to oscillate axially in antiphase. The piston and stack dimensions are generally less than a wavelength, and if all of the element were in contact with the water, the radiation from some parts would cancel that from others and a low output would result. The most important acoustic requirement is to prevent radiation from the rear face of the piston itself, which would be of equal amplitude to that of the front face, and in antiphase to it, so that the cancellation effect would be very strong. This is usually avoided by mounting the element in a watertight and air-filled case, so that only the front face of the piston is in contact with the water, whilst the remaining surfaces of the element itself are surrounded by the air in the casing. Thus, only the front face of the piston can radiate into water, the other surfaces radiating only weakly into the low impedance of the air. This arrangement also reduces the electrical insulation problems, by keeping the sea-water away from the stack. In practice, the fundamental acoustic requirement of low radiation from the rear of the piston is fairly easy to achieve, but maintaining good electrical insulation of the stack is one of the most difficult design requirements to guarantee, and leakage of water into the housing is probably the most common reason for failure of transducers in service.

It is often supposed that making an enclosure watertight is much easier than making it vacuum-tight, and that the accepted standards of care needed for the latter are not necessary for the former. Let us, however, estimate what the maximum tolerable leakage rate for a transducer might be. Problems with low insulation resistance across the ceramic could be expected if the gas around it became saturated with water vapour. The mass of water which saturates $1m^3$ of air at 10°C is approximately 10g.

Gay-Lussac's Law states that 18kg of water would occupy 22.4m³ if it existed as a vapour at normal temperature and pressure. Thus 10g would occupy about 12×10^{-3}m³. In a volume of 1m³, this mass of gas would produce a pressure of 1200Pa ($\simeq$ 10torr). Suppose that the leakage rate is just sufficient to cause saturation in three years (approx 10^8sec). Then the rate of rise of pressure would be about 1.2×10^{-5}Pa/s ($\simeq 10^{-7}$torr/s). A typical transducer volume may be taken as 2 litres (eg 0.1m × 0.1m × 0.2m), and the leakage rate to cause saturation of this volume in three years is therefore of order 2×10^{-5}Pa-L/s ($\simeq 2 \times 10^{-7}$torr-L/s). This leakage rate is typical of that required for good quality vacuum equipment. The standards needed in transducer assembly, and the testing methods, are thus comparable to those involved in good quality vacuum equipment.

It is the seal around the radiating piston which poses the most difficult problems, since it needs to be flexible, low loss, and reliably watertight. It must not have a low acoustic impedance, since that would allow the acoustic pressure produced at the piston face to be dissipated partly in the seal. Probably the simplest solution is the use of O-ring seals around the piston, as in *Figs 8.8a* and *b*. O-ring seals are widely used for making watertight seals, both static and dynamic, and some aspects of their use are discussed in later paragraphs. Although simple, they may introduce significant motional losses and tolerance problems, and require appreciable mass around the edges of the piston. A metal diaphragm around the piston, as in *Fig 8.8c*, offers a solution with very low internal damping, but if made very flexible it is likely to have a low acoustic impedance. A thin diaphragm also has mechanical disadvantages in difficulty of machining, and in being susceptible to damage in operation. For these reasons, and those mentioned in Section 8.5, diaphragm seals are therefore not commonly used.

Bonded rubber seals, as shown in *Figs 8.16a* and *b*, represent another approach to the problem. The design in *Fig 8.16a* effectively replaces the compressed O-ring with an annular rubber seal which is moulded into the gap between the piston and its surround. The seal is made by moulding the rubber around the metal components at an elevated temperature, but unfortunately the rubber shrinks as it cools, so that it tends to pull away from the surrounding metal. It is therefore necessary to ensure that the rubber is well bonded to the metal parts, and that it will remain bonded even after prolonged exposure to sea-water at high pressure. Unless very carefully manufactured, the bond can be susceptible to progressive failure

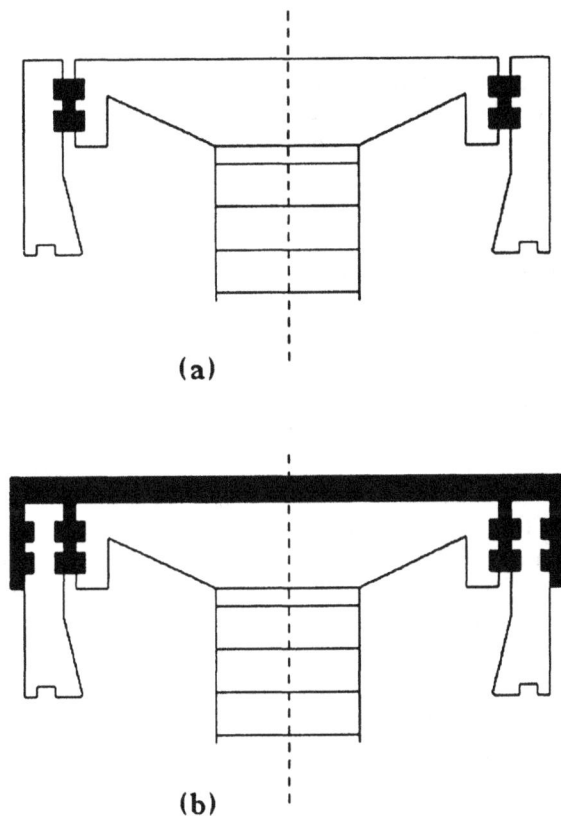

(a)

(b)

Fig 8.16   Piston designs using bonded rubber seals.

from the edge exposed to the sea–water, and it is desirable for
the design to make the potential leakage paths as long as
possible, including the addition of corrugations in the actual seal
area as shown in *Figs 8.16*. These corrugations may also
possibly provide a mechanical seal in the event of failure of the
chemical bond. The seal should be formed of a material with a
low dynamic loss, and preferably with a loss which does not
vary too much with temperature. Natural rubber has a low loss
factor, but suffers from the effects of oxidation and stress
concentrations, and a nitrile compound is generally preferred.

The design in *Fig 8.16b* takes the bonded rubber right
across the front face, thus reducing the number of exposed bond
edges, but it requires very reliable bonding over the whole

piston area, since even tiny air bubbles under the rubber can have dramatic effects on the acoustic output. It was observed in Section 8.5 that such a seal had considerable advantages in widening the choice of possible materials for the piston. These bonded seals avoid the compression of the rubber and the scratch-free surfaces which are necessary in O-ring designs; they thus offer potentially better acoustic performance, and greater reliability, provided that they can be consistently produced to high quality.

It was noted above that O-ring seals may be used as a dynamic seal around the piston, and they are commonly used also as a simple and effective means of making static watertight joints. When axial vibration of a piston is needed, an arrangement incorporating O-ring seals between the piston and its surrounding cylindrical housing, as in *Figs 8.8a* and *b*, can be used. Where there is no requirement for any motion, face or corner seals as illustrated in *Fig 8.17* are usually to be preferred, as they can be more accurately controlled. Guidance in the application of O-ring seals is given in British Standards (BS 1806 and BS 4518), and in manufacturers' literature. The surface finish of the housings should not be worse than 16 μinch (0.4 μm), and any machining marks should preferably lie around the seal – ie they should not traverse the seal. Attention to detail in the design can often reduce the possibility of scratches occurring across the seal face during assembly. A little Silicone Grease may be used where necessary as a lubricant. The O-ring should be supported against the direction of the applied pressure; thus, for a face seal exposed to an external pressure, the inside diameter of the O-ring should be supported by the inner diameter of the groove in the housing. This prevents too much movement of the O-ring, and reduces the risk of any consequent "pumping" action from any pressure cycling. For static seals which do not need to be disassembled, it is possible to use a little flexible epoxy resin adhesive in the groove with the O-ring to improve reliability. The moulding process involved in manufacturing O-ring seals sometimes produces "flash" rings around their inner and outer diameters, and any excessive flash represents a potential source of leakage, especially for piston-type seals.

Experimental studies have shown, however, that even carefully assembled single O-ring seals may often have a leak rate exceeding 10^{-5}Pa-L/s, and two O-seals in series should be used wherever possible. Experience has also shown that lower leak rates are obtained by increasing the compression applied to the O-ring above that specified in the British Standards, and the following design guides for seals between

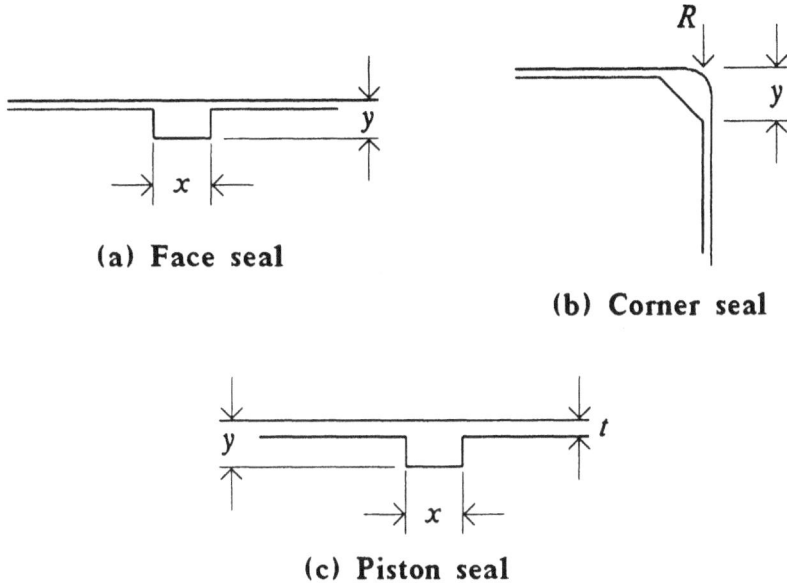

(a) Face seal

(b) Corner seal

(c) Piston seal

Fig 8.17   Types of O-ring seals.

about 50 and 100mm in overall diameter are recommended (where d is the diameter of the cross-section of the unstrained O-ring):-

Face seals (*Fig 8.17a*):
    $y = 0.75d$ $x = 1.34d$ ie nominal compression = 25%

Corner seals (*Fig 8.17b*):
    $y = 1.43d$ $R = 0.043d$
(for 45 degree chamfer angle)

Piston seals (dynamic) (*Fig 8.17c*):
    $y = 0.85d$ $x = 1.34d$ ie nominal compression = 15%
    $t = 0.09$mm for pressures above 3.5MPa (500psi); t may be increased for lower pressures.

    The small annular gap (t) for the piston-type seal is necessary mainly to prevent the O-ring being extruded into the gap as the external pressure is raised, which would result in a marked increase in the stiffness and losses of the seal. It also serves to reduce the damage which would be caused to the seal by repeated pressure cycling, with a consequent loss of reliability. Anti-extrusion rings fitted behind the O-rings are sometimes recommended for static applications, but they generally introduce rather high mechanical losses and are not commonly used around transducer pistons. However, designs with a narrow gap between the piston and the housing do introduce potential difficulties in achieving adequate control of the mechanical tolerances of the assembly to maintain the gap at its desired value. In many designs, the applied pressure forces the piston inwards against the spring supporting the element, and the O-seal must accomodate this displacement as well as the alternating vibrations. If the displacement is large (eg comparable with the O-ring diameter), it is possible for dirt to be introduced into the gap and to score the seal surfaces. The dangers of leakage from this mechanism can be minimised by using two O-seals in series and by reducing the displacement of the piston as much as possible. Apart from this effect, there is little difference between the leakage at low pressure and that at high pressure, and it is certainly unsafe to assume that a reliable seal for use at only shallow depths can be achieved with less care than for higher pressures.

Although the seal around the vibrating piston poses the most difficult basic design problems, the seal around the cable entry is also a common source of leakage. This may often arise because the cable is connected to the transducer only after it has been installed in its operating position, probably in rather poor conditions and without the opportunity for subsequent testing. If these circumstances cannot be avoided, it is essential to ensure that the design of the connector is suitable for such conditions, and that assembly is simple. A cable gland moulded on to the end of the cable and connected to the transducer housing by means of a double O-ring seal is generally satisfactory, if assembled with care.

It is worth repeating that water leakage is probably the most common cause of failure for transducers in service, and experience suggests that many of the failures can be traced to careless assembly in which the surfaces become scratched or dirt is introduced into the seal. The best precautions appear to be careful design to minimise risk, good quality control during assembly, and adequate testing during manufacture.

8-8   GENERAL ARRANGEMENTS

We now reach the stage of considering some of the arrangements for mounting the element in its housing. This is a stage which greatly involves engineering judgement, and it would be difficult to propose any rules which could be used to lay down optimum designs over a range of applications. Instead, the arrangements indicated in *Fig 8.18* will be used to illustrate some of the main features which should be considered in deciding on a suitable design in any particular case. Although only one element is shown in each example, it is assumed that it forms part of an array in which it is necessary to maximise the packing fraction of active/total area. The dead space between the pistons should therefore be kept to a minimum. All of the examples shown are for essentially similar element characteristics, and all use O-ring seals around the front piston; bonded seals would involve similar types of considerations, though differing in detail. The main features which influence the design are the search for watertightness, ease of machining, and the need for making electrical connections.

Fig 8.18a uses a diaphragm as a nodal mounting; as illustrated it is near the middle of the element – ie the element is nearly balanced – but this type of mounting may also be used where the tail mass is larger than the piston and the node is nearer the tail. The element is inserted into its case from the front, and the diaphragm is fixed into place by bolts from the rear. A tail cover is then fitted to complete the watertight enclosure around the element. Since the diaphragm has to be fixed from the rear, the tail mass must be inserted through the mounting flange and this makes it inconvenient for designs requiring heavy rear masses, which would need to be rather long. The flange on the tail cover also tends to be wasteful of space between elements. The shoulder mounting the diaphragm must be accurately machined, so that the tolerances of the assembly can be controlled well enough to ensure that the gap around the piston is within its permissible range. Careful attention to tolerances and design details is necessary at this stage. For example, the diaphragm needs to fit snugly into the housing in order to locate it precisely, but the corner of the diaphragm outer ring should be relieved to allow for the inevitable radius on the corner between the cylinder bore and the mounting shoulder. If this is not done it is quite probable that the piston will be canted over sufficiently to rub against the housing and perhaps even score the surface used for the O-seal. Electrical connections can be soldered to the element during assembly, whilst the diaphragm is behind the holes carrying the connecting leads through the housing, but before the piston enters the housing; however the gap for a soldering iron is likely to be small at this stage of the assembly for this design.

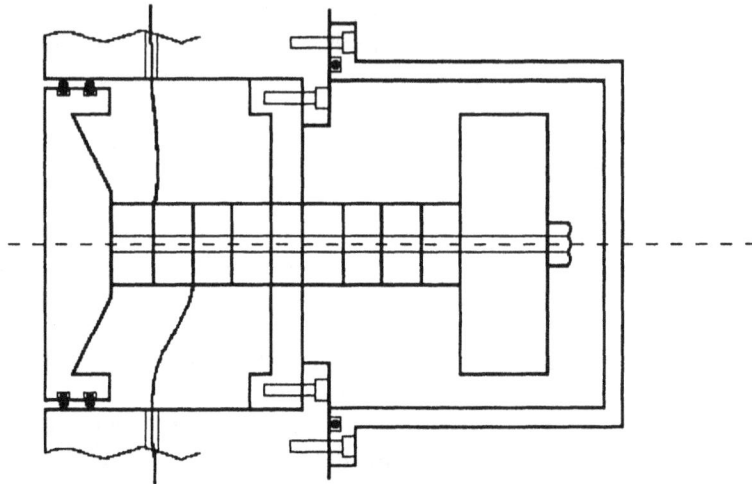

(a)

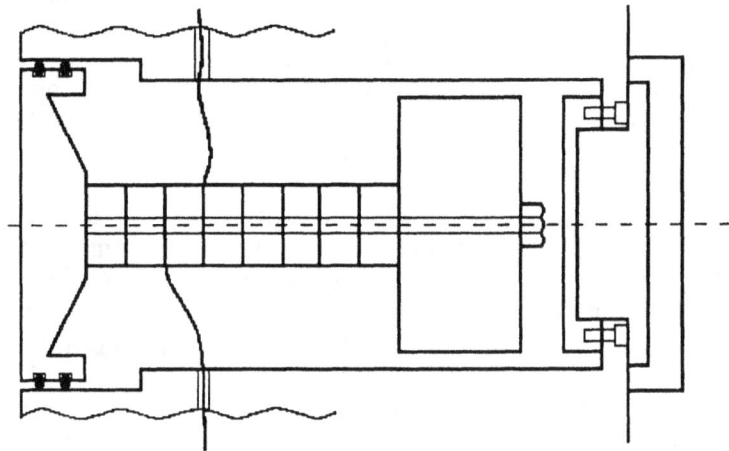

(b)

Fig 8.18 Typical   arrangements   for   low   frequency
piezoelectric transducers.

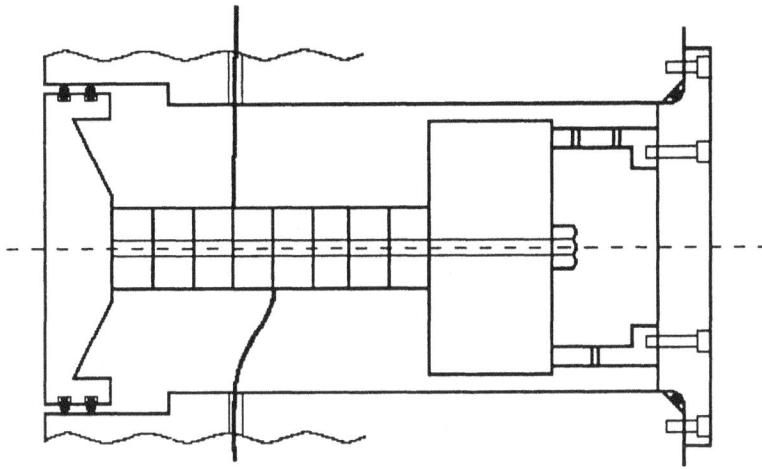

(c)

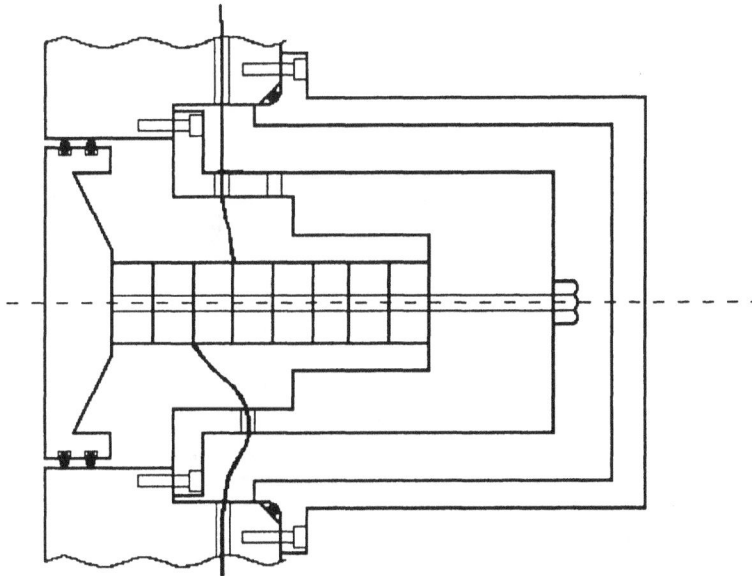

(d)

Fig 8.18   contd.

The arrangement shown in *Fig 8.18b* transfers the mounting diaphragm to behind the rear mass. This is of course most appropriate for an unbalanced element where the tail mass is much heavier than the piston. It has the advantage of relaxing the restriction on the diameter of the tail mass, which can now be as large as the piston diameter. The diaphragm is again bolted into place from the rear, and a cover plate is used to complete the housing. Making the electrical connections is much easier than in the previous example, since a larger gap is possible for inserting a soldering iron. However, a much thicker casing is now required, and it is necessary to machine the supporting flange accurately at the bottom of a deeper bore. The build-up of tolerances is also likely to be worse than in (*a*).

Fig 8.18c shows an attempt to overcome the problem of accurate machining at the bottom of a cavity by instead supporting the element on the tail cover. In the example illustrated, a slotted spring is used instead of a diaphragm to mount the rear mass. It is unnecessary to machine the whole length of the bore to the surface finish needed for the O-seal, and in this example the bore is machined out at the front to give the O-seal finish. In this case, the element must be inserted from the front and the fixing bolts attach through the cover plate. This would clearly be undesirable, as it would introduce further possible water entry paths which would be difficult to make reliably watertight. This example is included partly to illustrate how overcoming one problem can easily lead to others, and in particular how it can be very tempting to solve design difficulties by adding further potential leakage paths. As a general rule, the most important criterion in reducing the chances of leakage is to keep the number of possible water leakage paths to a minimum. In this particular example the build-up of tolerances would also be likely to cause difficulties in maintaining concentricity of the piston.

It is possible to place the slotted spring support at the front of the tail mass instead of at the rear, and such an arrangement is illustrated in *Fig 8.18d*. This design is effective in using the space around the stack to achieve a heavy tail mass, and brings the mounting position towards the front, so that the thickness of the housing can be reduced. The element is inserted from the rear, connections being made before it is fully engaged in its housing. A tail cover is needed over the rear. In this design, the spring is most conveniently made of the same material as the countermass, and if the spring is required to withstand high pressures this limits the choice of metal. Tungsten alloy and spent uranium have been used for tail masses when large mass in a small volume is wanted, but these are unlikely to be suitable if they are to be employed also as springs subjected to high loads. Brass is another, and more common, tail mass material, but this is also not very suitable for a highly loaded spring, and steel is therefore the most likely choice for this type of design.

It is evident that many alternative arrangements are possible to overcome the various problems in achieving a reliable design which meets particular requirements at the lowest cost. This section has merely touched on some of the main factors which should be considered; the resolution of the difficulties and conflicting demands is where the designer's engineering skills should be brought into play.

REFS Chapter 8

References

8.1 Bowden,F.P., and Tabor,D., *The Friction and Lubrication of Solids*, Clarendon Press, 1950. (Chap 1.)

8.2 Frost,N.E., Marsh,K.J., and Pook,L.P. , *Metal Fatigue*, Clarendon Press, 1974. (Chap 3.)

8.3 Lord Rayleigh, *The Theory of Sound*, McMillan & Co, 1926. See also Rijnja, H.A.J., "The design of sensitive hydrophones", *Acustica*, 33, 1-9, (1975).

8.4 Evans,D.J., "Calculation of the flapping frequency for a piston of linearly varying thickness driven at the centre", Admiralty Underwater Weapons Establishment, AUWE Tech Note 334/69, (1969).

8.5 Timoshenko,S., and Woinowsky-Krieger,S., *Theory of Plates and Shells*, McGraw-Hill, (2nd Ed), 1959, (p61.)

8.6 Roark,R.J., *Formulas for Stress and Strain*, McGraw-Hill, 1965.

8.7 Higgs,R.W., and Eriksson,L.J., "Acoustic decoupling properties of onionskin paper", *J Acoust Soc Am*, 46, 211-215, (1969).

8.8 Higgs,R.W., and Eriksson,L.J., "Acoustic decoupling properties of Corprene DC-100", *J Acoust Soc Am*, 46, 1254-1258, (1969).

8.9 Higgs,R.W., D'Amico,P.M., and Speerschneider,C.J., "The acoustical and mechanical properties of Sonite", *J Acoust Soc Am*, 50, 946-954, (1971).

8.10 Haan,D.E., Higgs,R.W., and Eriksson, L.J., "Performance degradation of deep-ocean transducers using onionskin paper for acoustic decoupling", *J Acoust Soc Am*, 51, 290-294, (1972).

8.11 Behrens,J.A., Blankenship,G.W., and Stokes,R.H., "An onionskin composite as a transducer backing material", *J Acoust Soc Am*, 65, 1562-1567, (1979).

8.12 Woollett,R.S., "Ultrasonic transducers: 2. Underwater sound transducers", *Ultrasonics*, 3, 243-253, (1970).

Chapter 9
TESTING

The testing of transducers is a vital element in confirming that they are acceptable for their intended purpose. The tests are generally of three types; those which confirm the acoustic performance of the completed transducer, those which are applied during manufacture for quality assurance purposes, and those which are used during the development phase to confirm the basic design.

The acoustic performance requirements must be agreed between the user and designer, and should then be described in a specification. It is important that this is realistic in the limits it lays down, and in the parameters it specifies, since it is easy inadvertently to specify tests which are difficult and expensive to carry out. It is for example tempting to specify a minimum value of the transducer's efficiency which may involve measurements of the directivity patterns in water, and these are usually difficult and time-consuming to achieve with the accuracy which may be necessary. Attention should therefore be paid to what properties genuinely need to be specified, and how they should be measured. Design requirements such as those affecting operating life are particularly difficult to specify in a way which can be rigorously tested, and these are usually established partly during the development stages. The "quality assurance" tests are generally the concern of the designer rather than the user; they are imposed to ensure that the component parts are satisfactory, and that assembly methods are adequate. These tests help to avoid waste in assembling unsatisfactory parts, and are sometimes necessary to confirm characteristics which cannot be measured on the completed transducer.

The status of the tests laid down in a specification can give rise to considerable discussion or argument. From one point of view, they may be regarded as essential requirements, with failure to meet the specified limits indicating total unacceptability of the device. In this case, it is clearly of vital importance to have precisely defined limits and accuracy in the measurements. At the other extreme, the specification is sometimes regarded merely as a target to be aimed at, with considerable latitude in its interpretation. In many cases, an intermediate position may be adopted, in which certain of the characteristics may be regarded as critical, whilst others may be treated as desirable aims which it is not essential to meet fully. It is not an aim of this book to judge between these positions, but it is important that the user and designer agree what status the specification does have for any particular application.

In this chapter, a number of tests will be described which may be performed on a transducer during and after its assembly, starting with tests on the ceramic and progressing through the stages of the assembly. Not all of the tests described will be necessary for all applications, and other tests may sometimes be appropriate. Those actually needed for any particular transducer depend of course on its design and the stringency of its requirement, but the measurements of resonance frequency and coupling coefficient (for projectors), and some form of watertightness test, should be omitted only in exceptional circumstances.

9-1 CERAMICS

Measurements of the fundamental parameters of piezoelectric ceramics are commonly carried out on thin discs of the material, and are described in some detail in the appropriate IRE Standard [9.1]. For our purposes, a simpler set of measurements is normally adequate, the basic techniques being the determination of the resonance and antiresonance frequencies for the sample, together with the low frequency capacitance. The capacitance of a sample may be measured by any good commercial capacitance

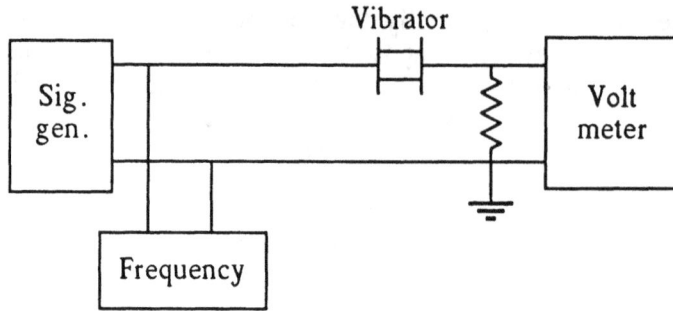

Fig 9.1 Basic circuit for measurement of f_m and f_n.

bridge operating at a frequency well below any resonances of the sample – a frequency of between 1 and 2kHz is commonly used. The bridge usually gives a value for the loss factor ($\tan\delta$) at the same time. Both measurements are made at a relatively low field, depending on the characteristics of the bridge.

A circuit as in *Fig 9.1* is used to determine the resonance and antiresonance frequencies, by measuring the variation with frequency of the modulus of the admittance of the sample. This circuit in practice determines the frequencies of minimum and maximum admittance, but these are sufficiently close to the resonance and antiresonance frequencies to give accurate enough values of the ceramic parameters for lightly loaded samples with a reasonably high coupling. The signal generator should have an output impedance lower than the minimum impedance of the specimen, so that the applied voltage is approximately independent of the frequency. The series resistor, which is used to measure the current through the specimen, should also be small compared with the lowest impedance of the specimen. A value of about 1 ohm is usually satisfactory. The frequency of the applied signal is varied until a maximum current is indicated (ie the voltage across the series resistor is a maximum), this being the frequency of minimum impedance (or maximum admittance), f_m. The frequency is then increased until the current is a minimum, thus indicating the frequency of maximum impedance (or minimum admittance), f_n. Since this impedance may be high, some care may be needed to minimise stray capacitance across the sample, but this is not usually a serious problem. A more common source of difficulties, because of its effect on reproducibility, is the mounting arrangement for the disc. Electrical connections must be made to the sample whilst allowing it to resonate freely. A simple and effective means for doing this is to mount the specimen horizontally, between two lightly sprung point contacts at the centre of the disc. Too tight clamping, or clamping off-centre, may cause errors in the measurements which lead to inconsistency from sample to sample.

The IRE Standard [9.1] gives expressions for the errors involved in measuring f_m and f_n instead of the true resonance and antiresonance frequencies, and for calculating the planar coupling coefficient of the material. The latter corrects for the non-uniform stress distribution within the disc, and results in a quite complex expression to derive the true coupling coefficient of the material. Further corrections may be applied if the thickness of the disc is greater than one-tenth of the diameter. For our purposes, it is usually satisfactory, and much simpler, to define the effective coupling coefficient by equation (4.5), ie

$$k_e^2 = 1 - (f_s/f_p)^2$$

which may be approximated, for high kQ_M samples, by

$$k_e^2 \simeq 1 - (f_m/f_n)^2 \qquad\qquad (9.1)$$

This value of k_e differs from the true coupling coefficient of the material, since it makes no allowance for the stress distibution through the sample; however, for thin discs the correction depends only on the radial stress distribution, and is independent of the actual dimensions of the disc. The use of this definition thus allows simple comparisons between samples, without introducing uncertainties about what formula has been used, or whether the formula is accurate for the dimensions of the particular sample. It also simplifies the use of these measurements for quality control purposes on batches of ceramic.

The piezoelectric stack in a Tonpilz transducer of the type described in Chapter 8 is generally assembled using ceramic rings instead of discs, and measurements on rings accentuate the problems of mounting the samples without interfering with the resonance. Some convenient method of lightly clamping the ring in a way which is symmetrical about the axis is desirable, and this may be effected by supporting the ring on three point contacts, and placing on it a light mass with three point contacts for the electrical connections. Trial and error is needed to confirm whether the arrangement gives consistent results. These ceramic rings are often thicker than the samples used for measurements of ceramic properties, and it may therefore be preferable to measure the resonance and antiresonance frequencies for their thickness mode instead of the radial mode. For this to be reliable, any harmonics of the radial resonance, or any other resonances, must be adequately separated in frequency from the thickness mode resonance. Such a thickness mode measurement should be regarded as a convenient method of confirming the consistency of the rings rather than as a method of establishing the basic characteristics of the material. These measurements are particularly useful for eliminating batches of rings which differ too much from the mean, or for detecting rings with cracks, which change the resonance frequencies appreciably.

Another test which can sometimes be useful, especially for samples where the dimensions make it difficult to separate the resonances, is a direct measurement of the pressure sensitivity of the sample. If the stress applied to a surface of a piezoelectric ceramic is changed, an output charge or voltage is generated and may be measured as an indication of the activity of the specimen [9.2]. The most straightforward way of applying a force to the sample is to use a known weight. It is not easy to place a weight on a sample smoothly, and it is therefore generally preferable to apply a weight, discharge the voltage across the sample, and then remove the weight. The aim is to measure the charge generated by this removal of the force, but this may be made difficult by the leakage of the charge through the input resistance of the equipment or through the ceramic itself. The measurement may be greatly improved by connecting a large capacitor across the sample, thus increasing the time constant of the combination. The generated charge is shared between the capacitors, and is measured by a high impedance voltmeter. If the stress is applied along the poling axis, the generated voltage is given by

$$V' = \frac{d_{33}T_3A_c}{C' + C_{LF}}$$

where T_3 is the change in stress, A_c is the cross-sectional area of the sample, C_{LF} is the low frequency capacitance of the ceramic, and C' is the added parallel capacitance. If $C' \gg C_{LF}$, this reduces to

$$V' \simeq d_{33}F_3/C'$$

ie $$d_{33} \simeq C'V'/F_3 \qquad (9.2)$$

where F_3 is the force applied to the sample. This method thus gives a measure of the piezoelectric d-constant of the ceramic sample, and can be useful particularly for ensuring consistency in samples for hydrophone applications. Unfortunately, piezoelectric ceramics are also strongly pyroelectric, and even small temperature changes – such as can arise from handling the specimens – can generate charges which are large enough to spoil the accuracy of this method (Section 3.5). As a general rule, it is preferable to use the measurements of maximum and minimum admittance as indications of acceptibility whenever possible. The ceramic pieces should be marked to indicate their polarity (ie their positive and negative faces during poling), and it is sometimes a wise precaution to check this marking, before assembling the pieces into a stack, by simply compressing the sample and confirming the sign of the voltage produced.

Measuring the changes of ceramic parameters under stress or temperature changes is a specialised subject which will not be discussed in any detail here. Care is needed, as for many experimental measurements, to ensure that the effects are correctly associated with the controlling causes. For example, when measuring the changes in parameters due to increasing the alternating electric field, care should be taken to avoid the temperature rise which is a natural consequence of the increased internal losses. The effects of applying mechanical stress are particularly difficult to measure in a consistent and reliable manner. One difficulty is the complicated nature of the effects, in which the behaviour during the first pressure cycle differs from that during succeeding cycles and also has a time dependence, so that it is not easy to specify and control all the relevant conditions. Another difficulty arises from the practical problems of applying a simple one–dimensional stress to the ceramic. If a load is applied to the faces of a disc, any constraints on sideways motion of the end faces may have a significant effect on the results, and some care is needed to make these end clamping conditions reproducible and consistent with the assumptions. In realistic conditions, the stress distribution may again differ from the ideal, and the effects of stress on a transducer stack in operation may therefore differ from those measured on the test samples.

One other characteristic of the ceramic pieces which is important for some designs is the adhesion strength of the silvered electrodes. This may be measured by cementing ceramic samples between metal end pieces and subjecting them to stress in a tensile testing machine. However, a much simpler qualitative test of the electroding may be applied by attaching adhesive tape to the silvering and peeling it back at an angle of nearly 180°. Good silvering should withstand this test without peeling off, and this is a good indication that the many hazards of silvering have been satisfactorily avoided.

9-2   TRANSDUCER IN AIR

Various tests may be applied to a transducer during its assembly, their main purpose being to confirm that the manufacture is proceeding satisfactorily. Since considerable effort is put into making good joints in a transducer, and they are therefore not easy to break down to recover the parts, it is often worth carrying out these tests in order to avoid the waste which would ensue from continuing with defective sub-assemblies. The decision of what tests should be carried out will depend on a comparison of the potential wastage compared with the costs of actually carrying out the tests, and may thus differ for each design.

The first test after assembling a stack, but before curing the joints, is to confirm the polarities of the ceramic rings. These are normally assembled with alternating polarities along the stack, and it is simple to check that a compression of the stack produces a voltage of the correct sign as connections are made to each ring in turn along the stack. The value of this test arises from the ease with which a ring can be inserted with the wrong polarity during assembly of the stack, and the subsequent wastage which could ensue in completing the transducer. A simple measure of the stack capacitance may also be advantageous, to confirm that it lies between preset limits.

The stack may be assembled separately and then attached to its front piston and tail mass, or it may be assembled complete with the front and tail masses. After curing, the resulting assembly constitutes the basic resonating element of the transducer, and several measurements on it are desirable. The element in air should have very low radiation loading, and the internal losses are also usually low, so that a motional Q-factor of several hundred is typical. Measurements appropriate to a high-Q_M resonator can therefore be applied. The element should be mounted to minimise its mechanical coupling to its surroundings, for example by suspending it by a flexible wire, preferably at its nodal position. Although the acoustic coupling in air should be small, it may occasionally be significant, and it is

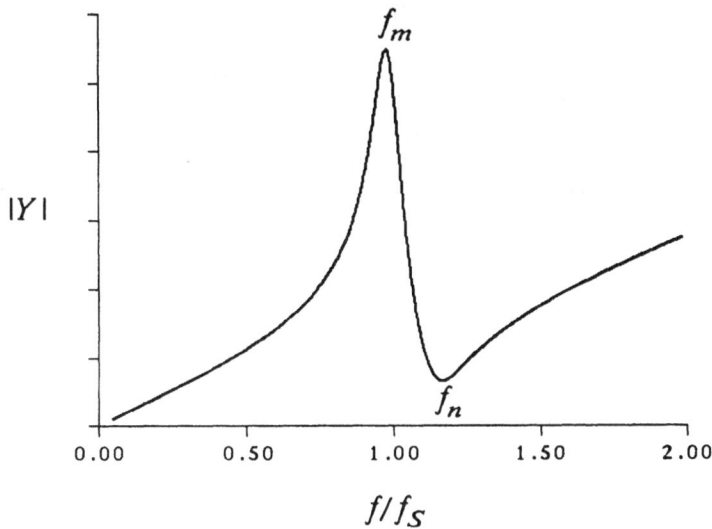

Fig 9.2 Variation of admittance magnitude with frequency around resonance for simple resonator.

best to avoid any reflecting surfaces near to the radiating piston. O-rings should not be fitted around the piston for these tests, since they introduce damping and tend to worsen the reproducibility of the measurements.

A circuit as in *Fig 9.1* may be used to measure f_m and f_n. The requirements for the circuit are similar to those for the measurements on ceramic specimens, except that the actual values to be measured will differ because the capacitance and resonance frequency of an element are markedly different from those of a single ceramic piece. The main condition is that the output impedance of the generator, and the series resistance, should be small compared with the impedance of the element at resonance. By increasing the applied frequency through the band around resonance, the circuit indicates the variation of the magnitude of the admittance, and a smooth curve as in *Fig 9.2* should be seen, thus confirming that only one resonance is present in the band. Provided that this is the case, the method gives a measurement of f_m, which is approximately the resonance frequency. By measuring also the frequency of minimum admittance (f_n), the effective coupling coefficient (k_e) can be calculated from (9.1). These two parameters (f_m and

k_e) are good indicators of the quality of the assembled element, since both are very dependent on achieving good joints.

The value of the admittance at f_m is also a useful check, since it is approximately equal to the conductance at resonance, and can be used, together with the value of low frequency capacitance, to give an approximate value of the motional Q–factor from equation (4.27); thus,

$$k_e^2 = \frac{G_{mS}}{\omega_S Q_M C_{LF}} \qquad\qquad (4.27)$$

$$\simeq 1 - (f_m/f_n)^2 \qquad \text{(from (9.1))}$$

If the maximum admittance value (at f_m) is denoted by Y_{max}, and this is approximately equal to the motional conductance at resonance (G_{mS}), this gives

$$Y_{max} \simeq \omega_S Q_M C_{LF}\{1 - (f_m/f_n)^2\}$$

ie $$Q_M \simeq \frac{Y_{max}}{\omega_S C_{LF}\{1 - (f_m/f_n)^2\}} \qquad\qquad (9.3)$$

Although the value of Q_M is likely to vary appreciably, even for well constructed elements, an especially low value indicates higher than expected internal losses, and hence suggests poor assembly. It is not of course necessary to calculate Q_M for all elements as a production test; specifying a lower limit for the value of the maximum current (as indicated by the peak voltage across the resistor R_T) should be an adequate check, and thus allow rejection of elements in which the internal losses are too high.

If the element is satisfactory, the next stage is to assemble it into its housing. It is then likely to have some sort of rubber seal around its piston, and probably some other mounting arrangement for the element, and the losses in these components will reduce the value of Q_M, typically to a value between 20 and 50. The simple circuit of *Fig 9.1* may again be used to give an approximate value of f_S, or at least to confirm whether f_m lies within specified limits, and also the value of the admittance at f_m. This is a very simple and useful test to show, for example, that the element has been inserted correctly into its housing, and that the piston is not touching the sides of the surrounding bore. In sweeping through the frequencies around resonance, the current should show a single peak and single trough, thereby confirming that no unwanted resonances are present. Alternatively, an automatic admittance bridge can be used to plot the admittance curve of the transducer in air, and thus confirm more clearly that only a single resonance occurs within the frequency band of concern. Such a curve can be used to measure the resonance frequency of the transducer, and by measuring the frequencies at which the conductance is half its peak value the motional Q_M can be derived. An approximate value of the coupling coefficient can be calculated by using equation (4.27) in the form

$$k_e^2 = G_{mS}/\omega_S C_{LF} Q_M$$

One method of defining the acceptable limits for the behaviour of the admittance is to draw limiting curves around the nominal curve and reject any transducers which have admittance values outside these limits. This technique of plotting the admittance curve can be very useful in giving a qualitative picture of the behaviour of the transducer, and for investigating unexpected results, but it is not a particularly quick test to perform or easy to specify precisely.

Other methods have been devised to permit simple measurements of the coupling factors of transducers with moderate values of Q_M. The equivalent circuit of a piezoelectric transducer is as shown in *Fig 4.5*, and when the dielectric losses are small the loss resistor R_e tends to infinity and may then be omitted. The capacitance of this circuit as a function of frequency is described by equation (4.20) and shown in *Fig 4.4*. At low frequencies the capacitance (C_{LF}) becomes $C_0 + C_1$, and at high frequencies C_{HF} tends to C_0. Since the coupling coefficient k is related to these capacitors by

$$k^2 = C_1/(C_0 + C_1) = (C_{LF} - C_{HF})/C_{LF}$$

it is clear that k could in principle be determined by measuring the capacitance of the transducer at very low and very high frequencies. However, whilst C_{LF} is easily measured, it is not usually practicable to obtain a reliable value for C_{HF} because higher frequency resonances cause further variations in the capacitance curve.

Fig 9.3 shows idealised capacitance curves for a transducer with a large value of Q_M (dashed), and for one with a lower Q_M (solid curve). Both transducers have the same coupling coefficient, so C_{LF} and C_{HF} would be the same for both if no other resonances occurred. The figure shows the curves only at

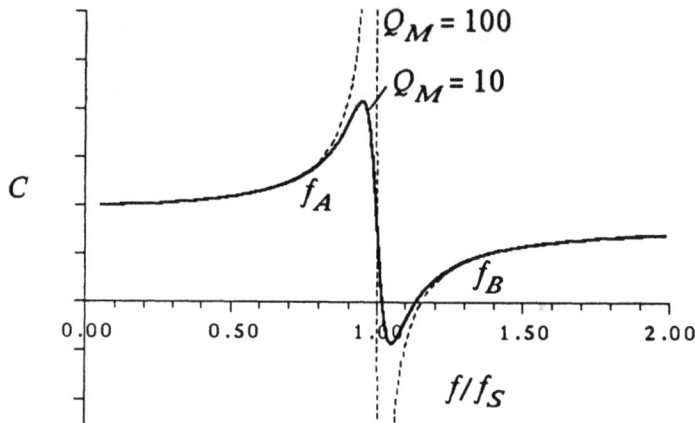

Fig 9.3   Variation of capacitance with frequency around resonance for simple resonator.

frequencies around resonance, and assumes that no other resonances interfere in this band. At frequencies near resonance, the curves obviously differ in shape, but as the frequency diverges from resonance the curves become closer together as they tend towards the high-Q_M curve. For a resonator with an infinite Q_M, the curve could be constructed (and hence k deduced) if we knew the resonance frequency (f_S) and the values of the capacitances at any two frequencies. If Q_M is finite, it is possible to measure the capacitances at frequencies such as f_A and f_B, where the capacitance curve is very close to the infinite-Q_M curve, and thus deduce the shape of the idealised curve. From this can be calculated the value of the coupling factor. The condition for this method to be applicable is that no other resonances should interfere over the frequency band between these two measurement frequencies. It is particularly suitable for transducers having Q_M between about 10 and 100; for higher values, the simpler f_m, f_n method for determining k_e is usually adequate.

The method is based on the expression (4.20) for the frequency dependence of the capacitance of a transducer which can be represented by the simple equivalent circuit of *Fig 4.5*, ie

$$C = C_0 - \frac{\omega_S C_1 Q_M^2 \Omega}{\omega(1 + Q_M^2 \Omega^2)} \qquad (4.20)$$

For frequencies at which $Q_M{}^2\Omega^2 \gg 1$, this expression may be written as

$$C = C_0 - \frac{\omega_S C_1 Q_M^2 \Omega}{\omega Q_M^2 \Omega^2}$$

$$= C_0 - C_1/(x^2 - 1) \qquad (9.4)$$

where $x = f/f_S$. (ie $\Omega = x - 1/x = (x^2 - 1)/x$.) Choose frequencies f_A, f_B such that

$$(f_A/f_S)^2 \equiv x_A^2 = \alpha/(\alpha + 1)$$

and $\qquad (f_B/f_S)^2 \equiv x_B^2 = \alpha/(\alpha - 1)$

where α is a parameter to be selected as described below. The capacitances at these frequencies are then

$$C_A = C_0 + C_1(\alpha + 1) \qquad (9.5a)$$

$$C_B = C_0 - C_1(\alpha - 1) \qquad (9.5b)$$

Thus, $\qquad \dfrac{C_A - C_B}{C_A + C_B} = \dfrac{2C_1\alpha}{2(C_0 + C_1)}$

$$= \alpha k_e^2$$

ie $\qquad k_e^2 = \dfrac{1}{\alpha}\dfrac{C_A - C_B}{C_A + C_B} \qquad (9.6)$

where k_e is the effective coupling factor of the transducer. The condition that $Q_M{}^2\Omega^2 \gg 1$ is equivalent to the condition in selecting f_A and f_B that

$$Q_M^2(x^2 - 1)^2/x^2 \gg 1,$$

which corresponds approximately to $Q_M{}^2 \gg \alpha^2$. The condition on the choice of frequencies is therefore that

$$\alpha^2 \ll Q_M^2 \qquad (9.7)$$

A fuller treatment shows that the value of k_e derived from this method is in error by a factor of approximately $(1- \alpha^2/Q_M{}^2)$. An accuracy of about 1% is therefore achieved by using $\alpha < Q_M/7$.

The method can be implemented by using a capacitance bridge which can operate over the range of frequencies about resonance. The procedure is:–

a) Measure C_{LF} (eg at 1kHz)

b) Measure f_S. This may be done by measuring the frequency of maximum admittance, as described above, using the circuit of *Fig 9.1*. Alternatively, the bridge may be set to an estimated value for the capacitance at resonance – ie $C_{LF}(1 - k_e{}^2)$, assuming an estimated value of k_e, – and an estimated value of conductance G_{mS}. Leaving C untouched, adjust frequency and G to balance the bridge, and then measure the frequency. Since the capacitance varies rather rapidly near resonance, this can give quite an accurate value of f_S.

c) Choose a value of the parameter α. If the value of Q_M is known, or can be estimated, make $\alpha < Q_M/7$. Too low a value of α reduces the accuracy of the method, since it corresponds to a wider difference between f_A and f_B, and hence the value of $C_A - C_B$ is smaller and there is a greater chance of other resonances interfering. For typical values of Q_M between 20 and 50, a value of α between 3 and 7 would be suitable.

d) Calculate two frequencies f_A and f_B from the Table 9.1.

e) Measure the capacitances C_A and C_B at the two frequencies corresponding to x_A and x_B (where $x_A = f_A/f_S$ etc).

f) Then the coupling coefficient k_e is given by (9.6), ie

$$k_e^2 = \frac{C_A - C_B}{\alpha(C_A + C_B)}$$

g) It is advisable to confirm that no other resonances are interfering by repeating the measurement with a different value of α.

TABLE 9-1

α	2	3	4	5	6	7
x_A	0.8165	0.8660	0.8944	0.9129	0.9260	0.9354
x_B	1.414	1.225	1.155	1.118	1.095	1.080

Tuned transducer

If the transducer is fitted with a tuning coil, the admittance curve should be approximately symmetrical about the G-axis, rather like the curve for $\beta > 1$ in *Fig 5.2*, though with a large loop if the value of Q_M is high. When admittance measurements around resonance are made (using the circuit of *Fig 9.1*), there will therefore be two frequencies of minimum admittance, one on either side of resonance. These two frequencies should be approximately symmetrical about the resonance frequency, and the size of the admittance loop can be determined by evaluating the ratio of the admittance at resonance and at each of the two frequencies of minimum admittance.

It is possible to deduce an approximate value of the coupling coefficient from this admittance curve, provided the value of Q_M is high. From (5.2), it can be shown that the susceptance becomes zero when $\Omega = 0$, corresponding to the resonance frequency, and also at frequencies given by

$$C_0 - \frac{C_1 Q_M^2}{1 + Q_M^2 \Omega^2} = 0 \qquad (9.8)$$

For $Q_M^2 \Omega^2 \gg 1$, this gives $\Omega^2 = C_1/C_0 = W^2$. From (5.11), for small deviations from f_S, $\Omega \simeq \pm 2(f - f_S)/f_S$. Thus, if the two frequencies given by the solutions to (9.8) are denoted by f_1 and f_2, then

$$2(f_S - f_1)/f_S \simeq W$$

and

$$2(f_2 - f_S)/f_S \simeq W$$

Thus,

$$2(f_2 - f_1)/f_S \simeq 2W \simeq 2k_e$$

Hence,

$$k_e \simeq (f_2 - f_1)/f_S \qquad (9.9)$$

This method, although only approximate, can sometimes be useful for checking the coupling coefficient of a transducer which already has a tuning coil connected across it.

9-3   HIGH STRESS TESTS

The above tests are all carried out with a low applied field. For projector applications, tests should be applied to confirm that the transducer will also function satisfactorily at high power. Although high power tests in water are possible, they are generally time-consuming and inconvenient and, for those cases where cavitation prevents the full power being radiated near the surface, they can become difficult to perform. It is however possible to devise tests which can be carried out in air, and which impose on the transducer mechanical and electrical stresses corresponding to those during operation; the price to be paid for this simplification is that the stresses have to be applied separately. In order to ensure adequate reliability in service, the test stresses should be greater than the predicted operating stresses by a safety factor, which is often taken as a factor of two. The tests are of two types. In one, an electric field of twice the operating field is applied to the transducer, with low mechanical stresses; in the other, twice the operating mechanical stress is applied, with a low electrical field. The first is called an "over-voltage" test, and the second a "motional over-drive" test.

Specifying these tests depends upon an understanding of the admittance curves and the associated equivalent circuits for the transducer in air and water, as described in Chapter 4. The impedance at resonance in air is much lower than at resonance in water. If the full operating voltage were applied to the transducer at resonance in air, the input current would be much greater than the current in water. Since the input current corresponds approximately to the motional current, and hence to the piston velocity, the stresses in the ceramic stack would be much higher than when operating in water, and would very probably lead to failure of the element. However, if the input *current* in air is set to correspond to that during operation in water, it will reproduce the mechanical stresses in the element, whilst the applied field will be much lower than in service. In order to apply the operating voltage safely in air, it is necessary to move away from resonance, and carry out the test where the input impedance is high. For an untuned transducer, this could be done conveniently at a frequency well below resonance; for a transducer incorporating a tuning coil and transformer, the choice of frequency is more critical, and it is usual to choose one of the frequencies of high impedance on either side of resonance. With the appropriate choice of frequency, the test voltage can be applied with only a modest input current, and hence without producing large mechanical stresses in the element.

These two tests are relatively simple, and very suitable as production tests aimed at ensuring adequate power handling capacity; they will be discussed in more detail below. Other techniques have been studied, in order to be able to apply both electrical and mechanical stresses simultaneously without immersing the array in water. In these techniques, a mechanical load is applied to the transducer face to simulate the acoustic impedance which would be experienced in operational conditions [9.3]. The transducer under test is coupled to a second transducer, whose resonance frequency is usually close to the test frequency. Coupling between the transducers may be effected either by a direct mechanical attachment, if the piston is not subject to flexing, or by means of a fluid-filled tube. The second transducer is connected to an electrical network, the vibration of the piston thus being converted into electrical power by the second transducer and then dissipated in the electrical network. By varying the components of the network, the load on the test transducer can be controlled, and it is thus possible to simulate a range of acoustic loads. This range may be extended even further by electrically driving the second transducer, with a controllable amplitude and phase, instead of using a passive termination. These techniques need considerably more careful setting up than the over-drive and over-voltage tests, and are therefore less convenient for production tests, but they can be useful for specific tests during development, especially for measurements at high power.

Motional over-drive tests

This test is intended to confirm that the transducer can safely withstand the amplitude of vibration to be encountered during operation. The motional output power in water is given by (4.29)

$$W_m = i_m^2 R_1 = i_m^2 (R_i + R_r)$$

where R_i and R_r represent the internal and radiation losses (in electrical terms), as in *Fig 4.6*, and i_m is the current into the motional arm. The values of these resistors can be evaluated during the transducer design phase, and the value of i_m to produce the required operating power can thus be calculated. If a safety factor of two is to be used, a motional test current of twice this calculated value of i_m should be applied to the transducer at its resonance frequency in air. For a reasonably efficient transducer, the impedance of R_i is appreciably less than that of either R_e or the parallel capacitance C_0, and the motional current is therefore approximately equal to the input current, for a transducer at its resonance in air. This approximation is even better if the transducer has a tuning coil connected across the element, tuned to the operating frequency. For the test, an input current of twice the calculated motional current during operation is thus applied to the transducer and maintained for about 30 seconds; if it is applied for too long (eg longer than about 1 minute), overheating of the element may occur.

The test is best performed by adjusting the frequency to resonance at low drive and then increasing the input current to about 2/3 of the test current, readjusting the frequency as necessary to give the maximum current. The drive is then increased to the test current for the specified duration, noting the resonance frequency at over-drive and also the voltage needed to produce the test current. An oscilloscope should be connected across the drive terminals, to observe the input waveform. After the test, the transducer should be allowed to cool for some time, and the low power resonance tests repeated on it. A transducer fails this over-drive test if it cannot be driven up to the specified current, if the applied voltage needed is too high, if significant distortion of the waveform is observed, or if the resonance frequency changes too much (eg more than 1–2%) during or after this test. Failure of this test is usually due to faulty joints in the element, or to inadequate compression of the stack by the centre bolt.

The over-drive test generates high amplitudes of vibration, and hence may be dangerous to hearing. If the operating frequency is in the audio region, it will almost certainly be loud enough to cause complaint! This test should therefore be carried out with the transducer in a sound-proofed box.

Over-voltage tests

This test involves the application of high voltages, and suitable safety precautions should be taken. The test is intended to confirm that the completed transducer will withstand the necessary operating voltage, which may again be calculated from a knowledge of the transducer's impedance at resonance in water. In some cases, the applied voltage at the edge of the operating band may be greater than that at resonance, depending on the characteristics of the drive amplifier. However, it is usually adequate to evaluate the voltage from the expression that the motional power is equal to $V^2/(R_i + R_r)$. If the transducer contains no tuning coil, the test could be carried out at a frequency well below resonance, but this may incur difficulties in generating high enough voltages at very low frequencies. It is therefore sometimes preferable to conduct the test at the anti-resonance frequency of the element in air (ie its frequency of maximum impedance, just above the resonance). If the transducer contains a tuning coil, it is necessary to determine the frequencies either side of resonance at which the impedance is relatively high. One of these should be chosen for this test, taking into account the magnitude of the impedance at each point, and the sensitivity of the tuning coil losses to frequency, in order to minimise the load on the driver amplifier and to allow for the unrepresentative losses in the coil.

After setting the specified frequency at a low input voltage, the input voltage should be increased to the test value, and maintained for 10–15 seconds. The input current should be determined by measuring the voltage drop across a small series resistance. A transducer fails this test if the voltage cannot be driven up to the specified value (ie voltage breakdown occurs), or if there are fluctuations in the current, indicating incipient breakdown.

Problems may arise in applying this over-voltage test to transducers which contain a number of elements with a single tuning coil, because of the difficulty of tuning the system in air when the individual anti-resonance frequencies are not identical. In such cases, an over-voltage test should be carried out at low frequency on the untuned array, and the tested tuning coil then connected. A high voltage DC tester may then be applied to check the insulation between the element and the casing.

It is sometimes useful to carry out an over-voltage test at a low frequency on bare elements, in order to detect faulty elements at an early stage. This may even be carried out at DC, provided that the current is limited to avoid catastrophic breakdown.

9-4 WATERTIGHTNESS

Before the transducer is immersed in water, it should be submitted to tests to confirm its watertightness, since leakage of sea water into the housing may necessitate scrapping the transducer and complete wastage of the previous work. The simplest test is to pump a gas into the transducer housing to a pressure of about 10psi (7×10^4Pa) and check that the pressure does not decrease. Immersion in water permits detection of bubbles leaking out of the housing, and the application of soapy water solution around the seals allows a better localisation of any leaks. This technique can detect leaks down to about 10^{-3}torr–L/s, compared with the required sensitivity of 10^{-7}torr–L/s (Sect 8.7). It should therefore be regarded only as a preliminary test to detect gross leakage; it is nevertheless useful for eliminating major defects in assembly such as omitted O–seals.

Much better sensitivity can be obtained by using a commercial leakage tester based on detecting a trace gas. One of the most convenient systems uses a gaseous halide (eg Freon) as the trace substance. With this **halide leak detector**, a small amount of the halide is introduced into the transducer housing, with a small internal over–pressure. Leakage to the outside may be detected either by surrounding the whole transducer in an enclosure, or by "sniffing" around the transducer with a probe. Operation of the leak tester should conform to the instructions for the particular test device being used, and the following precautions should also be observed:–

a) Cigarette smoke, and some floor polishes and adhesives, may be detected and should be avoided in the test area.

b) Do not fill the transducer with its tracer gas in the same room as the test is carried out.

c) Since most tracer gases are denser than air, it is better to start probe testing at the top of the transducer and work downwards.

d) It is common to introduce the gas through a Schrader valve, and care must be taken to seal this valve fully, for example by fitting a cap on it, to prevent leakage from the valve spoiling the test.

e) The sensor head should be protected from exposure to high concentrations of the tracer gas, which will reduce its life and lower its sensitivity for a time.

f) The sensitivity of the leak tester should be checked regularly (eg weekly or even daily) against a standard leak.

With care, this method can detect leaks down to about 10^{-6}torr–L/s, which is generally adequate for a preliminary test. Even higher sensivity could be obtained by using a mass spectrometer, but this is rarely justified for transducer applications.

The great advantage of the methods above is that they do not expose the transducer to the dangers of water entry, but their major disadvantage is that they do not reproduce the operational conditions of stress on the seals. For example, any O-ring seals are subjected to low internal pressure, rather than external pressures which may be much higher and may cause significant displacement of the front piston. It is therefore necessary to subject the transducer to a pressure test in water to at least the operating pressure for some hours. A test pressure of 1.5-2 times the operating pressure is desirable. Measurement of the insulation resistance across the transducer stack, or from one side of the stack to earth, is usually the most sensitive and convenient indicator of any leakage. The use of a desiccator within the transducer housing is sometimes suggested, to mop up any small amounts of water which penetrate the seals, but this prevents any simple indication of the effects of leakage during the pressure test, and it is therefore doubtful whether the advantages outweigh the disadvantages. It is desirable to check the electro-acoustic characteristics of the transducer after the pressure test, in order to confirm that its performance has not been adversely affected by the high mechanical stress. A repeat of the low power test around resonance is usually sufficient.

The importance of these watertightness tests cannot be over-emphasised in view of the serious consequences of leakage and the frequency with which leakage problems are observed. All temptations to dispense with this test, in order to save time or expense, should be strongly resisted, as the consequences of yielding to the temptation will very often be regretted.

9-5 ACOUSTIC TESTS IN WATER

The final test of any transducer is to measure its performance in water in its operational conditions. In some cases, this may be very difficult to carry out, for example if the transducer forms a part of a large array or if it is to be used at great depth. But in most cases it is quite practicable to carry out some acoustic measurements in water, provided that the appropriate facilities are available. The techniques of underwater acoustic testing have been comprehensively treated by Bobber [9.4], and in this section we shall consider only the main requirements for such tests.

Testing can be performed most conveniently in an **acoustic tank**. However, problems can arise from the limited size of such tanks, primarily because of reflections of the acoustic energy from the tank walls. The situation is improved if the tank is fitted with an acoustically absorbing lining, but this is usually a relatively expensive undertaking, particularly if measurements at frequencies below 15kHz are needed. It is sometimes possible to use an unlined tank for impedance or admittance tests, since these are less susceptible to the effects of reflections than are direct measurements of the field such as beam patterns and sensitivities. Thus, if the admittance of a transducer is measured at several locations within the tank (avoiding positions too near the walls), and the variation in values is less than say 5%, then the tank may be used for comparative measurements to that accuracy. We shall see later that the minimum test distance for accurate sensitivity measurements increases as the frequency is reduced, if a pulse method is used. For measurements below about 10kHz, the size of tank needed may become impractically large, and recourse to a larger body of water may then become necessary.

Several techniques have been investigated to overcome these restrictions, at the expense of greater complexity in the measurement system. One method is to carry out the measurements with pulses which are short enough to permit separation between the direct pulse and the reflected pulses. If this separation is practicable, and the other conditions for accuracy can be satisfied, the technique can give satisfactory results, but involves considerable care in identifying and processing the correct pulses. If a resonant transducer is being measured, the number of cycles needed for the response to rise to 70% of the steady state value is approximately Q_M/π. Thus, for a transducer resonant at 10kHz with a Q_M of 10, about three cycles (ie 0.3ms) would be needed to reach a level suitable for measurement. A total pulse length of about 0.5ms might thus be acceptable, such a pulse having a spatial length of 0.75m. The separation between the direct and reflected pulses must be sufficient to allow measurements in the second half of the direct pulse. As the frequency is reduced, the required pulse length becomes longer and the limitations of the method therefore become more severe.

Another method depends on measurements of the sound field very near to the transducer face and subsequent calculation of the radiated far-field [9.5, 9.6]. Since the measurements are carried out very near to the transducer, the effects of reflections are much reduced and good results have been obtained. However, the method needs a good system for accurately positioning and scanning the measuring hydrophones in the near field, and facilities for computing the far field from the measured values. An alternative approach was proposed by Trott

[9.7], who used an array of sources to generate a plane wave field over a volume within the near–field of this array, and then carried out measurements on the transducer by placing it within this volume. Both of these techniques require specialist study, and the interested reader is recommended to consult the references.

The **minimum range** for conventional measurements in water is affected by a number of considerations [9.8]. It was noted in Section 6.4 that a complex and rapidly varying acoustic field exists near to the face of a radiating transducer, and that the field only settles down to its far–field pattern beyond the Fresnel range of L^2/λ, where L is the maximum dimension of the radiating face. Even for a point source, the acoustic field at very short ranges has a reactive component which approaches its far field value of zero only for ranges beyond about one wavelength. For simplicity, measurements of the acoustic characteristics of a transducer are normally made in the far field, and this implies that they should be made (a) at a range of at least one λ, and (b) beyond the Fresnel range of L^2/λ. Measurement of a transducer's beam pattern is usually made by rotating the transducer in the field generated by an acoustic source, and this is equivalent to the far field response only if the intensity generated by the source is approximately constant over the volume swept out by the transducer as it rotates. This requires that the range of the source is large compared with the length of the transducer in the plane perpendicular to its axis of rotation, so that the amplitude of the acoustic field at all parts of the transducer face remains constant as the transducer is rotated. In practice, this means that when measurements of beam patterns are required there is the further condition that the distance between source and receiver should be at least 10 times the dimensions of the transducer in the plane of the measurement. These conditions thus determine the minimum separation between source and receiver for reliable far–field measurements. For measurements in a tank, the limitations due to reflections from the walls must also be considered. However, too long a test range may introduce problems with reflections from the water surface.

The transducer characteristics which are most commonly specified for measurement in water are the admittance curves, the projector and receiver sensitivities, and the beam patterns. Measurements of the admittance curves are straightforward, using a commercial admittance bridge. They may be carried out either at successive frequency points or by sweeping the frequency smoothly through the band of interest, using a bridge

with automatic tracking of the admittance. The swept frequency
technique is quick and gives a good qualitative indication of the
behaviour of the transducer, showing for example if more than
one resonance is present. But it needs a frequency scale marked
around the admittance curve if results for a particular frequency
are wanted, or if a value of the resonance frequency is needed.
Details of the precautions needed to obtain good accuracy are
described by Bobber [9.4]. Probably the one most easily
forgotten is the need to ensure that the transducer face is fully
wetted. When a transducer is immersed in water, it is common
for air bubbles to adhere to the surfaces, especially if they are
at all greasy. These air bubbles can have a very marked effect
on the acoustic characteristics, and are a frequent cause of
inconsistency in the measurements. It is desirable to clean the
radiating face with a detergent before immersion, but even so it
is often necessary to wait for some hours for the measured
values to become consistent. **Wetting** is often faster in sea
water than in fresh, and it may be assisted by driving the
transducer to near cavitation.

The commonest method of measuring the receiving
sensitivity is to place the transducer in a known acoustic field,
which is generated by a calibrated source. The acoustic far field
produced by this source in the absence of any obstacles may be
derived from its calibration data. The hydrophone is then placed
in this field at a known range from the source (beyond the
minimum range), and its output voltage is measured. Although
the presence of the hydrophone may have distorted the actual
field, the output voltage is related to the calculated field, – ie
the field which existed before introducing the hydrophone. This
is then known as the "**free field sensitivity**" of the
hydrophone. For a receiver which is more than a small fraction
of a wavelength across, the actual pressure field over the face
may show considerable variation, and it would thus be difficult
to relate the output voltage to any particular pressure value. For
a small hydrophone, however, the actual acoustic pressure over
the face is reasonably constant, and the output may be
expressed in terms of the actual pressure, giving the "**pressure
sensitivity**". It is the pressure sensitivity which is calculated in
the idealised expressions (eg in Chapter 10) derived by
multiplying the piezoelectric g-coefficient of the ceramic by the
applied stress. A small hydrophone having a high mechanical
stiffness (eg well below resonance) will in practice cause only a
small disturbance of the acoustic field, and in that case the two
hydrophone sensitivities are almost equal. For sound incident
normally on a large rigid baffle, the acoustic pressure at the

face is approximately doubled, and the free field sensitivity of a
hydrophone in the baffle would be about twice that of its
pressure senstivity. In most cases, the free field sensitivity is
therefore between one and two times the pressure sensitivity.

If the free field sensitivity and the admittance of a
transducer have been measured, there is no need for a separate
measurement of the projector sensitivity. This is because there
is a relationship which applies between the two sensitivities and
the admittance of a linear transducer; this is commonly known
as the **"reciprocity"** relationship, and takes the form

$$\hat{M} = J\hat{S}_I \qquad (9.10)$$

where $\hat{M}$ is the transducer's hydrophone sensitivity in its linear
form, $\hat{S}_I$ is its projector sensitivity referred to its input current,
and J is a "reciprocity parameter" [9.4]. If the separation (d)
between the projector and receiver is large enough to consider
the energy to be diverging spherically (ie $d > L^2/\lambda$), the
reciprocity parameter has the value

$$J = 2\lambda d/\rho c = 2d/\rho f \qquad (9.11)$$

The projector sensitivity is referred to a nominal range of 1
metre, so, although the measurements may be made well beyond
1 metre, a nominal value of $d = 1$ may be assumed when
referring to the corrected sensitivities. Converting the sensitivities
to their more common logarithmic form, and adopting the usual
practice of referring them to 1µPa instead of 1Pa, equation
(9.10) becomes

$$S_I = M + 20\log F + 354 \qquad (9.12)$$

where S_I is the projector sensitivity in dB re 1µPa/A at 1
metre, M is the hydrophone sensitivity in dB re 1V/µPa, and F
is the frequency in kHz. S_I may readily be converted to the
voltage projector sensitivity (S_V) by noting that, if the
transducer's electrical input impedance is Z, a voltage of |Z|
would be needed to produce an input current of 1 ampere. We
thus obtain the relationship

$$S_V = M + 20\log F - 20\log|Z| + 354 \qquad (9.13)$$

It is common practice to measure the receiving sensitivity M
of a transducer, and its admittance (or impedance), and use
these to deduce the value of S_V or S_I.

Measured **beam patterns** are especially sensitive to reflections, particularly in their effects on the side lobes of the radiation pattern. It is thus quite difficult to draw up a meaningful and precise specification for the beam pattern, although it is a very useful indicator of the performance of a transducer array and often has a significant influence on the operational performance. The features which are most readily specified are the angular width of the main beam to some level such as the −3dB points, and the maximum height of the side lobes. For a symmetrical transducer array in its broadside beam, the side lobes should also be symmetrical, and any significant deviation should be investigated. If the beam is steered, some asymmetry will develop, but this should be readily understood from the discussion of Chapter 6. As already noted in Chapter 2, the theoretical level of the first side lobe for an unshaded linear (or rectangular) array is −13dB relative to the main lobe level. Any significant increase on this level is a strong indication that the transducer is not behaving correctly. If an array has shading applied to give much lower side lobes, however, it becomes much more susceptible to the effects of tolerances, and deviations from say −30dB side lobe levels are of less concern. Indeed, the achievement of side lobe levels below −30dB requires considerable care in the quality control of the array elements, particularly if the array is operating at the resonance frequency of the elements. The interpretation of the beam patterns thus depends on the stringency of the requirements.

Several different ways of presenting beam pattern data are used, and these can affect their ease of interpretation. As an illustration, *Fig 9.4* shows a variety of presentations of the beam patterns for the same transducer. A polar plot of the pattern, as in *Fig 9.4a*, gives a good qualitative picture of the response, but it is usually easier to take measurements of beam width or side lobe levels from a rectangular plot as in *Fig 9.4b*. A logarithmic (dB) scale for the sensitivity is most commonly used, in order to give a clear indication of the side lobe levels. Using a linear scale of acoustic intensity, as in *Fig 9.4c*, makes it difficult to read off the side lobe levels precisely, although it is sometimes worth noting how low the power is in directions away from the main lobe. It is also worth noting that curves for hydrophone or projector sensitivity may be plotted against either a linear or a logarithmic scale of frequency. The logarithmic scale is convenient especially for displaying hydrophone response, where it emphasises the low frequency, constant sensitivity region, which is often of most interest.

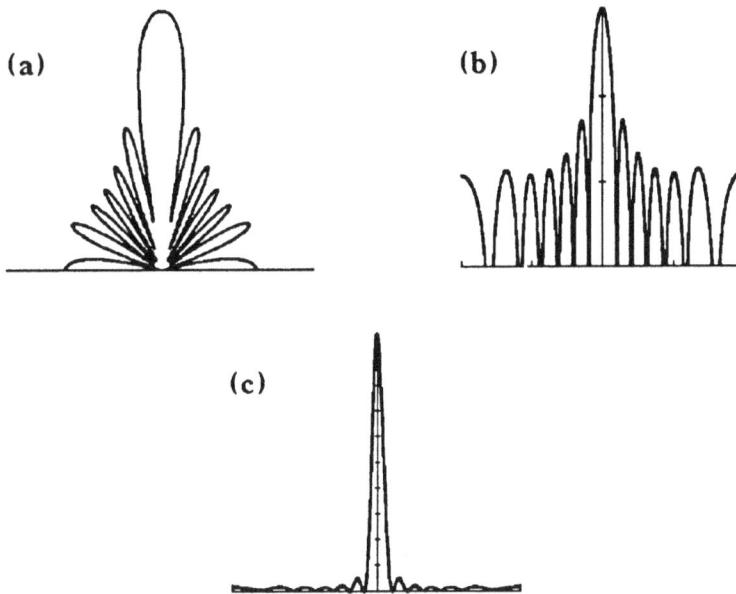

Fig 9.4 Beam   pattern   plots: (a)   Logarithmic,   polar;
         (b) Logarithmic, Cartesian; (c) Linear (intensity),
         Cartesian.

The directivity index of a transducer is often used in assessing its performance in a system. However, its measurement involves the determination of the total power radiated, and this requires measurements of the directional curves in several planes. It is therefore not an easy parameter to measure with any great accuracy, and it is generally preferable to specify the main features of the beam pattern, including a qualitative check that the beam pattern is behaving generally as expected, and then to estimate the DI from the relationships in Chapter 2. The overall efficiency of a transducer is also a characteristic of some interest, which may be derived from the measured DI by using equation (2.8). Since this again involves the measurement of the DI, it is a relatively lengthy process, which is subject to considerable error, and should be avoided whenever possible. A more practical estimate of the efficiency may be made for many transducers by determining the components of the equivalent circuit from measurements of the admittance in air and water.

Performance in operational conditions may in many cases be confirmed by driving the transducer at its working power in water. Although useful as a demonstration to the user that the desired source level can be achieved, this does not provide the confirmation of the safety factors in the design which can be established by the high power tests in air described in Section 9.3. Another use of a full power test in water can be to show whether the transducer cavitates when operating near the surface. For transducers which are required to work at great depth, it is often necessary to carry out some acoustic tests over the range of working pressures. This necessitates the use of a pressure vessel which is large enough to allow acoustic measurements in it. From the comments above about the size of vessel required for such measurements, it is clear that such tests need special facilities and are generally difficult to perform accurately because of the restricted space available in practical pressure vessels. For measurements at lower frequencies, where the restrictions due to the size of practical pressure vessels become most serious, it is possible to contain the transducer within an acoustically transparent pressure vessel (eg made of glass-reinforced plastic) which is itself immersed in a larger body of water. The standard source may then be placed at a suitable measurement range, outside the pressure vessel, and conventional calibration techniques used. This method can give reasonable accuracy if the acoustic loss through the pressure vessel is low, and is particularly useful in measuring any changes of transducer characteristics as the pressure is changed.

It is also possible to carry out tests with the transducer radiating into high pressure gas. Because the radiation loading of the gas is much smaller than for water, the effects of reflections from the walls of the pressure vessel are much less significant, and it is for example possible to measure any changes in resonance frequency or internal losses as the gas pressure is raised. Such measurements can give some confidence that the characteristics of the transducer do not change too much under pressure, although they are not a direct measure of the performance at pressure. The dangers of carrying out such tests with high pressure gas must however be recognised, and the need for suitable precautions cannot be over-emphasised.

In principle, the acoustic pressures may be applied in air as well as in water, and this is used in the **"air pistonphone"** method of calibration [9.9]. In this method, alternating pressures are applied to the transducer face in a small air-filled chamber, thus avoiding the need for tests in water. However, the technique is limited to low frequencies, generally below about

200Hz. It thus provides a very useful test method for measuring pressure sensitivity at low frequencies without recourse to in-water facilities; indeed, it is often the most reliable absolute calibration method at such low frequencies.

Low frequency measurements have also been made by generating the alternating pressure in a small water-filled tube, in which the transducer is immersed. A pressure wave is generated in the test vessel by a source at the bottom of the tube, the pressure at the hydrophone being calculated from the drive signal and geometry. The diameter of the tube must be small compared with a wavelength of sound in water at the test frequency, and the walls must be effectively rigid, to satisfy the conditions assumed in the calculations. This method can provide a convenient bench test for comparing the pressure sensitivities of small hydrophones up to about 3kHz [9.10].

Pick-up of electrical or magnetic fields at the hydrophone or along its cable is a potential source of error when measurements of sensitivity are carried out in air or in fresh water, although in sea water the transducer is usually adequately screened by the sea water itself. It is therefore advisable to check that pick-up is not introducing errors – for example by observing the time delay associated with the acoustic transmission between source and receiver. When pick-up is a problem, it may be reduced by introducing electromagnetic screening around the hydrophone and cables, and by extra care in earthing the measurement system to avoid "earth loops".

REFS Chapter 9

References

9.1 IRE Standards on Piezoelectric Crystals: Measurements of Piezoelectric Ceramics, 1961. (IEEE Standard 179-1961) (Reprinted in Jaffe et al, Ref [3.5])

9.2 IEEE Standards on Piezoelectricity, *IEEE Trans on Sonics and Ultrasonics*, SU-31, March 1984.

9.3 Woollett,R.S., "Power limitations of sonic transducers," *IEEE Trans on Sonics and Ultrasonics*, SU-15, 218-229, (1968).

9.4 Bobber,R.J., *Underwater Electroacoustic Measurements*, US Government Printing Office, 1970.

9.5 Baker,D.D., "Determination of far-field characteristics of large underwater sound transducers from near-field measurements." *J Acoust Soc Am*, 34, 1737-1744, (1962).

9.6 Bobber,R.J., op. cit., Chapter 4, Sections 4.2, 4.3.

9.7 Bobber,R.J., op. cit., Chapter 4, Section 4.4.

9.8   Bobber,R.J., op. cit., Chapter 3, Section 3.4.
9.9   Bobber,R.J., op. cit., Chapter 2, Section 2.5.
9.10 Bobber,R.J., op. cit., Chapter 2, Section 2.5.2.

Additional Reading

1   Van   Buren,A.L.,   Luker,L.D.,   Jevneger,M.D.,   and
     Tims,A.C.,   "Nearfield   calibration   arrays   for   transducer
     evaluation," *Proc IoA*, <u>6</u>, Part 3, 16–23, (1984).
2   Albers,V.M.(Ed), *Underwater Acoustics Handbook - II*,
     Pennsylvania State University Press, 1965, Part 4.
3   Gale,R.J., "Sonar Transducer Test Methods," 2nd Ed,
     AUWE Publication 32273/R1, 1977.

Chapter 10
HYDROPHONES

It has already been remarked that resonant transducers which have been designed primarily as projectors may also be used as receivers, and this is commonly done in active sonar systems, where only moderate frequency bands are involved. In such cases, the incorporation of a parallel tuning coil across the element to improve the matching to its drive amplifier also affects its hydrophone response; the variation with frequency of the receiving sensitivity of a tuned transducer is calculated in the first section of this chapter.

For many applications it is desirable to achieve a response which is independent (or almost independent) of frequency, and in such cases it is necessary to avoid the variations around resonance. The use of an untuned piezoelectric hydrophone well below resonance produces a substantially flat frequency response, and is the usual way of achieving a uniform sensitivity over a wide frequency band. The frequency response of such a hydrophone is calculated in Section 10.2, and the conversion to absolute sensitivity levels in Section 10.3. The remainder of the chapter deals with the various sources of noise in hydrophones, and their influence on the design of broad-band hydrophones, including spherical, cylindrical, and pressure-gradient designs.

10-1   HYDROPHONE RESPONSE OF TUNED TRANSDUCER

The receiving response of a piezoelectric transducer may be calculated by considering the behaviour of an equivalent circuit as in *Fig 10.1*, which is similar to *Fig 5.1* except that the driving voltage is made zero, and the voltage generated in the ceramic by the incident pressure is represented by e_p in series with R_1, L_1, and C_1 in the motional arm. For generality and convenience we include the three electrical components C_0, L_0, and R_0, and use the subscript 0 for all three. The output voltage from the hydrophone due to the incident pressure is then the voltage (V) across the resistor R_0.

The impedance of the $R_0 L_0 C_0$ combination (Z_0) is given by

$$\frac{1}{Z_0} = \frac{1}{R_0} + \frac{1}{j\omega L_0} + j\omega C_0$$

Thus,
$$Z_0 = \frac{j\omega L_0 R_0}{j\omega L_0 + R_0 - \omega^2 L_0 C_0 R_0}$$

The voltage output across R_0 is then

$$\frac{V}{e_p} = \frac{Z_0}{Z_0 + R_1 + j\omega L_1 + 1/j\omega C_1}$$

After some further working, and making the following substitutions,

$$\omega_1^2 = 1/L_1 C_1 \quad \text{(ie } \omega_1 \text{ is the motional resonance frequency.)}$$

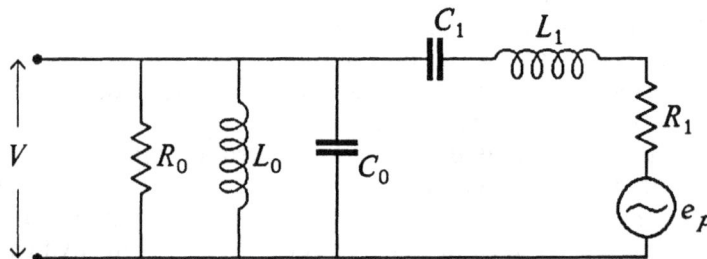

Fig 10.1   Equivalent circuit of piezoelectric hydrophone.

$\omega_0^2 = 1/L_0 C_0$ (ie ω_0 is the electrical resonance frequency.)

$Q_1 = \dfrac{1}{\omega_1 C_1 R_1} = \dfrac{\omega_1 L_1}{R_1}$ (ie Q_1 is the motional Q-factor)

$Q_0 = \omega_0 C_0 R_0 = \dfrac{R_0}{\omega_0 L_0}$ (for parallel combination)

$x_1 = \omega/\omega_1$

$x_0 = \omega/\omega_0$

$W^2 = C_1/C_0$ (as in equation 5.3),

this gives the result

$$\frac{V}{e_p} = \frac{x_0^2 W^2}{x_0^2 W^2 (1 + R_1/R_0) - (1 - x_1^2)(1 - x_0^2) \atop \qquad - j\{x_1(1 - x_0^2)/Q_1 + x_0(1 - x_1^2)/Q_0\}} \qquad (10.1)$$

The magnitude of the output voltage is then given by

$$\left|\frac{V}{e_p}\right|^2 = \frac{(x_0^2 W^2)^2}{\{x_0^2(W^2 + x_1/x_0 Q_1 Q_0) - (1 - x_1^2)(1 - x_0^2)\}^2 \atop \qquad + \{x_1(1 - x_0^2)/Q_1 + x_0(1 - x_1^2)/Q_0\}^2} \qquad (10.2)$$

and the phase angle (θ) of the output voltage relative to the incident pressure, represented by e_p, by

$$\tan\theta = \frac{x_1(1 - x_0^2)/Q_1 + x_0(1 - x_1^2)/Q_0}{x_0^2 W^2 + x_1 x_0/Q_1 Q_0 - (1 - x_1^2)(1 - x_0^2)} \qquad (10.3)$$

These equations represent the general solutions for the output voltage generated by an incident alternating pressure. They may be reduced to simpler expressions for particular cases, some of which are discussed below.

A) $\omega_1 = \omega_0$

The electrical resonance frequency is usually intended to be equal to the motional resonance; in that case, $\omega_1 = \omega_0$, so that we may write $x_1 = x_0 = x$, and the expressions then become

$$\left|\frac{V}{e_P}\right|^2_A = \frac{(x^2W^2)^2}{\{x^2(W^2+1/Q_1Q_0)-(1-x^2)^2\}^2 + \{x(1-x^2)(1/Q_1+1/Q_0)\}^2} \tag{10.4}$$

$$\tan\theta_A = \frac{x(1-x^2)(1/Q_1 + 1/Q_0)}{x^2(W^2 + 1/Q_1Q_0) - (1-x^2)^2} \tag{10.5}$$

A1) $\omega_1 = \omega_0$ and power matched.

The analysis of Section 5.2 in terms of filter networks showed that maximum power transfer over the widest possible band is achieved when $R_0 = R_N/\beta$ and $R_1 = R_N\beta$, so that $R_1 = \beta^2R_0$. This is equivalent to making $Q_1 = Q_0 (= Q)$, and the expressions then become

$$\left|\frac{V}{e_P}\right|^2_{A1} = \frac{(W^2)^2}{\{(W^2+1/Q^2)-(x-1/x)^2\}^2 + \{(x-1/x)(2/Q)\}^2} \tag{10.6}$$

$$\tan\theta_{A1} = \frac{(2/Q)(x - 1/x)}{(x - 1/x)^2 - (W^2 + 1/Q^2)} \tag{10.7}$$

These may be expressed in terms of the parameters used in Section 5.2, ie

$$y = \frac{1}{W}(x - 1/x) \qquad \text{(as in equation 5.4)}$$

and

$$\beta = \frac{R_1}{R_N} = \frac{1}{QW}$$

Then,

$$\left|\frac{V}{e_P}\right|^2_{A1} = \frac{1}{(\beta^2 + y^2 - 1)^2 + 4\beta^2} \tag{10.8}$$

$$\tan\theta_{A1} = \frac{-2\beta y}{1 + \beta^2 - y^2} \tag{10.9}$$

These expressions are similar in form to those derived from the filter methods of Section 5.2, and appear somewhat simpler than (10.6) and (10.7); they are also symmetrical about $y = 0$. However, their use is restricted to the cases for which $\omega_1 = \omega_0$, and for general purposes it is more useful to use the expressions in terms of the (non-symmetrical) frequency parameter x_1.

A2) $\omega_1 = \omega_0$, high impedance input.

It is quite common for the tuned hydrophone to be connected to a high input impedance amplifier, especially for measurements. In this case, $R_0 \rightarrow \infty$ (ie $Q_0 \rightarrow \infty$) and the expressions tend to

$$\left|\frac{V}{e_p}\right|^2_{A2} = \frac{(x^2W^2)^2}{\{x^2W^2-(1-x^2)^2\}^2 + \{x(1-x^2)/Q_1\}^2} \tag{10.10}$$

$$\tan\theta_{A2} = \frac{x(1-x^2)/Q_1}{x^2W^2-(1-x^2)^2} \tag{10.11}$$

Some examples.

Examples of curves for the hydrophone response of a tuned transducer feeding into a high impedance receiver are shown in *Fig 10.2*. The curves are for a transducer having $\omega_1 = \omega_0$ and for which $W^2 = 0.3$, which is a typical value for a lead zirconate titanate transducer. For $Q_1 > 2$, two peaks in the response curve are evident, becoming higher as Q_1 increases. When Q_1 is high, the separation in x_1 between the two peaks is approximately equal to W, ie approximately equal to the coupling coefficient. As Q_1 is reduced to a value near to $1/W$ the two maxima coalesce to form a single broad peak. Note that in these figures the V/e_p scale is linear (not logarithmic), and the peaks are thus exaggerated to some extent. Q_1 has been used in this section to indicate the Q-factor of the $L_1C_1R_1$ combination, and this is equivalent to the motional Q-factor of the transducer. If the fractional bandwidth of a resonant transducer is taken as approximately $1/Q_1$, it will extend in *Fig 10.2* from about $x_1 = 1/(1 + 1/2Q_1)$ to $x_1 = 1 + 1/2Q_1$, and this should be remembered in studying the curves in *Fig 10.2*. For example, the bandwidth for $Q_1 = 4$ extends approximately from $x_1 = 8/9$ to $x_1 = 9/8$, whilst for $Q_1 = 2$ the bandwidth would extend from $x_1 \simeq 4/5$ to $x_1 \simeq 5/4$.

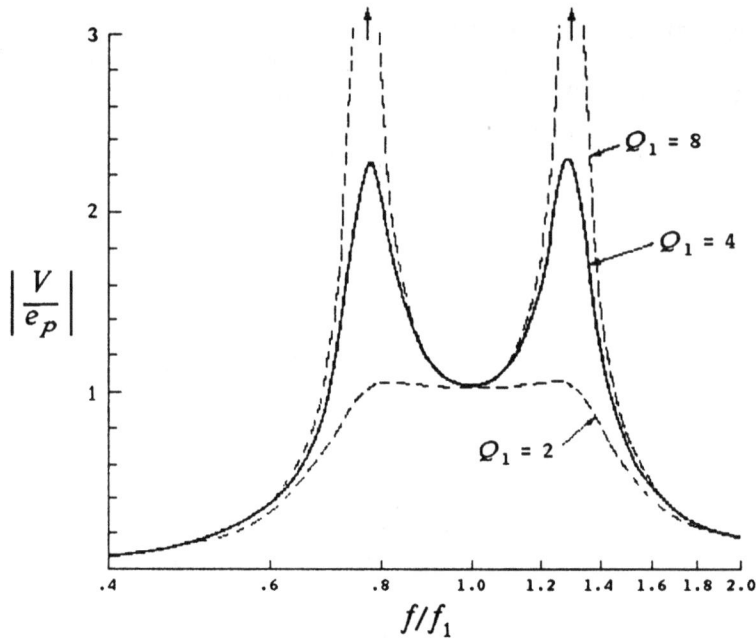

Fig 10.2 Hydrophone response of tuned transducer: effect of varying Q_1. ($W^2 = 0.3$, $R_0 = \infty$, $\omega_0 = \omega_1$)

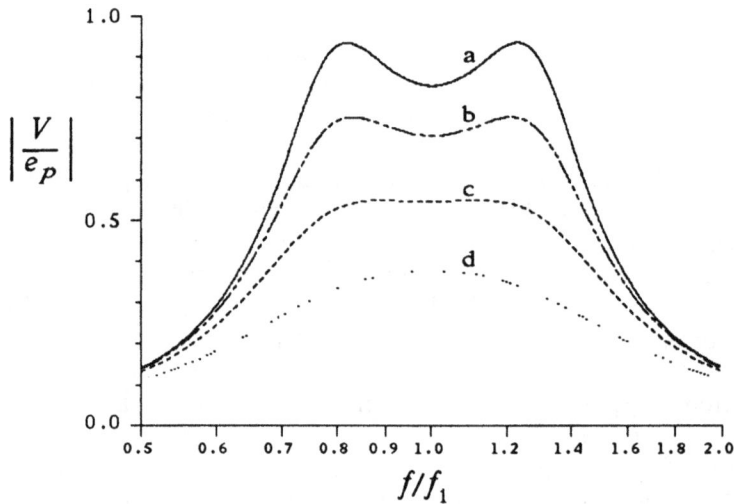

Fig 10.3 Hydrophone response of tuned transducer: effect of varying input resistance. ($W^2 = 0.3$, $\omega_0 = \omega_1$, $Q_1 = 2$) (a) $Q_0 = 8$; (b) $Q_0 = 4$; (c) $Q_0 = 2$; (d) $Q_0 = 1$.

The effects of varying the input impedance of the amplifier are illustrated in *Fig 10.3*. As the input resistance is reduced (ie Q_0 is reduced) from a high value, the two peaks reduce in size, forming a reasonably square-topped response when $Q_0 = Q_1$, the matched case. If Q_0 is reduced even further, a single peak results, rather narrower in bandwidth. This variation in response can be significant, because it is not uncommon for the hydrophone response of a transducer to be measured when it is connected to a high impedance calibration instrument, whilst in operation it is correctly connected into a matched impedance.

Equations (10.2) and (10.3) can be used to investigate the effects of deviations from exact tuning of ω_0 and ω_1, which can arise for example from changes in operating conditions. As might be expected, in such cases the response curves lose their symmetry, to a degree depending on the extent of the deviation from the ideal conditions.

Series tuning.

One important characteristic of the curves for tuned transducers is that the response falls progressively as the frequency is decreased towards zero. This fall-off in sensitivity usually prevents the use of a tuned transducer at frequencies well below resonance. There may nevertheless be some applications where a transducer is required to be used both as a transmitter at resonance and also as a receiver at low frequencies. One solution to this problem is to use a tuning inductor in series with the element instead of connecting it in parallel, thus avoiding the shorting effect of the parallel inductance at low frequencies. In practice, however, this is often less convenient than using a parallel inductor, which offers more flexibility in allowing the tuning inductor to act as a transformer to control the impedance of the transducer; parallel tuning also ensures that the impedance becomes low at high frequency, which is often desirable to minimise harmonic distortion in the drive signal. Another possible technique is to use diodes in series with the parallel inductor, so that it is open-circuited for small signals but effective for the larger voltages applied during transmission.

10-2 HYDROPHONE RESPONSE OF UNTUNED TRANSDUCER

B) The response of an untuned hydrophone is readily calculated from (10.2) and (10.3) by allowing L_0 to become infinite. This also implies $\omega_0 \to 0$, and $x_0 \to \infty$ except at $\omega = 0$. Also, $x_0 Q_0$ ($= (\omega/\omega_0)\omega_0 R_0 C_0$) becomes $\omega R_0 C_0$. Then,

$$\left|\frac{V}{e_P}\right|^2_B = \frac{(W^2)^2}{\{W^2+(1/Q_1\omega_1 R_0 C_0)+(1-x_1^2)\}^2 + \{(1/\omega R_0 C_0 x_1)(1-x_1^2)-x_1/Q_1\}^2} \tag{10.12}$$

$$\tan \theta_B = \frac{(1-x_1^2)/(\omega R_0 C_0) - x_1/Q_1}{W^2 + (x_1/Q_1\omega R_0 C_0) + (1-x_1^2)} \tag{10.13}$$

B1)   Untuned, high impedance input.

For the case where the hydrophone feeds into an infinitely high input impedance (ie $R_0 \to \infty$), these expressions become

$$\left|\frac{V}{e_P}\right|^2_{B1} = \frac{(W^2)^2}{\{W^2+1-x_1^2\}^2 + \{x_1/Q_1\}^2} \tag{10.14}$$

$$\tan \theta_{B1} = \frac{-x_1}{Q_1(W^2 + 1 - x_1^2)} \tag{10.15}$$

Some examples.

Examples of the response curves are shown in *Fig 10.4*. The solid curves are for untuned hydrophones having $W^2 = 0.3$ and various values of the motional Q–factor, Q_1, feeding into an infinite impedance. The effect of the Q–factor is restricted to the band around resonance. The most important frequency band for these untuned hydrophones is generally the region well below resonance, where the response is almost independent of frequency (eg the rise in sensitivity is not more than 4dB from low frequency up to $x_1 = 0.7$). In practice, the resistance R_0 cannot be infinite, since the ceramic itself has a finite leakage resistance, and the input resistance of any practical amplifier must also have a finite value. The effects of R_0 show up at low frequencies, as indicated by the dashed curve of *Fig 10.4*.

The response curves show a peak at a value of x_1 just above $x_1 = 1$, – ie at a frequency just above the motional resonance ω_1. From (10.14), the peak will occur when the denominator is a minimum. By differentiating, it is easy to show that this occurs when

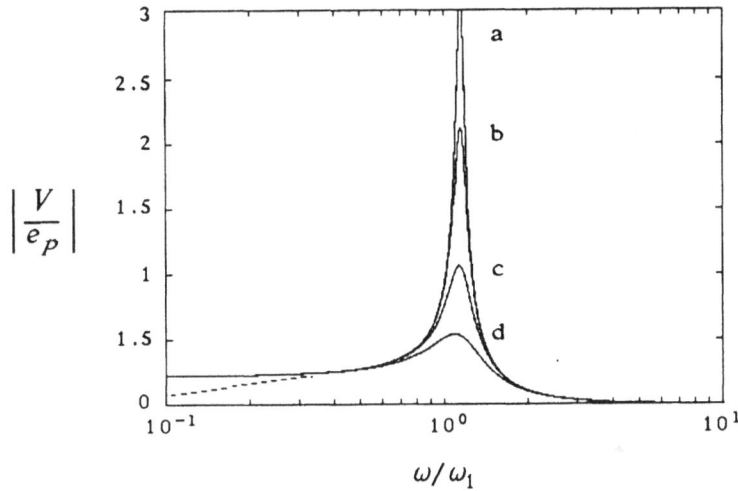

Fig 10.4 Hydrophone response of untuned transducer.
($W^2 = 0.3$, $R_0 = \infty$) (a) $Q_1 = 12$; (b) $Q_1 = 8$,
(c) $Q_1 = 4$, (d) $Q_1 = 2$.

$$x_1^2 = 1 + W^2 - 1/2Q_1^2 \quad (= x_{1m}^2) \qquad (10.16)$$

The frequency of maximum hydrophone response is thus slightly higher than ω_1, provided $2Q_1^2W^2 > 1$, as is usually the case. For a high Q_1, the peak response is at $x_{1m} \simeq (1 + W^2)^{1/2}$. Since W is related to the coupling coefficient (k) by $W^2 = k^2/(1 - k)^2$, the peak response frequency may be expressed in terms of k as

$$x_{1m}^2 = \frac{1}{1 - k^2}$$

x_{1m} therefore corresponds to the frequency of maximum resistance (the anti-resonance frequency) f_p in equation (4.5). The ratio of the sensitivity at the peak to that at low frequencies is given by

$$\frac{|V|_m}{|V|_{LF}} \simeq Q_1(1 + W^2)^{1/2}$$

$$= \frac{Q_1}{(1 - k^2)^{1/2}} \qquad (10.17)$$

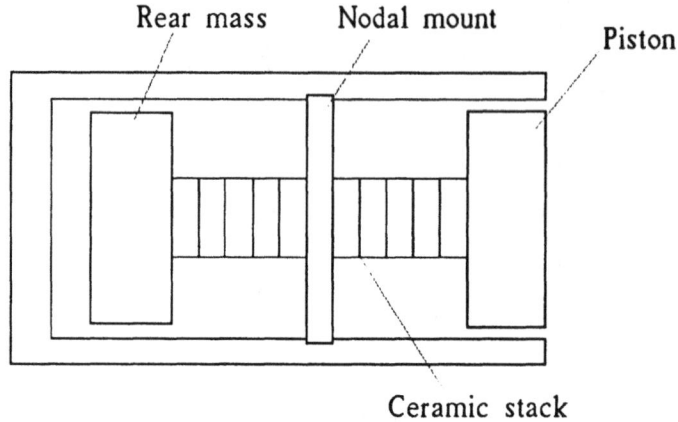

Fig 10.5   **Balanced hydrophone element.**

provided that $Q_1^2 \gg 1/4(W^2 + 1)$, ie $Q_1^2 \gg (1 - k^2)/4$, which is normally the case.

10-3   HYDROPHONE SENSITIVITY – ABSOLUTE VALUE

The above expressions apply to any transducer which can be adequately represented by the equivalent circuit of *Fig 10.1* (with or without the tuning coil), and show how the response varies with frequency. In order to transform the equations to give the absolute values of hydrophone sensitivity, the voltage source e_p must be expressed in terms of the incident acoustic pressure. This relationship depends on the design of the transducer. As a start, we can consider the simple case of a balanced element supported at its nodal mid-point, as in *Fig 10.5*, and calculate the response at a frequency well below the resonance of the element, but not so low that the response is significantly reduced by the low frequency roll-off due to the finite value of the parallel resistance (R_0) across C_0.

The acoustic pressure acts on the front piston, which transfers the force to the piezo-ceramic stack. If the area of the piston is A_p, and the cross-sectional area of the ceramic stack is A_c, the stress on the ceramic due to a low frequency acoustic pressure p is given by

$$T = p.A_p/A_c \qquad (10.18)$$

The electrical field generated in the ceramic is (Section 3.2)

$$E = gT$$

where g is the appropriate piezoelectric g-coefficient. In the usual case, the polarisation of the ceramic is parallel to the axis of the stack, and the relevant coefficient is g_{33}. Thus, if the thickness of each piece of ceramic is t, the output voltage is given by

$$V = g_{33} t p A_p / A_c$$

If the ceramic pieces are connected in parallel, the resultant output voltage is the same as for a single piece. This voltage is generated only in the front half of the element, since that is the only part of the stack which is subjected to the acoustic pressure. The use of the rear half in reducing the acceleration response of the element will be discussed later; for the moment, we merely note that the output voltage is halved by the effect of the capacitance of the rear half of the stack connected across the front half. The resultant output voltage of the hydrophone at low frequency is thus

$$V = \frac{1}{2} g_{33} \frac{A_p}{A_c} t p \qquad (10.19)$$

The value given by (10.14) for the response of an untuned hydrophone at low frequency $(x_1 \to 0)$ is

$$V = e_p \cdot \frac{W^2}{W^2 + 1} \qquad (10.20)$$

Thus, equating values for the low frequency response from these two equations gives the expression for e_p,

$$e_p = \frac{1}{2} g_{33} \frac{A_p}{A_c} t p \cdot \frac{(W^2 + 1)}{W^2} \qquad (10.21a)$$

$$= \frac{1}{2} g_{33} \frac{A_p}{A_c} \frac{t p}{k^2} \qquad \text{since } W^2 = k^2/(1 - k^2) \qquad (10.21b)$$

This may now be substituted into the previous expressions, to give the variation of the absolute hydrophone sensitivity with frequency. For example, for an untuned hydrophone feeding into an infinite resistance, (10.14) becomes

$$\left|\frac{V}{p}\right| = \frac{g_{33}t}{2(1-k^2)} \frac{A_p}{A_c} \frac{1}{\{(W^2+1-x_1^2)^2 + (x_1/Q_1)^2\}^{1/2}} \qquad (10.22)$$

This expression, and others derived using (10.19), refer to the pressure sensitivity of the hydrophone, relating the output voltage to the acoustic pressure at the transducer face. The hydrophone sensitivity of a transducer is however often interpreted as the free-field sensitivity, ie the output related to the acoustic pressure which existed at the position of the transducer face before the transducer was introduced into the field. For a small transducer, the difference between the free-field and the pressure sensitivities is very small, but if the size of the transducer is increased to $\lambda/2$ or more, the acoustic pressure at the face may become significantly different from its undisturbed value. For example, if the hydrophone face is large in terms of wavelengths, and its acoustic impedance is high, the pressure at the face is approximately twice its value in the undisturbed field. In that case, the free-field hydrophone sensitivity would be nearly twice the pressure sensitivity (an increase of nearly 6dB). If a small hydrophone is fitted in a larger baffle, the effects of the diffraction field due to the baffle may cause the pressure at the hydrophone face itself to be either larger or smaller than the undisturbed value, depending on the geometry of the arrangement in terms of wavelengths at the frequency concerned, and in that case the sensitivity curve may deviate appreciably from its ideal smooth shape.

It is worth noting also that the expressions have been derived on the assumption that the simple equivalent circuit is a valid representation of the transducer behaviour. In particular, this assumes that the components in the circuit remain independent of frequency. In practice this is not exactly true; for example, the value of the radiation impedance is a function of frequency, as discussed in Chapter 6. For greater accuracy in the calculations, more accurate values of radiation resistance and reactance could be used in the equations for any particular frequency. However, this is unlikely to be justified in general, since the values are significant only near resonance, and it is therefore usually acceptable to use values appropriate to the resonance frequency, and assume them to remain constant over the whole frequency band. The remaining circuit elements are generally almost independent of frequency.

10-4   BROAD BAND HYDROPHONES

Many applications need a hydrophone which has a constant sensitivity over a wide frequency band, and it is clear from the previous section that this is conveniently effected by using a piezoelectric transducer well below resonance. The hydrophone itself may take a variety of forms, such as a piston type of design, or a cylindrical or spherical element, depending on the particular requirements. Although the absolute value of the pressure sensitivity is determined by the detailed hydrophone construction, the frequency dependence of the pressure response is generally as calculated in the previous section, provided only that the electrical input admittance is satisfactorily described by the simple equivalent circuit. However, further variations in the free-field sensitivity may be caused by diffraction effects if the hydrophone dimensions exceed about a half wavelength. In most cases where a uniform broad band response is needed, the hydrophone dimensions are therefore kept small compared with a wavelength. If this is not possible, it is usually necessary to accept the consequent variations in response and correct for them by measuring the actual reponse curves. In the more usual case where the hydrophone can be made sufficiently small, the main aim of the design is to ensure that all resonances are outside the band of interest. In practice, this means making the element resonance frequency above the operating band, and avoiding any other resonances in the band. This will be considered in more detail in the following sections.

Another major aim of the design is to ensure that the inherent **self noise** of the hydrophone and its input amplifier is low enough to permit measurements down to the acoustic levels required, over the whole frequency band. Contributions to the self noise arise from

    (a)   thermal noise.
    (b)   amplifier noise.
    (c)   noise due to vibration of the mounting.

These need to be compared with the output due to the ambient acoustic noise levels, which constitute the background limiting the basic performance. The significance of (a) and (b) on the design will be discussed in Sections (10.5) and (10.6).

The importance of the contribution (c) due to mounting vibration depends on the nature of the application – in particular, the magnitude of the vibrations of the mounting. The hydrophone response to these vibrations, – often called its **"acceleration response,"** – may be reduced by two approaches. One method is to use a "balanced" design, as in *Fig 10.5*, in which the element is supported at its node at its mid-point. At low frequencies, only the front half of the ceramic stack is thus exposed to the acoustic pressure, and the rear half contributes no output voltage; indeed, as noted in Section 10.3, it reduces the output voltage by half since it is in parallel with the active portion of the stack. However, if an oscillatory force is applied to the nodal mounting (along the axis of the element), one half of the stack will be compressed whilst the other half will be stretched. The voltages generated in each half will thus oppose each other, and the resulting output due to the mounting vibration should ideally be zero. This technique is commonly adopted to reduce acceleration sensitivity, and designs of this type will be assumed in most of the calculations in this chapter. As is evident from the discussion in Section 7.2, the use of a balanced element design produces an element with a higher Q_M than for an unbalanced design. This is not usually important where the hydrophone is to be used only well below resonance, but the response at resonance should be taken into account in assessing performance of the whole system.

The second method of reducing the acceleration sensitivity is to decouple the element from its housing by means of a compliant mounting, as indicated diagrammatically in *Fig 8.10*. If the resonance of the element mass on its supporting spring is well below the operating frequency band, the element will be mechanically decoupled from the vibrations of the housing, and the acceleration sensitivity thus reduced. It may be difficult to make the supporting resonance sufficiently low to give satisfactory isolation down to low operating frequencies, and this method is therefore not particularly suitable for wide band applications. However, it can be useful when an asymmetrical element is necessary to give acceptable performance at resonance, and low acceleration sensitivity is also wanted at lower frequencies. It is of course possible to combine the two techniques by supporting a balanced element on a compliant nodal mounting, thus combining the acceleration cancelling and the decoupling effect. Because any nodal support in practice has a finite compliance, it is necessary to avoid its resonance with the element mass falling within the operating frequency band. Equations describing the response of such systems are derived in Section 10.9.

Expressions for the pressure sensitivity of piston hydrophones are also derived in Section 10.9, to permit comparison between the acceleration response of a hydrophone and its acoustic pressure response. These expressions show in more detail than in Sections 10.2 and 10.3 the voltages generated in a piezoelectric ceramic Tonpilz transducer, allowing for the effects of the mounting compliance.

The assumption that a hydrophone is small compared with a wavelength normally implies also that it should be virtually omni-directional. If a small hydrophone is fitted into a large rigid baffle, it should respond of course only to signals from in front of the baffle, and in an ideal case this response (for a small hydrophone) should be independent of angle over the front half-space. Real baffles are neither infinitely large nor infinitely stiff, and this may cause the actual measured response to show some directionality, which should be taken into account.

A common requirement is for a small omni-directional hydrophone which can be used to act as a probe in an acoustic field. This is usually satisfied by a small spherical or cylindrical hydrophone. If the element is small enough compared with a wavelength, its exact shape has no effect on its directional pattern, which should be virtually omni-directional. This statement is true if the hydrophone senses pressure only; if however it has any sensitivity to acoustic velocity, its response will be directional. Indeed, it is possible to design a hydrophone which senses pressure gradient, or acoustic velocity, and in that case it should have a "figure-of-eight" (or sinusoidal) dependence on angle. (See Section 10.10.) In designing an omni-directional probe hydrophone, it is therefore important to avoid any response to the directional velocity field. This type of response can arise, for example, if the hydrophone is mounted in such a way that it has a transverse resonance within the band of interest. If such a resonance is set up, it may generate signals as a result of the acceleration sensitivity of the element, and hence cause the response to be directional. To avoid this, care is necessary to make sure that no resonances due to the design of the hydrophone support can occur in the operating band. The directionality associated with these resonances usually gives rise also to irregularities in the hydrophone sensitivity curve. In fact, irregularities in the sensitivity curve in a band where it should be flat are generally a strong indication of unwanted resonances in the overall mechanical system.

10-5   THERMAL NOISE

Any resistor at a finite temperature generates voltage fluctuations which are known as thermal noise. If a hydrophone is immersed in the sea, it responds to pressure fluctuations in the medium which arise similarly from thermal agitation of the medium itself. Mellen [10.1] has shown that this thermal noise pressure in a 1 Hz band is given by

$$p_T^2 = \frac{4\pi K_0 T \rho c}{\lambda^2} \qquad (10.23)$$

where K_0 = Boltzmann's constant (1.38×10^{-23} J/deg K)

         T = temperature in degrees Kelvin

and ρ, c and λ are the density, sound speed, and wavelength in water at the frequency being considered.

It is of interest to express this in terms of the voltage produced by the hydrophone when subjected to these pressure fluctuations. To do this, we need a value for the pressure sensitivity of the hydrophone, and this can be derived by first calculating the projector sensitivity. The acoustic pressure generated by an omni-directional source of power output W is given by (2.6), ie

$$p^2 = \rho c W / 4\pi r^2 \qquad (2.6)$$

in which p is the acoustic pressure at range r from the source. The power input to the transducer may be expressed as $i^2 R$, where i is the input current, and R is the resistive component of the input impedance. The output power is then given by $W = \eta_{ea} i^2 R$, where η_{ea} is the electro-acoustic efficiency. The current projector sensitivity ($\hat{S}_I$) referred to range r, is therefore given by:-

$$\hat{S}_I^2 = \frac{p^2}{i^2}$$

$$= \frac{\rho c R \eta_{ea}}{4\pi r^2} \qquad (10.24)$$

By the reciprocity theorem, the hydrophone sensitivity $\hat{M}$ is related to the projector sensitivity (from (9.11)) by

$$\frac{\hat{M}}{\hat{S}_I} = \frac{2r}{\rho f}$$

where f is the frequency (in Hertz). Thus, combining these expressions,

$$\hat{M}^2 = \frac{c R \eta_{ea}}{\pi \rho f^2}$$

$$= \frac{R \eta_{ea}}{\pi \rho c} \lambda^2 \qquad (10.25)$$

(a)

(b)

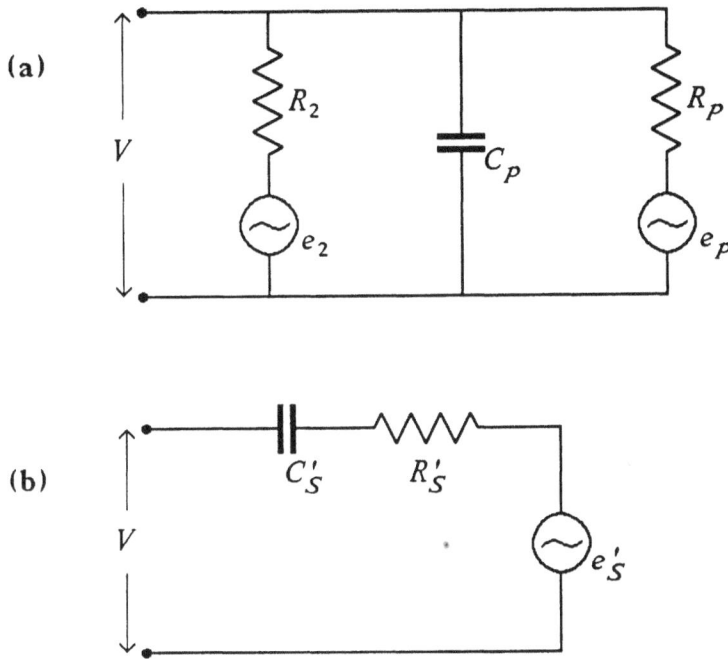

Fig 10.6   Equivalent circuit of hydrophone: (a) showing thermal noise sources; (b) converted to series form.

The mean square output voltage generated (in a 1 Hz band) by the pressure fluctuations of (10.23) is therefore

$$e_T^2 = \hat{M}^2 p_T^2$$

$$= 4K_0 TR \eta_{ea} \qquad (10.26)$$

For an ideal hydrophone having $\eta_{ea} = 1$ (ie 100% efficiency), the mean square output noise voltage is thus $4K_0 TR$, which is just what would be expected for the noise from a resistor R. We thus see that, for an ideal hydrophone, the electrical equivalent component R correctly represents the thermal noise due to the water as well as its acoustic impedance. For a non-ideal hydrophone, having $\eta_{ea} < 1$, additional electrical noise is contributed by the other dissipative elements, the effects of which may be calculated by representing the hydrophone as in *Fig 10.6a*. R_p and C_p

represent the ideal hydrophone, with its equivalent acoustic noise source e_p, and any additional losses are represented by the parallel resistor R_2, with its associated electrical noise voltage e_2. This circuit may be converted to its series form, as in *Fig 10.6b*, by first combining R_p and R_2 in parallel to give $R'_p = R_p R_2/(R_p + R_2)$; then

$$R'_s = \frac{R'_p}{1 + \beta'^2} \tag{10.27}$$

where $\beta' = \omega C_p R'_p$. The output noise voltage is then given by

$$V^2 = 4K_0 T R'_s$$

Using equation (10.25) for the hydrophone sensitivity, but with R'_s substituted for R, gives an expression for the equivalent noise pressure, p_{TR}

$$p_{TR}^2 = 4K_0 T R'_s \cdot \frac{\pi \rho c}{R'_s \eta_{ea} \lambda^2}$$

$$= \frac{4\pi K_0 T \rho c}{\eta_{ea} \lambda^2} \tag{10.28}$$

The equivalent acoustic noise power is thus $1/\eta_{ea}$ times that of the ideal hydrophone given in equation (10.23).

The variation of noise power with frequency can be deduced by using (4.24) for the variation of η_{em} with frequency.

$$1/\eta_{em} = 1 + \frac{\omega}{\omega_S} \frac{\tan\delta}{k^2 Q_M} (1 + Q_M^2 \Omega^2)$$

where $\tan\delta = 1/\omega C_{LF} R_e$; ie $\omega\tan\delta = 1/C_{LF} R_e = 1/C_p R_2$ if the circuit of *Fig 10.6a* represents a hydrophone at low frequency. Then, assuming $\eta_{em} \simeq \eta_{ea}$

$$p_{TR}^2 \simeq \frac{4\pi K_0 T \rho c}{\lambda^2} \left\{ 1 + \frac{1 + Q_M^2 \Omega^2}{\omega_S k^2 C_{LF} R_2 Q_M} \right\} \tag{10.29}$$

$$\simeq \frac{4\pi K_0 T \rho f^2}{c} \left\{ 1 + \frac{Q_M \Omega^2}{\omega_S k^2 C_{LF} R_2} \right\} \tag{10.30}$$

if $Q_M^2 \Omega^2 \gg 1$ (ie not too near resonance).

If R_2 is a constant, independent of frequency, which represents for example the input resistance of an amplifier, then we can define an angular frequency

$$\omega_C = 1/C_{LF}R_2 \qquad (10.31)$$

which is the **roll-off frequency** due to R_2 in parallel with C_{LF}. Then, (10.30) may be written as

$$p_{TR}^2 = \frac{4\pi K_0 T\rho f^2}{c}\left\{1 + \frac{Q_M}{k^2}\frac{\omega_C}{\omega_S}\Omega^2\right\} \qquad (10.32)$$

Substituting $K_0 = 1.38 \times 10^{-23}\text{J/°K}$, $T = 280°\text{K}$, $\rho = 10^3\text{kg/m}^3$, $c = 1.5 \times 10^3\text{m/s}$, this gives the rms noise pressure in a 1Hz band as

$$p_{TR} = 1.8 \times 10^{-7}F\left\{1 + \frac{Q_M}{k^2}\frac{F_C}{F_S}\Omega^2\right\}^{1/2} \text{ (Pa)} \qquad (10.33a)$$

where the angular frequencies have been converted to frequencies in kHz; ie F_S is the resonance frequency, F_C the roll-off frequency, and F the frequency being considered, all in kHz. In logarithmic terms, this becomes

$$P_{TR} = -15 + 20\log F + 10\log\left\{1 + \frac{Q_M}{k^2}\frac{F_C}{F_S}\Omega^2\right\} \qquad (10.33b)$$

in dB re 1 µPa in a 1Hz band.

At frequencies well below resonance, when $F \ll F_S$, so that $\Omega \simeq -F_S/F$, and if also $F^2 \ll (Q_M/k^2)^2 F_C F_S$, then (10.33a) simplifies to

$$p_{TR_o} = 1.8 \times 10^{-7}(F_C F_S Q_M/k^2)^{1/2} \qquad (10.34a)$$

and in logarithmic terms

$$P_{TR_o} = -15 + 10\log(F_C F_S Q_M/k^2)$$

$$\text{(dB re 1µPa in 1Hz)} \qquad (10.34b)$$

Examples of noise levels derived from these equations are shown in *Fig 10.7*. The effects of varying F_C and Q_M are illustrated in *Fig 10.7a*, all the curves being for a hydrophone having $k = 0.5$ and a resonance frequency of 100kHz. For an ideal hydrophone of 100% efficiency, $F_C \rightarrow 0$ and (10.33b) becomes

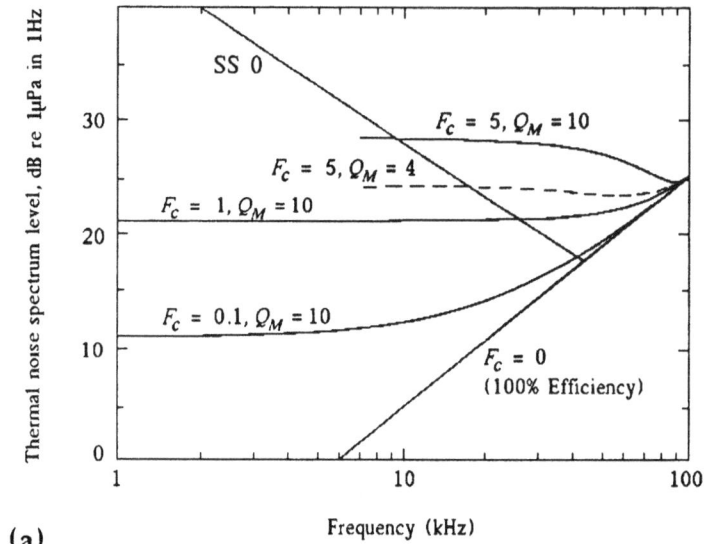

(a)

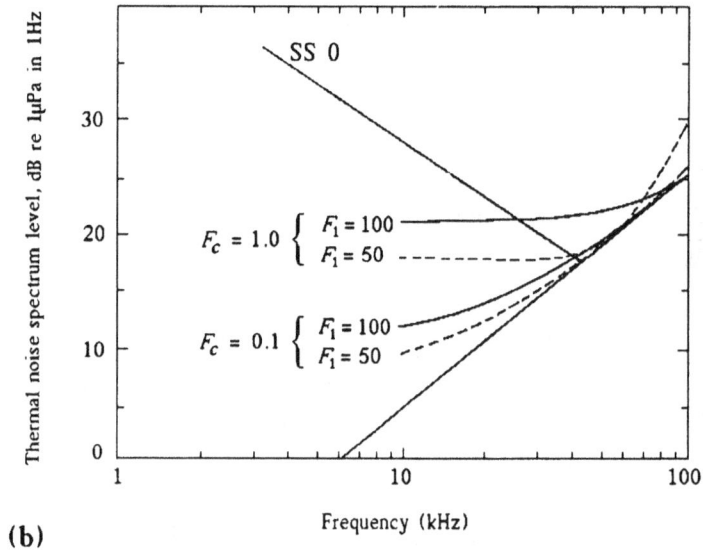

(b)

Fig 10.7   Equivalent thermal noise spectra for hydrophone:
(a) Effect of varying F_C (F_1 = 100kHz, k = 0.5);
(b) Effect of varying F_1 (Q_M = 10, k = 0.5).

$$P_{TR} = -15 + 20\log F$$

This is shown in *Fig 10.7a* as the line sloping upwards to the right; it corresponds to the expression given by Mellen [10.1, 10.2]. The curves corresponding to non-zero values of F_C show how the hydrophone's noise level is increased by the electrical loss factors, generally due to the input resistance of the preamplifier, and expressed in terms of the roll-off frequency F_C. These hydrophone self-noise levels need to be compared with the ambient noise levels expected in the sea, and these are indicated by the line sloping downwards to the right. This line represents the noise levels given by Knudsen as being typical for ambient noise in deep water at sea state zero (SS 0). Urick [10.2] has described the various sources of noise in the sea, including consideration of the frequency bands within which they are applicable and the effects of increasing sea state. For any particular application, it may be appropriate to select a different ambient noise curve, but the Knudsen curve for SS 0 is used for the examples in this section, as it represents a reasonable limiting value for the frequency band from about 500Hz to 50kHz. The Knudsen values are described approximately by

$$P_N = 45 - 17\log F \quad \text{(dB re 1}\mu\text{Pa in 1Hz band)} \qquad (10.35)$$

The curves in *Fig 10.7a* show how the increased self-noise levels due to a parallel resistor may prevent measurements down to SS 0 in the region around 30-60kHz. The effects of varying the resonance frequency F_S are shown in *Fig 10.7b*, to illustrate how reducing F_S to 50kHz can improve the thermal self-noise, especially in this band near resonance. However, it must be remembered that such a move will introduce the peakiness of the response around the resonance, and is therefore usually undesirable.

The parallel resistor R_2 is in practice a combination of a fixed resistor representing the input resistance of the pre-amplifier and a resistor representing the dielectric loss factor of the ceramic. The dielectric loss resistor varies inversely with frequency, thus keeping the loss factor ($\tan\delta$) constant (4.9). If the input resistance of the pre-amplifier is sufficiently high, and the value of $\tan\delta$ relatively high, so that the $\tan\delta$ effect dominates, the expression for the equivalent thermal self-noise in the hydrophone becomes

$$P_{TR} = -15 + 20\log F + 10\log\left\{1 + \frac{Q_M\tan\delta}{k^2}\frac{F}{F_S}\Omega^2\right\} \quad (10.36)$$

In this case, the self-noise curve becomes asymptotic at low frequencies to a straight line which is linearly dependent on frequency, and described by

$$P_{TR_o} = -15 + 10\log F + 10\log(F_S\tan\delta Q_M/k^2) \quad (10.37)$$

instead of the constant value which results from a constant R_2 (10.34).

Implications for hydrophone design.

Some guidance for the design of broad band hydrophones may be deduced from the above relationships. If the operating band does not extend above 30kHz, the resonance frequency of the hydrophone need not be above 50kHz, to maintain a reasonably constant sensitivity over the desired band. It is then evident from *Fig 10.7* that the thermal self-noise of the hydrophone is generally well below the ambient noise levels for *SS* 0, unless the efficiency of the hydrophone at resonance is very poor, or the value of the roll-off frequency is made unusually high. Thus, for operating bands up to about 30kHz, it is generally simple to design a system which can adequately measure acoustic levels down to deep sea state zero.

If a flat response to frequencies above 50kHz is wanted, however, more care is needed to avoid excessive thermal noise. As is clear from the curves of *Fig 10.7*, Q_M, F_S and F_C should be made as low as possible, and k as high as possible. The first step is therefore to make the resonance frequency as low as is consistent with the need for a smooth response, and to choose a design having a high k and low Q_M. It is then necessary to make F_C as low as possible. Although this may be achieved by making either C_{LF} or R_2 high, we shall see later that increasing C_{LF} is often associated with a reduction in sensitivity, which may lead to problems with amplifier noise, and it is therefore usually preferable to obtain a low value of F_C by making R_2 as high as is practicable. If the required value of R_2 becomes very large, the dielectric losses in the ceramic may become significant, and a ceramic with a low $\tan\delta$ should then be chosen. This may not be easy when balanced against the probably stronger need for a high coupling coefficient, since most of the ceramics intended for hydrophone use have relatively high values of tanδ, and a careful analysis of the predicted performance should be made in choosing the material for the most demanding applications.

10-6 AMPLIFIER NOISE

Any pre-amplifier connected to the hydrophone is a source of further noise, and the power from this must be added to the thermal self-noise of the hydrophone itself. Amplifier noise may be attributed to a combination of a voltage source across the input, and a current source in series with the input. It may then be characterised by four parameters:- S_e, the power spectral density of the voltage source; S_i, the spectral density of the current source, and two parameters representing the real and imaginary parts of the correlation coefficient between S_e and S_i. Woolett [10.4] showed that the correlation coefficient had only little influence on the noise characteristics, and it will therefore be assumed to be zero in this treatment. For the case of an amplifier with a piezoelectric ceramic hydrophone across its input, at a frequency well below resonance, the spectral density of the resultant amplifier noise may then be represented by a single source (S_A) in series with the input. The strength of this source depends on the impedance of the hydrophone, which may be taken as $1/\omega C_{LF}$ for low frequencies. Then

$$S_A = S_e + S_i/(\omega C_{LF})^2 \qquad (10.38)$$

Since the hydrophone thermal noise and the amplifier noise are uncorrelated, the total receiver noise is the sum of the two noise powers, provided that they are expressed in the same units – ie both in electrical or both in acoustic terms.

Typical curves for the equivalent input noise of a good quality pre-amplifier are shown in *Fig 10.8*. The values represent the electrical noise power in a 1 Hz band, referred to the input, for a range of capacitances across the input. The high capacitance curve represents the voltage source S_e. In practice, the noise of an amplifier may be determined by measuring its noise output as a function of frequency when it has a (noiseless) capacitor, equal to that of the hydrophone, connected across the input in place of the hydrophone. The measured output may be referred to the input by dividing by the gain of the amplifier, to give the noise spectrum level e_F in dB re 1V in 1Hz. This noise voltage may then be converted to its acoustic equivalent by dividing by the pressure sensitivity of the

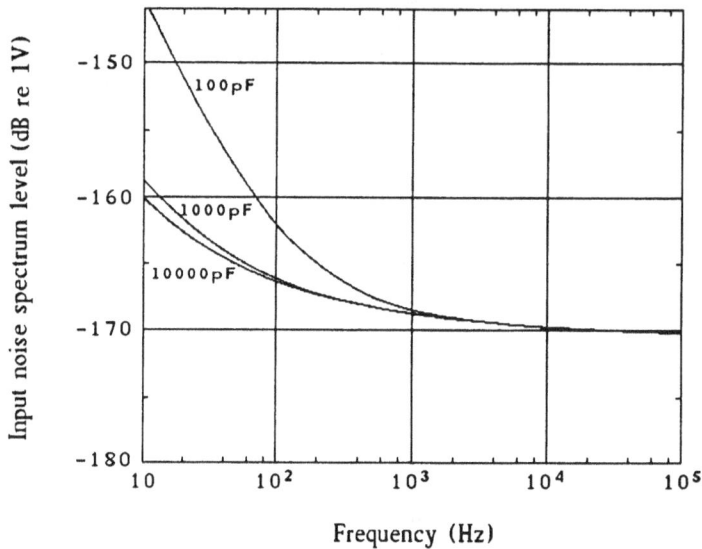

Fig 10.8 **Typical spectra for equivalent input noise of good quality hydrophone pre-amplifier, for various source capacitances.**

hydrophone. Thus, if we denote this amplifier noise spectrum level, expressed in equivalent pressure terms, by P_E (dB re 1μPa in 1 Hz),

$$P_E = e_E - M$$

In order to achieve a high input impedance for the pre-amplifier, as well as a low noise figure, a Field Effect Transistor is commonly used for the first stage of amplification.

Implications for hydrophone design.

Suppose that a hydrophone is wanted to measure acoustic levels down to *SS* 0 over a band extending up to 50kHz. Accurate measurements can be made only if its hydrophone sensitivity is high enough for the output due to the acoustic pressures to be significantly above the output due to the amplifier noise itself. Taking the minimum acoustic pressures as P_N (as given by (10.35)), this may be expressed as

$$e_E - M < P_N = 45 - 17\log F$$

ie
$$M > e_E - (45 - 17\log F) \qquad (10.39)$$

If the amplifier noise spectral level is assumed to have the approximate value of -169dB re 1V (for a good quality pre-amplifier), then this gives

$$M > -214 + 17\log F \qquad \text{(dB re 1V/}\mu\text{Pa)} \qquad (10.40)$$

as a typical minimum hydrophone sensitivity which would permit measurements down to deep sea state zero. To give an adequate margin over the amplifier noise, the sensitivity should be 5dB or more higher than the value given by (10.40). Ambient sea noise levels given by Wenz [10.3] are lower than those of Knudsen, and a requirement to measure down to the Wenz curves would thus increase the necessary hydrophone sensitivity. Conversely, if it is not necessary to measure down to *SS* 0, the sensitivity may be reduced. The simple assumption of a constant pre-amplifier noise may not be justified for frequencies below 1kHz if the hydrophone capacitance is below about 1000pF, and curves such as those in *Fig 10.8* should then be used for e_E.

Some of the equations derived earlier for elements satisfying the lumped-mass approximation can be used to obtain expressions for the hydrophone sensitivities of transducers. For a balanced element, as in *Fig 10.5*, with its ceramic rings connected electrically in parallel, the low frequency pressure sensitivity of the hydrophone is deduced from (10.19) as

$$\hat{M} = \frac{1}{2} g_{33} \frac{A_p}{A_c} t \qquad (10.41)$$

Also, from (7.7),

$$C_{LF} = \varepsilon A_c n / t$$

These equations apply for hydrophones where the ceramic rings are all connected in parallel. If the front and rear halves of the stack were connected electrically in series instead of in parallel, the sensitivity would be doubled, but the capacitance reduced by a factor of four. The factor $\hat{M}^2 C_{LF}$ thus remains constant, and this would be true also for other possible rearrangements of the electrical connections. We thus have a range of possible choices of $\hat{M}^2$ and C_{LF}, subject to the restriction that $\hat{M}^2 C_{LF}$ cannot exceed a value which depends on the overall element design. This factor is given by

$$\hat{M}^2 C_{LF} = \frac{1}{4} g_{33}^2 \left(\frac{A_p}{A_c}\right)^2 t^2 \; \varepsilon A_c n/t$$

$$= \frac{1}{4} g_{33}^2 \left(\frac{A_p}{A_c}\right)^2 \varepsilon V_c \qquad (10.42)$$

where $V_c = ntA_c$ is the volume of the ceramic. For geometrically similar stacks (ie having the same value of A_p/A_c), the parameter $\hat{M}^2 C_{LF}$ is thus proportional to the volume of ceramic, and does not depend on the particular values of A_c, t or n. For a given ceramic volume, the parameter depends on the value of the hydrophone figure of merit of the ceramic itself, $g_{33}^2\varepsilon$, (as in Chapter 3).

We would expect the ceramic volume to be related to the resonance frequency of the element, a lower resonance being associated with a larger element, and a useful relationship can be derived by using (7.4)

$$\frac{A_c}{l_c} = \frac{M_1 \omega_S^2}{E_e(1 + M_1/M_2)}$$

Thus, for a balanced element,

$$A_c = M_1 \omega_S^2 l_c / 2E_e \qquad (10.43)$$

which, with (10.42), gives

$$\hat{M}^2 C_{LF} = g_{33}\varepsilon E_e \frac{A_p^2}{2 M_1 \omega_S^2} \qquad (10.44)$$

From (7.3),

$$Q_M = \frac{\omega_S \cdot 2M_1}{r}$$

where r is the motional resistance of the hydrophone at resonance. Hence, for a balanced element,

$$\hat{M}^2 C_{LF} = g_{33}^2 \varepsilon E_e \frac{A_p^2}{\omega_S r Q_M}$$

$$= g_{33}^2 \varepsilon E_e \frac{A_p}{\omega_S \rho c \tilde{R} Q_M} \qquad (10.45)$$

where $\tilde{R}$ is the normalised radiation resistance (ie $r \simeq \rho c A_p \tilde{R}$).

Approximate values may be substituted in (10.45) for some of these factors, assuming the stack to use LZT II ceramic, viz:-

$$g_{33} = 28 \times 10^{-3} \text{ Vm/N}$$

$$\varepsilon = \varepsilon_{33}\varepsilon_0 = 1700 \times 8.85 \times 10^{-12} = 1.5045 \times 10^{-8} \text{ F/m}$$

$$E_e = 5.2 \times 10^{10} \text{ Pa (ie 90\% of Young's Modulus of LZT II,}$$
to allow for the effects of joints and spacers.)

$$\rho c = 1.5 \times 10^6 \text{ kgm}^{-2}\text{s}^{-1}$$

If the piston diameter is at least 0.4λ at resonance, then $\tilde{R}$ will be between 0.6 and 1.3. Thus, assuming $\tilde{R} = 1.0$, (10.45) becomes

$$\hat{M}^2 C_{LF} \simeq 6.51 \times 10^{-11} \frac{A_p}{F_S Q_M} \qquad \text{(V}^2\text{F Pa}^{-2}) \qquad (10.46a)$$

where A_p is in m² and F_S is in kHz. The equation remains the same if the hydrophone sensitivity is expressed in V/µPa and the capacitance in pF (instead of V/Pa and Farads). If A_p is in mm², the equation is

$$\hat{M}^2 C_{LF} \simeq 6.51 \times 10^{-17} \frac{A_p}{F_S Q_M} \qquad (10.46b)$$

When considering a range of geometrically similar designs of differing resonance frequencies, it is reasonable to assume that their dimensions scale in proportion to the wavelength at the resonance λ_S. The ratio of piston diameter ($2a$) to λ_S may thus be treated as a parameter; then,

$$A_p = \pi a^2 = \frac{\pi c^2}{4 \times 10^6 F_S^2} \cdot (2a/\lambda_S)^2$$

$$= \frac{1.77}{F_S^2} (2a/\lambda_S)^2 \qquad \text{(m}^2) \qquad (10.47)$$

Substituting in (10.46a) gives

$$\hat{M}^2 C_{LF} \simeq \frac{11.5 \times 10^{-11} \, (2a/\lambda_S)^2}{F_S^3 Q_M} \qquad (10.48)$$

For example, if $2a/\lambda_S = 0.5$,

$$\hat{M}^2 C_{LF} \simeq \frac{2.88 \times 10^{-11}}{F_S^3 Q_M} \qquad (10.49)$$

Values of Q_M for balanced hydrophones with reasonably lightweight pistons are typically between 10 and 20. Thus, assuming $Q_M = 15$ leads to the rough approximation,

$$\hat{M}^2 C_{LF} \simeq \frac{2 \times 10^{-12}}{F_S^3} \qquad ((V/\mu Pa)^2 pF) \qquad (10.50)$$

The parameter $\hat{M}^2 C_{LF}$ representing the maximum power output from the hydrophone is thus inversely proportional to the cube of the resonance frequency, and it is therefore desirable to make F_S as low as possible. A_p should also be made as large as is consistent with the other requirements, and Q_M as small as practicable. Equation (10.45) shows the significance of choosing a material with a high value of the parameter $g_{33}^2 \varepsilon$, as was noted in Chapter 3. If the piezoelectric ceramic is used in a mode other than the 33-mode, the appropriate g-coefficient must of course be used, and the corresponding relationships derived. The most important factor is the resonance frequency, which must be high enough to give a flat response curve over the operating band, but should not be made unnecessarily high if the ceramic volume, and hence the value of $\hat{M}^2 C_{LF}$, is to be maximised.

The values of hydrophone sensitivity $\hat{M}$ and capacitance may then be selected, subject to the limitation on $\hat{M}^2 C_{LF}$. The sensitivity may be made high, in order to increase the response to acoustic signals above amplifier noise, but only at the cost of a corresponding reduction in capacitance. This may be a disadvantage for two reasons:-

(a) For a given input resistance, the roll-off frequency will be raised, which will increase the thermal noise, as indicated in *Fig 10.7*.

(b) Any capacitance across the hydrophone – such as cable capacitance – causes a reduction in sensitivity, which becomes significant if the hydrophone capacitance falls to a value comparable with that of the cable.

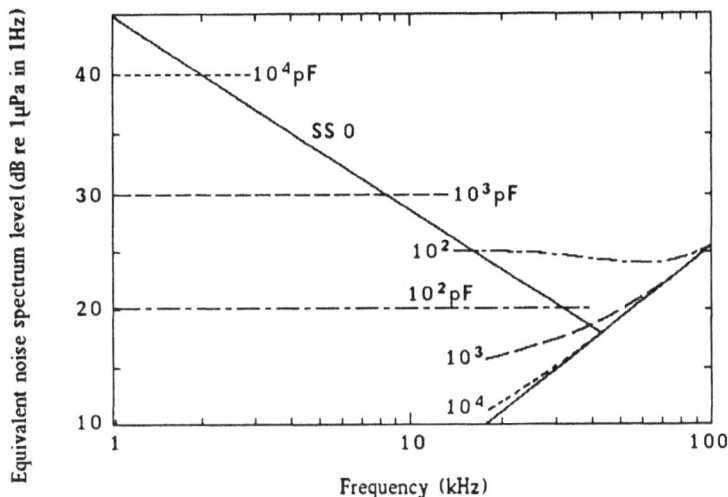

Fig 10.9   Hydrophone self-noise levels: effect of varying C_{LF} (F_S = 100kHz, Q_M = 15, k = 0.5)

The principles involved may be illustrated by considering an example. Suppose that we need a hydrophone with a piston diameter of 20mm, and a resonance frequency of 100kHz, to give an operating band up to 50kHz. Assuming a lead zirconate titanate hydrophone with the parameters above, and assuming Q_M = 15, (10.46b) gives $\hat{M}^2 C_{LF} \simeq 1.36 \times 10^{-17}$ (V/µPa)^{2}pF. Suppose that the capacitance is chosen as 1000pF. Then

$$\hat{M} = (1.36{\times}10^{-17}/10^3)^{1/2} = 1.17{\times}10^{-10} \text{ V/µPa}$$

$$\text{ie, } M = 20\log\hat{M} \simeq -199\text{dB re } 1\text{V/µPa}$$

The amplifier noise levels may now be converted to their equivalent pressure levels by dividing the noise voltages by the hydrophone sensitivity; in dBs, this is effected by subtracting M from the value e_E = −169 dB re 1V used in deriving (10.40), giving the value (30dB) indicated by the dashed line on the left hand side of *Fig 10.9*. (This assumes M to be constant for frequencies well below resonance.) For this case, the amplifier noise is seen to be below *SS* 0 up to about 8kHz.

Suppose also that the resistance (R_2) in parallel with the hydrophone is 1MΩ. Then, F_C = $10^{-3}/2\pi{\times}10^{-9}{\times}10^6$ = 0.16kHz. This value, together with appropriate values of F_S, Q_M, and k, may be substituted in (10.33b) to calculate the variation with frequency of the hydrophone's thermal noise. For example, assuming F_S = 100kHz, Q_M = 15, and k = 0.5, gives the dashed line at the higher frequency side of *Fig 10.9*. It is clear that in this case it is the amplifier noise which is the dominant factor limiting the noise performance above 8kHz. If the hydrophone is to be used only below 8kHz, its self-noise should be well below *SS* 0 levels. The low frequency response falls by 3dB at 160Hz (F_C), and continues to fall as the frequency is reduced. If constant sensitivity to frequencies below 160Hz is wanted, the roll-off frequency may be reduced by increasing R_2; however, this implies an input resistance for the pre-amplifier in excess of 1MΩ, and a correspondingly high leakage resistance across the ceramic and cable. In practice, values above 1MΩ become increasingly difficult to realise, and considerable care is necessary to achieve and maintain a value of R_2 above 10MΩ.

An alternative way of reducing F_C would be to increase the capacitance C_{LF}, for example by dividing the stack into more (and thinner) rings. Suppose, therefore, that we choose to make C_{LF} = 10,000pF instead of the 1000pF assumed above. Then, in order to keep M^2C_{LF} constant, the hydrophone sensitivity is reduced by 10dB, and the equivalent amplifier noise pressure is increased correspondingly, as indicated by the dotted line in *Fig 10.9*. In this case, the amplifier noise is below *SS* 0 levels only up to about 2kHz, and measurements above that frequency will be corrupted by amplifier noise. It is obviously possible to choose other values of C_{LF}, and to evaluate their effects on the response curves. The best choice depends on the operating band, and on the lowest ambient noise conditions to be measured. In the cases considered above, for example, C_{LF} = 1000pF is preferable (though not ideal) if measurements are wanted down to *SS* 0 over the band from 500Hz to 5kHz. If the range of interest is from 50 to 500Hz, however, C_{LF} = 10,000pF is a better choice, in order to make it easier to achieve the low roll-off frequency needed.

Conversely, the capacitance could be reduced below the 1000pF of the example by dividing the stack into fewer rings, or by connecting some rings in series instead of in parallel. In this case, the sensitivity of the hydrophone is increased. An increase in sensitivity is of little benefit at low frequencies, where the main effect is to depress the amplifier noise even further below the ambient noise levels, whilst the associated decrease in capacitance makes it more difficult to achieve a low roll-off frequency. If F_C is raised, it also increases the thermal noise from the hydrophone in the range around 40kHz. An increase in sensitivity at the expense of capacitance is therefore not necessarily beneficial, if measurements are wanted only at low frequencies. However, if the measurement range extends above 8kHz, it can be seen from *Fig 10.9* that a capacitance between 100 and 1000pF gives a better compromise between the thermal and amplifier noise limits.

Another disadvantage of reducing the hydrophone capacitance may arise if it becomes comparable with any **parasitic capacitance** which is in parallel with the hydrophone. The most probable source of such parasitic capacitance is the connecting cable, particularly for long cable lengths, since coaxial cables may have typical capacitances of around 100pF/m. The effect of a parasitic capacitance C_k is to reduce the sensitivity by a factor $(1 + C_k/C_{LF})^{-1}$, the reduction becoming significant when C_k is comparable to, or greater than, C_{LF}. If the capacitance of a particular hydrophone is reduced in order to increase the sensitivity, the effect of any cable capacitance becomes more important, and Rijnja [10.5] showed that the maximum sensitivity was obtained for $C_{LF} = C_k$. Except for those cases in which it is necessary to maximise the sensitivity, it is normally desirable to ensure that the hydrophone capacitance is significantly greater than the cable capacitance. In this way, the effects of any variations in cable length or characteristics may be minimised.

10-7   SPHERICAL HYDROPHONES

The relationships in the preceding sections apply to piston hydrophones, which are commonly used in baffles to measure the acoustic pressure at the surface of the baffle. A common requirement is for a hydrophone to measure the pressure at a point in an acoustic field. This implies that the sensor is small and omni-directional, and this is often realised by using a spherical hydrophone which is small compared with a wavelength at the highest frequency of interest. Considerations similar to those above may be applied to these hydrophones, but first we must derive the expressions for their characteristics.

Equations have been derived by Anan'eva [10.6] for the parameters of spherical shell receivers which are silvered over the inner and outer surfaces, and radially poled, viz:-

The hydrophone sensitivity (below resonance) is given by

$$\hat{M} = \frac{b}{2(1+\beta+\beta^2)} \{g_{33}(\beta^2+\beta-2) - g_{31}(\beta^2+\beta+4)\} \quad (10.51a)$$

in which b is the external radius of the shell, and β is the ratio of internal to external radii. For a thin shell, for which $\beta \to 1$, this simplifies to

$$\hat{M} = -bg_{31} \qquad (10.51b)$$

The low frequency capacitance is given by

$$C_{LF} \simeq \frac{4\pi\varepsilon b\beta}{(1-\beta)} = \frac{4\pi\varepsilon b^2}{t}(1-t/b) = 4\pi\varepsilon b\left(\frac{b}{t} - 1\right) \qquad (10.52)$$

where t is the wall thickness (and $t/b \ll 1$, corresponding to the thin-wall approximation). ε denotes the dielectric constant and g_{31} the appropriate piezoelectric constant for the ceramic, the dominant mechanical stress being circumferential, and hence normal to the poling direction. The hydrophone sensitivity is independent of the shell thickness, within the thin shell approximation. This occurs because the stress in the shell is greater than the external pressure by a factor proportional to b/t, whilst the voltage generated for a given stress is proportional to t. The output voltage, which depends on the product, is thus independent of t. The capacitance, however, reduces as the thickness is increased.

The resonance frequency is given by

$$f_S^2 = \frac{1}{(2\pi b)^2} \left(\frac{2E_{11}}{(1-\sigma)\rho_c}\right) \qquad (10.53)$$

where ρ_c is the density, σ the Poisson's ratio, and E_{11} the elastic modulus of the ceramic (in the plane normal to the poling direction, with the electrodes open-circuited).

Thus, for a spherical shell,

$$\hat{M}^2 C_{LF} = 4\pi\varepsilon g_{31}^2 b^3\{(b/t)-1\} \qquad (10.54)$$

Substituting typical values for a lead zirconate titanate hydrophone material (LZT II), ie:-

$$\varepsilon = 1700 \times 8.85 \times 10^{-12} = 1.5045 \times 10^{-8} \ \text{F/m}$$

$$g_{31} = -12.0 \times 10^{-3} \ \text{Vm/N},$$

this gives

$$\hat{M}^2 C_{LF} = 2.72 \times 10^{-11} b^3\{(b/t)-1\} \quad \text{(V/Pa)}^2\text{F} \qquad (10.55a)$$

for b in metres, or

$$\hat{M}^2 C_{LF} = 2.72 \times 10^{-20} b^3 \{(b/t) - 1\}$$

$$(V/Pa)^2 F \text{ or } (V/\mu Pa)^2 pF \qquad (10.55b)$$

when b is expressed in mm.

Substituting typical values for LZT II in (10.53) gives the relationship between resonance frequency and shell radius; ie, using

$$\varrho_c = 7.7 \times 10^3 \text{ kg/m}^3$$

$$\sigma = 0.3$$

$$E_{11} = 6.3 \times 10^{10} \text{ Pa}$$

gives $f_S = 770/b$ Hz, for b in metres   (10.56a)

and $F_S = 770/b$ kHz, for b in mm.   (10.56b)

Combining (10.55) and (10.56b) results in

$$\hat{M}^2 C_{LF} = \frac{12.4 \times 10^{-12}}{F_S^3} \{(b/t) - 1\} \qquad (10.57)$$

For example, a thin-walled spherical shell hydrophone of LZT II ceramic with a mean diameter of 20mm and a wall thickness of 1mm would have a resonance frequency of 77kHz. Its capacitance would be 17,000pF, and hydrophone sensitivity 12×10^{-5}V/Pa, ie -198.4dB re 1V/μPa; its value of $\hat{M}^2 C_{LF}$ would therefore be 2.45×10^{-16} $(V/\mu Pa)^2 pF$.

As for the piston designs, some choices may then be made between $\hat{M}$ and C_{LF}, to obtain the best bandwidth and noise performance, according to the principles described in Section 10.6. This may be achieved most simply by varying the wall thickness of the ceramic, but it is also possible to divide the electrode surfaces into separate areas and connect these in series instead of in parallel. Thus, for example, it is reasonably straightforward to connect the two halves of a sphere in series and hence double the sensitivity whilst reducing the capacitance by a factor of four compared with the usual arrangement of the two halves in parallel. Further subdivisions of the surface have also been used (eg Lipscombe [10.7]). Such configurations allow greater freedom in optimising sensitivity, but involve more care in manufacture, and some degradation can arise from inter-sector capacitances.

For a wall thickness which is one-tenth of the shell radius, as in the example above, (10.57) may be approximated by

$$\hat{M}^2 C_{LF} = 110 \times 10^{-12}/F_S^3 \qquad (10.58)$$

whilst for a balanced Tonpilz design, (10.50) gave

$$\hat{M}^2 C_{LF} = 2 \times 10^{-12}/F_S^3$$

Some of this 50-fold difference in output capability is due to the reduction in output caused by the acceleration-cancelling rear half of the balanced design, but the major effect arises from the magnification of acoustic pressure which occurs in the spherical shell. This stress magnification has a corresponding disadvantage in that the external hydrostatic pressure is also magnified, thus limiting the depth at which the hydrophone can be used.

This **depth limitation** is easily derived by recalling that for a thin walled spherical shell the stress in the shell is greater than the external pressure by a factor $2b/t$. If the maximum permissible stress in the shell is denoted by T_{max}, the external pressure (p_{max}) should thus not exceed

$$p_{max} = T_{max} \cdot (2t/b) \qquad (10.59)$$

We saw in Chapter 3 that stresses in lead zirconate titanate ceramics should generally not exceed 70MPa (10,000psi) to avoid serious changes in parameters, or even depoling. For $t/b = 0.1$, this corresponds to a maximum depth of operation of about 1200m (4000ft). To avoid problems with the effects of stress, if the hydrophone is to be used at considerable depth, it is therefore necessary to increase the ratio of t/b, which reduces the value of $\hat{M}^2 C_{LF}$ below that given by (10.58).

10-8   CYLINDRICAL HYDROPHONES

Cylindrical hydrophones are often used as small acoustic probes, since their use of a single piezoelectric tube instead of two hemispheres offers a potentially cheaper device than the spherical shell. Although the cylinder does not have the 3-dimensional symmetry of the sphere, it should nevertheless have an omni-directional response, provided that it is small compared with a wavelength in water at its highest frequency of interest. In this section, expressions are given for the main characteristics of cylindrical hydrophones and compared with those for spherical shells.

Piezoelectric tubes for such applications are generally silvered on the inner and outer surfaces and radially poled, and these will be considered first. Suppose that the inside radius of the tube is denoted by a, the outside radius by b, and the length by L. Assume also that the wall thickness t is small compared with the radius, so that we can when appropriate represent the mean radius by either a or b. The cylindrical tube has three basic modes of vibration associated with these dimensions.

a) The **radial mode** is one in which the radius expands and contracts. Its resonance frequency in air occurs when the circumference is equal to one wavelength of longitudinal vibrations in the ceramic, ie

$$f_{rr} \simeq c/2\pi a' \qquad (10.60)$$

where c is the velocity of longitudinal waves in the ceramic (normal to the poling direction), and a' is the mean radius.

b) The **length mode** is one in which the length expands and contracts. Its resonance frequency occurs when the length of the tube is half a wavelength of longitudinal vibrations in the ceramic, ie

$$f_{rl} \simeq c/2L \qquad (10.61)$$

c) The **thickness mode** is one in which the wall thickness expands and contracts. Its resonance frequency occurs when the thickness is a half wavelength of vibrations in the ceramic, but in this case it is the velocity (c') of thickness waves which is applicable, ie

$$f_{rt} \simeq c'/2t \qquad (10.62)$$

The resonance frequency of the thickness mode is higher than the others, and the lowest resonance arises from either the radial or length mode. The velocity is given by (elastic modulus/density)$^{1/2}$, and this may be used to express the equations in terms of the appropriate frequency constants for the particular modes and materials. Thus for LZT in the radial (or hoop) mode, a typical frequency constant is $N_3 \simeq 930$ Hz.m, and for the length mode the frequency constant is $N_1 \simeq 1430$. (More accurate values of the frequency constants for particular materials are quoted by the manufacturers.) Thus, in terms of the mean diameter, (10.60) becomes

$$f_{rr} \simeq 930/2a' \quad \text{(radial mode)}$$

$$\text{(10.60a)}$$

and (10.61) becomes

$$f_{rl} \simeq 1430/L \quad \text{(length mode)} \quad \text{(10.61a)}$$

These equations give the resonance frequency in Hz if the dimensions are in metres, and in kHz if the dimensions are in mm.

If a flat response curve is required, the first design criterion is (as usual) to make the lowest resonance frequency of the cylinder well above the highest receiving frequency. It is of interest to estimate the size of the cylinder compared with a wavelength in water. This may be done by recalling that the speed of sound in water is approximately 1500m/s, and its wavelength is thus given by $\lambda = 1500/f$. The diameter of the tube at its radial mode resonance frequency is therefore about 0.6λ. Over the flat frequency response region, well below resonance, the element is thus reasonably small compared with a wavelength in water. At the resonance frequency for the length mode, the length is almost equal to the wavelength in water, and significant directionality in the response pattern may therefore arise, even for appreciably lower frequencies. For an omni-directional probe hydrophone, it is therefore advisable to ensure that the length mode resonance is further above the operating band than the radial resonance. This is equivalent to a condition that the maximum length should be less than 1.5 times the maximum diameter of the tube.

In practice, some coupling always exists between these modes. For example, an externally applied pressure causes all of the dimensions to decrease. The piezoelectric output is then the sum of the voltages generated by the combination of radial and tangential stresses in the ceramic, involving both g_{33} and g_{31} coefficients. Since these have opposite signs, some cancellation occurs between the voltage generated by the stresses normal to the poling direction and that due to the change in wall thickness. The extent of this cancellation depends on the relative magnitudes of the stresses, which are functions of the frequency, the dimensions of the tube, and the basic design configuration. The coupling between the modes has a complicated dependency on frequency, involving the resonances of each mode, but for frequencies well below all the resonances, which is usually the region of interest, it is sufficient to consider only the low frequency response.

There are three basic design arrangements commonly considered for small cylindrical hydrophones. The fundamental design concept is that the acoustic pressure is applied only to the external surfaces, and that the inside diameter of the cylinder is shielded from the pressure. The method of achieving this shielding gives rise to three principal arrangements, which are:–

a) **End-capped**. In this case, the ends are sealed with end–caps, the interior of the cylinder usually being air–filled. The pressure on the whole area of the end–caps is applied longitudinally along the cylinder in addition to the circumferential stress.

b) **Exposed-ends**. This arrangement assumes that the inner surfaces are shielded, for example by means of cellular rubber, but recognises the difficulty of applying the shielding to the narrow ends of the cylinder walls. In this case, only the pressure on the ends of the cylinder wall itself is applied longitudinally along the cylinder in addition to the circumferential stress.

c) **Shielded-ends**. It is sometimes possible to shield the ends of the cylinder, and this condition envisages such an arrangement. Since this is however a difficult design to realise, and without any marked advantages, it will not be discussed further.

Expressions for the characteristics of such designs have been given by Langevin [10.8]. These were applied to LZT designs by Germano [10.9], and expanded to cover other possible arrangements by Wilder [10.10]. For radially poled ceramic tubes, Langevin's expressions are:–

a) **End-capped.**

Pressure Sensitivity (low frequency)

$$\hat{M} = b\left(g_{31}\frac{(2+\beta)}{(1+\beta)} + g_{33}\frac{(1-\beta)}{(1+\beta)}\right) \qquad (10.62a)$$

where $\beta = a/b$. For a thin–walled tube, for which $\beta \rightarrow 1$, this gives

$$\hat{M}_0 = \frac{3}{2}bg_{31} \qquad (10.62b)$$

Capacitance (low frequency)

$$C_{LF} = \frac{2\pi\varepsilon L}{\ell n\,(1/\beta)} \qquad (10.63a)$$

$$\simeq \frac{2\pi\varepsilon La}{t} \qquad \text{(for thin-walled tube)} \qquad (10.63b)$$

where $\varepsilon = \varepsilon_{33}{}^T \varepsilon_0$.

b) **Exposed-ends.**

Pressure Sensitivity (low frequency)

$$\hat{M} = b\left(g_{31}(2-\beta) + g_{33}\frac{(1-\beta)}{(1+\beta)}\right) \qquad (10.64a)$$

which becomes, for a thin-walled tube,

$$\hat{M}_0 = bg_{31} \qquad (10.64b)$$

Capacitance (low frequency):- as for the end-capped case, (10.63).

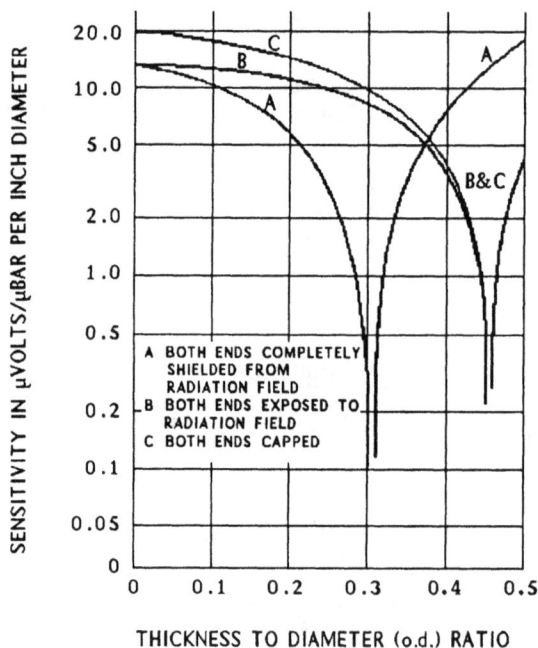

Fig 10.10 Hydrophone sensitivity of radially poled lead zirconate titanate (PZT-5) tubes; from *J Acoust Soc Am*, in *JASA*, <u>34</u>, 1139-1141, (1962).

Sensitivity curves for radially poled PZT-5 cylinders are shown in *Fig 10.10* (from [10.9]). These illustrate how the cancellation between the g_{33} and g_{31} modes may result in a very low sensitivity if the tube dimensions are badly chosen.

Although cylindrical elements are usually radially poled, with electrodes on the inner and outer surfaces, other arrangements have been described [10.10]. For example, it is possible to apply longitudinal stripe electrodes along the piezoelectric tube to divide the cylinder into an even number of curved segments. By applying a field in alternate directions between these electrodes, the ceramic may be poled tangentially around the walls. The tangential stress generated in the walls by an external pressure then produces a high sensitivity because the ceramic is being used in the 3-3 mode. The capacitance of such designs is however typically very low. Another possible configuration is to use a tube which has only the ends silvered, and hence uses longitudinal polarisation. Again, this can give a high sensitivity, associated with a low capacitance.

Comparison between types of designs.

A useful comparison between hydrophones is on the basis of their values of $\hat{M}^2 C_{LF}$. For thin-walled cylinders, radially poled, the expressions are:-

End-capped,

$$\hat{M}_0^2 C_{LF} = 2\pi g_{31}^2 \varepsilon \cdot \frac{9}{4} \frac{Lab^2}{t} \qquad (10.65a)$$

Exposed-ends,

$$\hat{M}_0^2 C_{LF} = 2\pi g_{31}^2 \varepsilon \cdot \frac{Lab^2}{t} \qquad (10.65b)$$

For comparison with a sphere, we make the resonance frequency of the sphere equal to that of the cylinder, since the value of resonance frequency is a basic design requirement. Assuming the resonance frequency of the cylinder to be that of the radial resonance, this implies (from (10.53) and (10.60)) that the radius of the sphere (a_s) is larger than that of the cylinder (a_c) by a factor

$$\frac{a_s}{a_c} = \left(\frac{2}{1-\sigma}\right)^{1/2} \tag{10.66}$$

$$\simeq 1.7 \quad (\text{for } \sigma = 0.3)$$

Substituting this into (10.54) allows the value of $\hat{M}^2 C_{LF}$ to be calculated for a spherical shell with the same resonance frequency as the cylinder. If the height of the cylinder is assumed to be equal to its diameter (ie $L = 2b$), the relative values of $\hat{M}^2 C_{LF}$ for elements with the same resonance frequencies are:-

for an end-capped cylinder;   2.25
for an exposed-end cylinder;   1.0
for a spherical shell;   4.36   (assuming $t/a=0.1$)
and for the balanced Tonpilz design; 0.08

The spherical shell thus has a significantly higher potential power output than the other designs.

Rijnja [10.5] observed that the value of $\hat{M}^2 C_{LF} f_s^3$ for hydrophones in their low frequency region commonly lay between 10^{-3} and 10^{-1}. (The square of his parameter K, capacitances being expressed in F and frequencies in Hz.) Expressed in dB, this implies that the value of $20 \log \hat{M} + 10 \log C_{LF} + 30 \log f_s$ is commonly between -30 and -10dB. For the examples above, the values are:-

Type	$\hat{M}^2 C_{LF} f_S^3$	$M + 10\log C_{LF} + 30\log f_S$
end-capped cylinder	5.75×10^{-2}	-12.4 (dB)
exposed-end cylinder	2.56×10^{-2}	-16
spherical shell	11.2×10^{-2}	-9.5
balanced Tonpilz	0.2×10^{-2}	-27

This relationship gives a useful means of predicting the main parameters of a hydrophone.

10-9   ACCELERATION SENSITIVITY

An acoustic pressure applied to a hydrophone produces a stress in the piezoelectric element, which generates the output voltage. However, any vibration of the hydrophone mounting will also produce a stress in the ceramic, in order to accelerate the end-masses in sympathy with the case vibration. The piezoelectric element of any transducer is sensitive to all the stresses in the ceramic, regardless of how they are caused, and the voltages resulting from these acceleration effects are therefore added to the acoustically generated voltages. They thus represent another source of unwanted noise in the hydrophone, and in some applications it is necessary to reduce this acceleration response, just as the other sources of noise need to be controlled.

One basic method of reducing acceleration response may lie outside the hydrophone itself, by reducing the vibration transmitted through the supporting structure to the actual hydrophone mounting. The importance of this should not be overlooked, and is another example of the need for adequate discussion between transducer designer and user. The mounting arrangement is particularly significant for spherical or cylindrical elements, since there is no immediately obvious satisfactory way of positioning them where they are wanted in an acoustic field. If they are rigidly attached to a supporting bracket, it is likely that vibration of the bracket will generate an acceleration response in the hydrophone, whilst if the mounting is very flexible it may be difficult to position the hydrophone accurately at the required point in the field. Vibration of the bracket may be caused by vibration transmitted directly through the structure, or by the action of the acoustic field itself. In the first case, the acceleration response generates an unwanted contribution to the background noise. In the second case, the acceleration sensitivity of the hydrophone may generate a response to the velocity components of the field and thus make the total output directional, instead of the ideal omni-directional pressure response. Some care is therefore needed to ensure that the hydrophone is mounted flexibly enough to avoid troublesome signals due to the mounting. The problem is best tackled by using engineering common sense backed up by adequate testing.

For piston-type hydrophones, it was explained in Section 10.4 how the acceleration sensitivity of the hydrophone itself could be reduced by adding a symmetrical rear half to the stack. This rear half is shielded from the acoustic field, but is subject to the same acceleration forces as the front half. However, when the front half of the stack is compressed by the acceleration, the rear is in tension, and by suitable electrical connection of the two halves, the acceleration voltages may be made to oppose each other and thus reduce the total acceleration

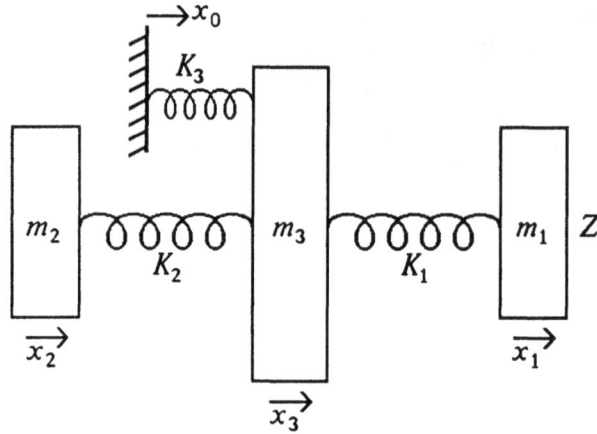

Fig 10.11 Mass-spring model of hydrophone with mounting.

response. Such designs are often described as "acceleration cancelling" designs, although in practice the acceleration signals rarely cancel each other exactly. Because the rear half is protected from the acoustic field, it does not generate any acoustic response at low frequencies; its capacitance does however reduce the output signal from the front half, and this is the price which is paid for the reduction in acceleration response.

In this section, the equations describing the response of a piston-type hydrophone to vibration of the casing are derived. This is done partly as an example of how the equations of motion can be applied to Tonpilz transducers, and the mathematical details may be omitted if only the results are of interest.

Acceleration response.

The hydrophone is represented by the system shown in *Fig 10.11*, which assumes a "lumped-mass" approximation. The mass m_1 represents the front piston, m_2 the rear mass, K_1 and K_2 the stiffnesses of the front and rear parts of the stack, m_3 any mass at the mounting position in the stack, and K_3 the stiffness of the mounting support. The radiation impedance of the load on the piston in its operating condition is denoted by Z, and any other losses are neglected. It is assumed that applied external forces cause a vibration (x_0) of the casing along the axis of the hydrophone, at an angular frequency ω. The consequent displacements of the masses are denoted by x_1, x_2, x_3, as indicated, and the resulting forces in the springs by F_1, F_2, and F_3 (where a positive value indicates a compressive force).

The basic equations of motion are then

For m_1, $\qquad m_1\ddot{x}_1 = F_1 - Z\dot{x}_1$ $\qquad\qquad$ (10.67a)

For m_2, $\qquad m_2\ddot{x}_2 = -F_2$ $\qquad\qquad\qquad$ (10.67b)

For m_3, $\qquad m_3\ddot{x}_3 = F_3 + F_2 + F_1$ $\qquad$ (10.67c)

where $\dot{x}$ and $\ddot{x}$ denote the first and second differentials of x with respect to time. The forces are given by

$$F_1 = K_1(x_3 - x_1) \qquad\qquad (10.68a)$$

$$F_2 = K_2(x_2 - x_3) \qquad\qquad (10.68b)$$

$$F_3 = K_3(x_0 - x_3) \qquad\qquad (10.68c)$$

Assuming the displacements to be sinusoidal functions of time, the relationships (1.5) between displacement, velocity, and acceleration may be used, to give

$$\frac{x_3}{x_2} = 1 - \frac{\omega^2 m_2}{K_2}$$

$$= 1 - n_2^2 \qquad\qquad (10.69a)$$

$$\frac{x_3}{x_1} = 1 - \frac{\omega^2 m_1}{K_1} + j\frac{\omega Z}{K_1}$$

$$= 1 - n_1^2 + jZ_1 \qquad\qquad (10.69b)$$

$$\frac{x_3}{x_0} = \frac{K_3}{K_1 + K_2 + K_3 - n_3^2 K_3 - \dfrac{K_1}{1 - n_1^2 + jZ_1} - \dfrac{K_2}{1 - n_2^2}} \qquad (10.69c)$$

where,

$$\omega_2^2 = K_2/m_2, \quad \text{and} \quad n_2 = \omega/\omega_2 \qquad (10.70a)$$

$$\omega_1^2 = K_1/m_1, \quad \text{and} \quad n_1 = \omega/\omega_1 \qquad (10.70b)$$

$$\omega_3^2 = K_3/m_3, \quad \text{and} \quad n_3 = \omega/\omega_3 \qquad (10.70c)$$

$$Z_1 = \omega Z/K_1 \qquad (10.70d)$$

The compression of the front part of the stack is given by

$$\frac{x_3-x_1}{x_3} = 1 - \frac{1}{1-n_1^2+jZ_1} \qquad (10.71a)$$

and that of the rear part by

$$\frac{x_2-x_3}{x_3} = \frac{n_2^2}{1-n_2^2} \qquad (10.71b)$$

The compression of the whole stack is thus given by

$$\frac{x_2-x_1}{x_3} = \frac{(1-n_1^2)(n_2^2-n_1^2)+Z_1^2+jZ_1(1-n_2^2)}{(1-n_2^2)\{(1-n_1^2)^2+Z_1^2\}} \qquad (10.71c)$$

These are all related to x_3; they may be related to the driving excitation x_0 by using (10.69c), which may be re-written as

$$\frac{x_3}{x_0} = \frac{(1-n_2^2)}{A^2+Z_1^2B^2} \left[\{(1-n_1^2)A-Z_1^2B\} \right.$$
$$\left. + jZ_1\{A+(1-n_1^2)B\} \right] \qquad (10.72)$$

where $A = (1-n_1^2)(1-n_2^2)(1-n_3^2) - n_1^2(1-n_2^2)K_{13}$
$$- n_2^2(1-n_1^2)K_{23} \qquad (10.73a)$$

$$B = (1-n_2^2)K_{13} - n_2^2K_{23} + (1-n_2^2)(1-n_3^2) \qquad (10.73b)$$

$$K_{13} = K_1/K_3, \quad \text{and} \quad K_{23} = K_2/K_3 \qquad (10.73c)$$

The compression of the front part is then given by

$$\frac{x_3-x_1}{x_0} = \frac{(1-n_2^2)}{A^2+Z_1^2B^2} \left[(-n_1^2A-Z_1^2B) + jZ_1(A-n_1^2B) \right] \qquad (10.74a)$$

and of the rear part by

$$\frac{x_2-x_3}{x_0} = \frac{n_2^2}{A^2+Z_1^2 B^2} \left[\{(1-n_1^2)A-Z_1^2 B\}\right.$$
$$\left. + jZ_1\{A+(1-n_1^2)B\}\right] \qquad (10.74b)$$

The springs K_1 and K_2 represent the front and rear parts of the piezoelectric stack, including appropriate allowances for the effects of joints and spacers as well as the true ceramic stiffness. If the front and rear parts are connected electrically in parallel, the resultant output voltage is calculated by summing the piezoelectric charges generated in the two parts and dividing by the total capacitance of the stack. The charge generated in each part may be calculated using the relationship (from (3.4))

$$\frac{Charge}{Force} = d$$

where the piezoelectric coefficient is d_{33} when the stack is built of rings poled along the length of the stack. Thus, the charge generated in the front part of the stack is proportional to $K_1(x_3-x_1)$, and for the rear part to $K_2(x_2-x_3)$. The total output voltage (V_a) due to the acceleration of the casing is thus given by

$$V_a = \frac{d_{33}}{C} \{K_1(x_3-x_1) + K_2(x_2-x_3)\} \qquad (10.75a)$$

where C is the total capacitance of the stack.

The behaviour of the resultant output voltage thus depends on the factor in brackets, which can be related to the compressions of the front and rear parts by defining a factor

$$\frac{\Delta t}{x_0} = \left(\frac{x_3-x_1}{x_0}\right) + \left(\frac{K_2}{K_1}\right)\left(\frac{x_2-x_3}{x_0}\right)$$

$$= \frac{\Delta f}{x_0} + \frac{K_2}{K_1}\frac{\Delta r}{x_0} \qquad (10.75b)$$

where $\Delta f = x_3-x_1$ = compression of front part
and $\Delta r = x_2-x_3$ = compression of rear part.

Then, $$V_a = \frac{d_{33}}{C} x_0 K_1 \left(\frac{\Delta t}{x_0}\right) \qquad (10.75c)$$

Element in air.

The behaviour of the element in air may be calculated approximately by allowing Z to tend to zero. Then, (10.74a and b) become

$$\frac{x_3-x_1}{x_0} = \frac{(1-n_2^2)}{A^2}\{-n_1^2 A\} = (1-n_2^2)\left(\frac{-n_1^2}{A}\right) \quad (10.76a)$$

$$\frac{x_2-x_3}{x_0} = \frac{n_2^2}{A^2}\{(1-n_1^2)A\} = (1-n_1^2)\left(\frac{n_2^2}{A}\right) \quad (10.76b)$$

and the total output voltage is given by

$$\frac{V}{x_0} = \frac{d_{33}}{AC}\{-K_1 n_1^2(1-n_2^2) + K_2 n_2^2(1-n_1^2)\} \quad (10.77a)$$

Since the magnitude of the acceleration is related to the displacement by $\ddot{x}_0 = \omega^2 x_0$, the voltage related to the acceleration of the housing in air is given by

$$\frac{V}{\ddot{x}_0} = \frac{d_{33}}{\omega^2 AC}\{-K_1 n_1^2(1-n_2^2) + K_2 n_2^2(1-n_1^2)\} \quad (10.77b)$$

The output due to the case accelerations can thus be made to vanish by making the factor in square brackets equal to zero – ie by making $K_1 = K_2$ and $\omega_1 = \omega_2$. This is realised by using a **balanced element**, in which the two halves are symmetrical, and this can in principle produce perfect cancellation of the acceleration response over most frequencies. In practice, the cancellation is never quite perfect, and the response may then become large at frequencies where the factor A in the denominator becomes zero. For example, for a balanced element, the expression for A (10.73a) can be simplified to

$$A = (1-n_1^2)\{(1-n_1^2)(1-n_3^2) - 2n_1^2 K_{13}\} \quad (10.78)$$

Thus, the frequencies for which $A = 0$ occur when $n_1 = 1$ (ie the resonance frequency of the element) and when

$$(1-n_1^2)(1-n_3^2) - 2n_1^2 K_{13} = 0$$

One solution to this equation is given approximately by

$$n_1^2 = \frac{1}{1 + N_{31}^2 + 2K_{13}} \tag{10.79}$$

where $N_{31} = n_3/n_1 = \omega_1/\omega_3$, and subject to the condition that $1 + N_{31}{}^2 + 2K_{13} \gg 2N_{31}$, which holds unless the supporting stiffness (K_3) is much greater than the stack stiffness (K_1). This value of n_1 is very close to that corresponding to the frequency at which the total mass $(m_1 + m_2 + m_3)$ resonates on the mounting spring (K_3), which is given by

$$n_r^2 = \frac{1}{N_{31}^2 + 2K_{13}}$$

The acceleration response of the element in air is thus expected to show peaks at the basic element resonance and near the resonance of the element on its mounting. This result, although not particularly surprising, emphasises the significance of controlling the mounting resonance, and this can be an important constraint on the design for a broad band hydrophone. For example, the usual condition would be to make the mounting resonance below the operating band. If this is intended to extend down to low frequencies, it implies a mounting of low stiffness, and this may introduce serious design conflicts if the hydrophone is also to operate at high pressures.

Imperfect cancellation.

The ideal cancellation envisaged above is spoilt by any deviations from perfect balancing. The first of these effects to be considered is that due to having one end of the stack radiating into water, in which case the value of the radiation impedance Z must be taken into account. This is most conveniently done by relating Z_1 to the mechanical Q-factor (Q_M). Assuming that the radiation impedance is entirely resistive, and noting that for a balanced element (from (7.3)),

$$Q_M = \omega_1 \frac{2m_1}{r}$$

then,

$$Z_1 = \frac{\omega Z}{K_1} \tag{10.70d}$$

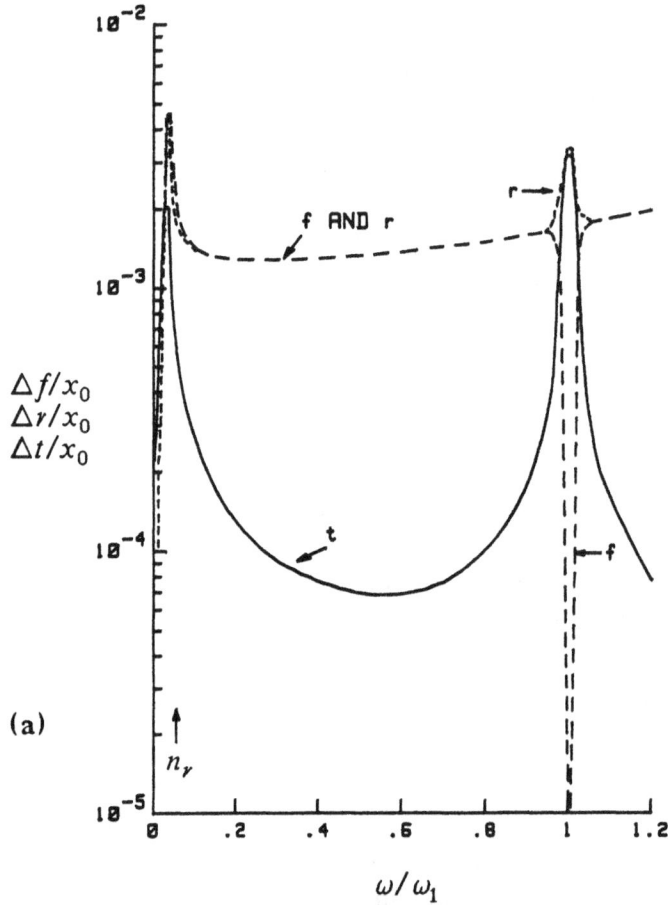

Fig 10.12 Acceleration response for nominally balanced elements, showing effect of radiation damping on

$$= \frac{\omega \omega_1 2 m_1}{K_1 Q_M}$$

$$= 2n_1/Q_M \qquad\qquad (10.80)$$

(since $\omega_1{}^2 = K_1/m_1$). This value may be substituted into (10.74) to calculate the resultant displacements, with results

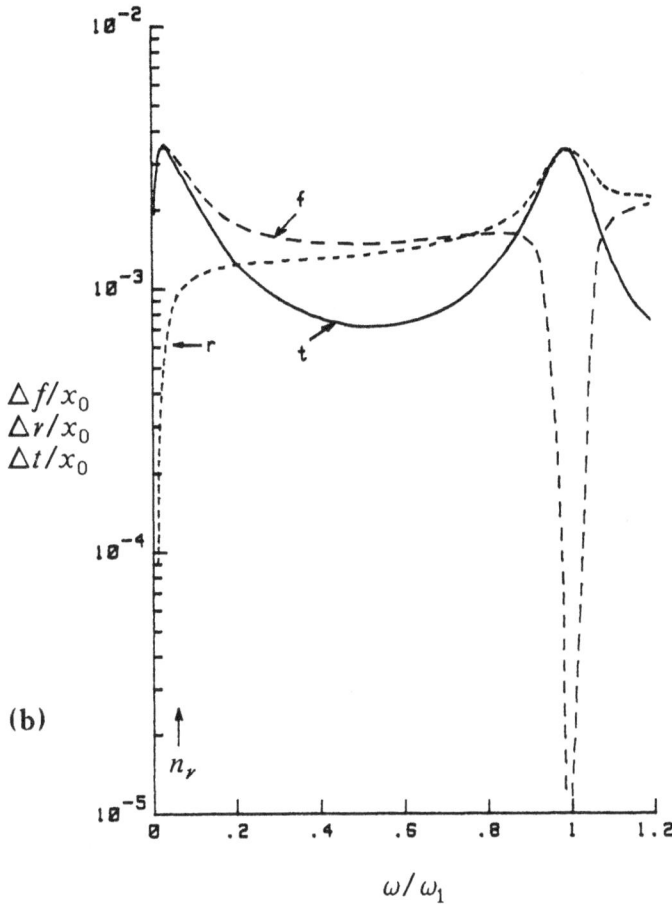

cancellation effect (from equations (10.74)): ($K_1 = K_2 = 300K_3$, $\omega_1 = \omega_2 = 15\omega_3$). (a) $Q_M = 100$; (b) $Q_M = 10$.

illustrated in *Fig 10.12a* for a balanced element with low radiation damping ($Q_M = 100$). This figure shows how the responses from the front and rear halves oppose each other, to give a total ouput voltage which is reduced by 20dB (compared with the front half alone) over a wide band between the resonances. For a hydrophone in water, when the radiation damping is generally considerably higher, the degree of cancellation is poorer, as is illustrated by the curves in *Fig 10.12b* for an element with $Q_M = 10$.

Imperfect cancellation may also arise from mechanical unbalance of the components of the stack, and these effects may be calculated from equations (10.74) and (10.75) by substituting appropriate values of K_1, K_2, ω_1, and ω_2. The results of such calculations show that a 10% error in resonance frequency between the two halves, corresponding approximately to a 20% error in mass or stiffness, reduces the cancellation appreciably for a high Q_M element, but produces only a relatively small degradation for an element with $Q_M = 10$. The effects of mechanical unbalance are therefore fairly easy to measure in air, and such measurements may be used to confirm that moderate mechanical unbalance should not cause any significant increase in acceleration sensitivity for elements which have a low Q_M in water.

Pressure sensitivity.

The absolute acceleration sensitivity of a hydrophone is not in itself important, since a high acceleration response may be acceptable if the pressure response is also high. In this section, expressions for the pressure response of a hydrophone are derived, for comparison with the acceleration sensitivity curves. We assume that the hydrophone can still be represented by the three mass and three spring system shown in *Fig 10.11*, but in this case the displacement (x_0) of the housing is zero, and the effect of the incident acoustic pressure is represented by a force F_0 acting normally on the front piston. The equations of motion for sinusoidal displacements may then be written:-

For m_1, $\qquad (K_1 - m_1\omega^2)x_1 = K_1x_3 - F_0$ $\qquad\qquad$ (10.81a)

For m_2, $\qquad (K_2 - m_2\omega^2)x_2 = K_2x_3$ $\qquad\qquad\qquad$ (10.81b)

For m_3, $\qquad (K_1 + K_2 + K_3 - m_3\omega^2)x_3 = K_2x_2 + K_1x_1$ $\qquad$ (10.81c)

Solution of these equations as before gives the results:-

For the front spring, the compression is

$$\frac{x_3 - x_1}{F_0} = \frac{1}{K_1 A}\left\{(1-n_2^2)(1-n_3^2) - K_{23}n_2^2\right\} \qquad (10.82a)$$

where A is as defined in (10.73a). For the rear spring, the compression is

$$\frac{x_2 - x_3}{F_0} = \frac{-n_2^2}{K_3 A} \qquad (10.82b)$$

The charge generated in the front part is proportional to the force (F_{pf}) in K_1, ie to

$$\frac{F_{pf}}{F_0} = \frac{K_1(x_3 - x_1)}{F_0}$$

$$= \frac{1}{A}\left\{(1 - n_2^2)(1 - n_3^2) - K_{23}n_2^2\right\} \qquad (10.83a)$$

Similarly, the charge in the rear part is proportional to the force (F_{pr}) in K_2, ie to

$$\frac{F_{pr}}{F_0} = \frac{-K_{23}n_2^2}{A} \qquad (10.83b)$$

The resultant output voltage is given by $d_{33}(F_{pr} + F_{pf})/C$ (as in (10.75)), and is thus proportional to

$$\frac{F_p}{F_0} = \frac{F_{pf} + F_{pr}}{F_0}$$

$$= \frac{1}{A}\left\{(1 - n_2^2)(1 - n_3^2) - 2K_{23}n_2^2\right\} \qquad (10.83c)$$

At low frequencies, when n_1, n_2, n_3 tend to zero, the force in the front spring $F_{pf} \rightarrow F_0$, and that in the rear spring $F_{pr} \rightarrow 0$, as would be expected. If the frequency is increased, some force is transmitted to the rear part, so that it makes a significant contribution to the total charge. For a balanced element, for which $K_1 = K_2$ and $\omega_1 = \omega_3$, (10.83c) simplifies to

$$\frac{F_p}{F_0} = \frac{1}{1 - n_1^2} \qquad (10.84)$$

This represents the simple behaviour of a mass-spring system, and is similar in shape to the curve given by (10.14) for an element with an infinite Q_M; in this ideal case, the stiffness of the mounting spring has no effect. For a more realistic case, where the balance is not perfect, F_p/F_0 becomes large when $A \to 0$. Thus, as for the acceleration sensitivity, the pressure response shows a peak approximately at the resonance frequency of the element on its mounting spring (10.79). More detailed calculations show that this is not a simple peak, but is a deviation on each side of the ideal curve. This behaviour again emphasises the importance of ensuring that the mounting resonance lies outside the operating band. There is a converse effect, in that if a hydrophone exhibits unexpected irregularities in its pressure response curve, it is a good indication that it may have some unpredicted mechanical resonances.

Relationship between acceleration and pressure sensitivities.

In order to compare the acceleration and pressure sensitivities, it is usual to divide the former by the latter − or in logarithmic terms, to subtract the pressure sensitivity from the acceleration sensitivity, both in dB, making sure that the units are consistent. This in effect expresses the acceleration sensitivity in terms of the pressure which would produce an equal output voltage. The absolute pressure sensitivity can be determined by expressing the force on the piston as $F_0 = A_p p$, where p is the acoustic pressure and A_p the area of the piston, and inserting this in (10.83c). Using the same method as for (10.75), this then gives the ratio of output voltage to applied acoustic pressure as

$$\frac{V}{p} = \frac{d_{33}A_p}{C} \frac{F_p}{F_0}$$

$$= \frac{d_{33}A_p}{AC} \left\{ (1-n_2^2)(1-n_3^2) - 2K_{23}n_2^2 \right\} \qquad (10.85)$$

The acceleration sensitivity in terms of its equivalent acoustic pressure ($\hat{M}_a$) is then given by the ratio of the acceleration sensitivity (from (10.77b)), to the pressure sensitivity (from (10.85)). For an element in air, the resulting expression is

$$\hat{M}_a = \frac{1}{\omega^2 A_p} \frac{\{-K_1 n_1^2(1-n_2^2) + K_2 n_2^2(1-n_1^2)\}}{\{(1-n_2^2)(1-n_3^2) - 2K_{23}n_2^2\}} \qquad (10.86)$$

In evaluating the suitability of a hydrophone, its acceleration sensitivity is normally measured in air, since as noted above this provides a convenient and sensitive test of its characteristics. Values of $\hat{M}_a$ for nominally balanced designs at low frequencies are typically in the range 1–10 Pa.s^2/m, a higher value indicating poorer acceleration cancellation. In logarithmic terms, this is equivalent to 120–140dB re 1μPa.s^2/m, or 140–160dB re 1μPa/g. With care, even better cancellation can be achieved.

It is possible to estimate the acceleration sensitivity which can be tolerated for a small hydrophone fitted in a large baffle. If the baffle is subject to vibrational forces, it will generate an acoustic pressure which will in turn produce an output voltage from the hydrophone. The required condition for the acceleration sensitivity is that any voltage due to the acceleration should be less than the output due to the *acoustic* field produced by the vibration of the baffle, since that will constitute the minimum acoustic background level. If the acceleration of the baffle normal to its face is denoted by $\ddot{x}_0$, then its velocity is $\dot{x}_0 = \ddot{x}_0/\omega$, and the pressure in front of the baffle is

$$p_0 = \rho c \ddot{x}_0 / \omega \qquad (10.87)$$

The acceleration sensitivity of the hydrophone casing will produce directly a voltage which is equivalent to a pressure given by $\ddot{x}_0 \hat{M}_a$, and the condition that the acceleration output shall be less than the acoustic signal is thus that

$$\ddot{x}_0 \hat{M}_a < \rho c \ddot{x}_0 / \omega$$

ie,
$$\hat{M}_a < \rho c / \omega \qquad (10.88)$$

For example, at 1kHz, $\hat{M}_a$ should be less than about 200 Pa.s^2/m to avoid corruption of the pressure signals, but there is little point in struggling to make $\hat{M}_a$ much less than (say) one tenth of this value.

For other types of hydrophones, there is no such simple indication of the maximum tolerable acceleration sensitivity, and in those cases it may be necessary to measure the acceleration levels of the mountings and hence deduce the required level of acceleration reduction. The general rule for minimising acceleration sensitivity is to ensure that the hydrophone element is as symmetrical as possible, since this tends to give a cancellation effect. Thus, a spherical hydrophone, or a cylindrical hydrophone mounted at its mid-point, should have a low acceleration response. In practice, however, deviations from the ideal configuration are caused by factors such as the need to take electrical leads into the sphere or cylinder, or inhomogeneities in the ceramic, and these produce an acceleration response which is virtually impossible to predict reliably.

In addition to the output due to accelerations along the axis of a piston-type design, which has been considered above, the response of the hydrophone to accelerations in other planes should also be measured, when the requirements justify it. For ideally balanced designs, the predicted response to such transverse vibrations is generally very small, because of the cancellation between parts of the element, and the actual measured output will depend on the uncontrolled deviations from ideal behaviour. The accelerations sensitivities of spheres and cylinders in all three orthogonal planes are similarly difficult to predict, and are best investigated experimentally.

It is worth noting that the equations used for analysing the acceleration sensitivity of a hydrophone could also be used as the basis for designing an accelerometer. In contrast to a hydrophone, where the aim is to minimise the acceleration sensitivity, the aim for an accelerometer is to achieve a good acceleration response whilst minimising any sensitivity to acoustic pressure. In practice, accelerometers often use the piezoelectric ceramic in a shear or bending mode, for which the design equations would differ in detail, although the fundamentals remain similar. The design of accelerometers is however a special subject of its own, and will not be discussed here.

Other mechanical sources of noise.

Several other sources of self-noise should be mentioned briefly.

a) If a hydrophone is suspended by a cable past which water is flowing, the vortices produced by the flow may generate **strumming** of the cable. The frequency of the vortex shedding is given by $f = Sv/d$, where v is the speed of the water past the cable, d is the diameter of the cable, and S is the dimensionless "Strouhal number", which has the value 0.18 over most of the relevant cable sizes and water speeds [10.11]. This effect may become particularly troublesome if the suspension system has a resonance frequency which coincides with the strumming frequency. The induced vibrations of the cable can produce an output from the hydrophone either through its acceleration response, or by an acoustic path. Strumming is generally reduced by using a faired cable, by introducing some mechanical isolation between the hydrophone and the cable, and by ensuring that the acceleration sensitivity of the hydrophone is low. Strumming may occur for hydrophones buoyed up from the sea bed, as well as for those suspended from the surface.

b) Motion of the cable tends to generate **triboelectric noise** (or **"microphony"**), which results from friction between the conductor and its screen when the cable is bent. The presence of this source of noise can sometimes be difficult to identify, although it can readily be observed by bending the cable in air. It can be reduced by paying attention to adequate clamping of the cable, and by using low-noise cables in which these triboelectric effects are minimised.

c) Water flowing past the hydrophone itself can generate turbulent fluctuations in pressure over the sensitive face of the hydrophone, and these produce an electrical noise output which is often described as **"flow noise."** This source of noise is observed especially in low frequency measurements. It can be tackled by increasing the area of the hydrophone to give some averaging of the fluctuations, and by using some form of dome or screen to separate the flow from the sensitive face of the hydrophone [10.12].

10-10 VELOCITY/PRESSURE GRADIENT HYDROPHONES

It has been remarked in the previous section that, if an omni-directional hydrophone response is wanted, it is important to avoid any sensitivity to acoustic velocity. For many applications, however, it would be very desirable to have a hydrophone which is both small and directional, and for such cases it is natural to consider a sensor which responds specifically to acoustic particle velocity. This is because the particle velocity in an acoustic wave is a vector quantity, having a maximum in the direction of propagation (and in the reverse direction), and with zero amplitude perpendicular to the propagation direction. This is in contrast with the acoustic pressure itself, which is a non-directional scalar. A hydrophone which is sensitive to acoustic velocity should in fact have a sinusoidal (or "figure of eight") directional response in a plane wave, with its axis along the direction of propagation. Unfortunately, despite the desirability for some applications of achieving directionality with a small hydrophone, their potential advantages are accompanied by some difficulties and disadvantages, and their main features will be discussed briefly in this section.

In practice, it is difficult to make a sensor which responds directly to acoustic particle velocity, and it is more common to make use of the fact that the velocity and pressure gradient are directly related. A measurement of acoustic pressure gradient should thus show the same directional characteristics as the velocity. It is simplest to envisage a pressure gradient sensor in which two small pressure probes are used to measure the pressure at two close points in the field; the gradient is then determined by subtracting their outputs. If a plane sound wave is incident at right angles to the line joining the pair of hydrophones, the signals at the two hydrophones are equal in both amplitude and phase, so that the difference between them is zero, assuming ideal behaviour. If a plane wave is incident along the line joining the pair, the phase difference between the signals at the two hydrophones is finite, and the output of the pair will therefore be finite, with a magnitude which depends on the difference in phase.

More precisely, if the hydrophone pair are assumed to be point receivers of equal (unit) sensitivity, with separation d, the directivity function can be calculated by substituting $n = 2$ and $\varphi = \pi$ in (6.28b), giving

$$F(\theta) = 2\sin\{(\pi d/\lambda)\sin\theta\} \qquad (10.89a)$$

If the sensitivity of each hydrophone is denoted by $\hat{M}$ the sensitivity of the doublet is thus

$$\hat{M}_D = 2\hat{M}\sin\{(\pi d/\lambda)\sin\theta\} \qquad (10.89b)$$

which, for $\pi d/\lambda \ll 1$, can be approximated by

$$\hat{M}_D \simeq (2\hat{M}\pi d/\lambda)\sin\theta \qquad (10.89c)$$

The $\sin\theta$ factor represents the figure–of–eight directional pattern, as in *Fig 10.13*, with maxima along the line joining the elements; this pattern is independent of frequency for these small values of d/λ, whilst the factor in brackets shows that the maximum output is proportional to d/λ. In order to maximise the output, it is therefore clearly advantageous to use as large a value of d/λ as is possible, within the limit assumed. If the separation approaches $\lambda/4$, the directional pattern deviates

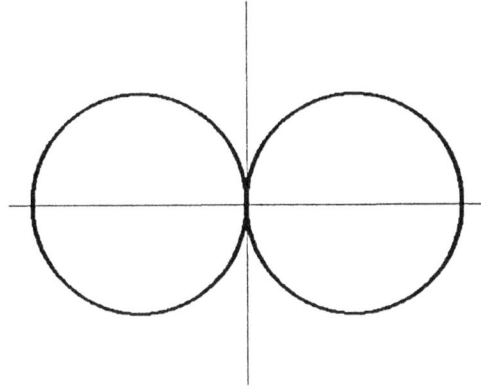

Fig 10.13 Dipole (figure-of-eight) directivity pattern.

significantly from the ideal figure-of-eight. If the response curve is to be independent of frequency, it is therefore desirable to restrict the separation to below about $\lambda/6$ at the highest frequency, and for a given separation, this sets the upper frequency limit. As the frequency is reduced, the sensitivity falls off linearly with frequency, and this limits the useful operating band to those frequencies where the output signal exceeds the self noise of the hydrophones. For large incident signals, the effective low frequency limit is thus lower than for small signals.

The inherent difficulties of these pressure gradient hydrophones are evident from this description of the basic mechanism. Firstly, the probes must be small, and hence are limited in output. Secondly, and more importantly, the subtraction of the signals leaves only a small difference, and the sensitivity is therefore much lower than for a conventional pressure hydrophone. Further, it is essential that the sensitivities of the two probes should be well matched if an accurate result of the subtraction is to be obtained. The effects of imperfect matching may be evaluated by assuming the two receivers to have sensitivities $\hat{M}_1$ and $\hat{M}_2$, and calculating the total output from the pair after subtraction of the signals at each hydrophone. If the average hydrophone sensitivity is denoted by

$$\hat{M}_0 = (\hat{M}_1 + \hat{M}_2)/2$$

and the difference between the two by

$$\Delta \hat{M} = \hat{M}_2 - \hat{M}_1$$

then the resultant output of the doublet is given by

$$\hat{M}_{D1} = \hat{M}_0 \left\{ \left(\frac{\Delta \hat{M}}{\hat{M}_0} \right)^2 \cos^2 \left(\frac{\pi d}{\lambda} \sin\theta \right) \right.$$

$$\left. + 4\sin^2 \left(\frac{\pi d}{\lambda} \sin\theta \right) \right\}^{1/2} \qquad (10.90a)$$

For small values of $(\pi d / \lambda)$, this approximates to

$$\hat{M}_{D1} \simeq \hat{M}_0 \left\{ \left(\frac{\Delta \hat{M}}{\hat{M}_0} \right)^2 + \left(\frac{2\pi d}{\lambda} \sin\theta \right)^2 \right\}^{1/2} \qquad (10.90b)$$

If the probes are exactly matched (ie $\Delta \hat{M} = 0$), this becomes the same as (10.89c), with a figure-of-eight response. As the frequency is reduced, so that d/λ becomes smaller, the amplitude of the sinusoidal component decreases. If the probes are not precisely matched, the $(\Delta \hat{M}/\hat{M}_0)$ term produces a residual pressure signal which becomes increasingly significant as the frequency reduces. When this happens, the directivity pattern reverts to a more nearly omni-directional shape. It is easy to estimate the tolerance limits for this effect. Suppose that the hydrophone pair is separated by 150mm, and is to be used at 100Hz, for which $\lambda = 15$m. Then the maximum output from the pair is smaller than the sensitivity of the individual hydrophones by a factor $2\pi d/\lambda = \pi/50 \simeq 0.06$. If the sensitivities of the two hydrophones differed by a factor of 1.06 (ie 0.5dB), the factor $(\Delta \hat{M}/\hat{M}_0)$ would be 0.06, and (10.90b) can then be used to show that the depth of the null in the pattern would be reduced to only 3dB. The difficulty of adequately matching the pairs of hydrophones thus represents a major problem for such designs.

Other designs have been developed which show a figure-of-eight response without specifically measuring the pressure at two separate points. For example, a watertight closed tube aligned along the propagation direction will have different acoustic pressures at each end, and will therefore be subjected to alternating forces acting along its axis. The resultant vibration of the tube may be measured by means of an accelerometer in the tube, thus giving an output which is proportional to the acoustic pressure gradient. If the accelerometer is sensitive only to axial accelerations, it will respond only to the acoustic field in that direction, and have zero response in the perpendicular plane. The accelerometer could be, for example, a piezoelectric ceramic bender element connected between the tube wall and an internal inertial mass, and such an arrangement would be free from the need for accurate matching of two hydrophones. It is not, however, easy to predict the sensitivity of such a system, or to ensure that its directivity curve has the desired shape.

Other designs have been developed in which the velocity field acts on a ceramic diaphragm or cantilever strip, producing a stress in the material which thus generates an electrical output voltage. Some of them have similarities with the velocity microphones which are quite commonly used in air. These designs are also generally much less sensitive to the effects of mis-match than are the direct gradient sensors which involve the measurement of the pressure at two separate points. Calculation of the output of all of these designs depends on a knowledge of the diffraction constants of the structures, and an experimental approach is usually adopted. If the element could be made to respond directly to the particle velocity, rather than to the pressure gradient, it should have a sensitivity which is independent of frequency, since the particle velocity in a plane wave is equal to the pressure multiplied by the acoustic impedance (Equation 1.8). In that case, the sensitivity would remain constant as the frequency was decreased, thus avoiding the low frequency self noise problem of the gradient designs. In practice, most of these small directional hydrophones have an output which falls as the frequency is reduced, and they are therefore more properly described as pressure gradient hydrophones than velocity hydrophones.

The figure-of-eight response curve has maxima in two directions, but the phase of the output for signals from one lobe is opposite to that for signals from the other. This can be used to generate a directivity pattern which has only one maximum. If the output from the doublet is added to the output from an omni-directional hydrophone of equal response, the resultant is a directional curve with a single maximum [10.13]. For small spacings (ie $d/\lambda \ll 1$) the directivity curve can be described by the equation

$$F(\theta) = (\pi d/\lambda)(\sin\theta - 1) \qquad (10.91)$$

which has a maximum response in one direction along the line of the hydrophone pair, when $\theta = 3\pi/2$, and a null in the opposite direction, when $\theta = \pi/2$. Such a pattern is known as

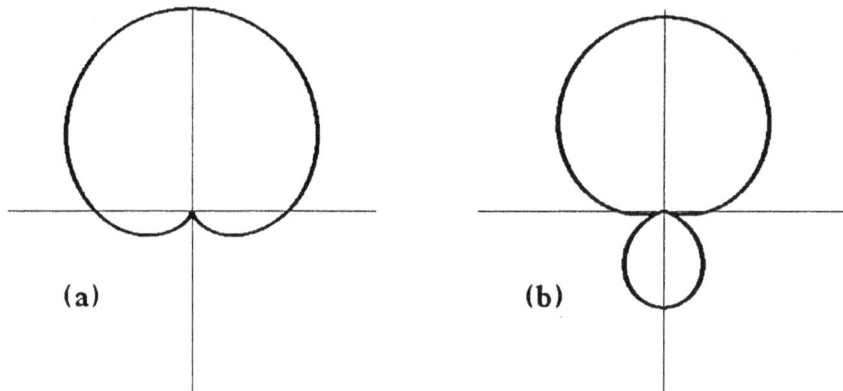

Fig 10.14 Unidirectional directivity patterns for small hydrophone pairs: (a) Cardioid; (b) Limaçon, $\delta = d/3$.

a "cardioid" pattern, as in *Fig 10.14a*, and has the great advantage for some applications of being uni-directional instead of having the ambiguity of the dipole response. It is however critically dependent on the amplitudes of the dipole and omni-directional sensors being equal; if they are not, the $(\sin\theta - 1)$ term has to be replaced by a factor $(\sin\theta - a)$, where a is a factor representing the relative sensitivities of the omni-directional and sinusoidal sensors, and in general this will not show the simple cardioid shape. The matching of the two responses is made more difficult by the fact that the omni-directional hydrophone is usually independent of frequency, whilst the dipole has a sensitivity which is approximately proportional to frequency. It is then necessary to match one to the other by means of a frequency-dependent correction network.

An alternative technique for producing a cardioid response is again to use two point hydrophones, but in this case to apply to one of them a time delay equal to the acoustic delay corresponding to the separation between the hydrophones. In one direction along the line of the pair, the output from one hydrophone plus electrical delay will exactly match that from the other hydrophone including the acoustic delay. If one output is subtracted from the other, the resultant will be zero. In the opposite direction, the delays will add, and the response will be a maximum. The response of such a system may again be calculated from (6.28b), using $n = 2$. For the cardioid response, the applied phase shift – which corresponds to the appropriate time delay – would be

$$\varphi = -(2\pi d/\lambda) + \pi$$

Then (6.28b) gives

$$\hat{M}_C = 2\hat{M} \sin\left(\frac{\pi d}{\lambda} (\sin\theta-1)\right) \qquad (10.92a)$$

which for small $\pi d/\lambda$ becomes

$$\hat{M}_{C1} \simeq \frac{2\hat{M}\pi d}{\lambda} (\sin\theta-1) \qquad (10.92b)$$

This has the cardioid pattern as in (10.91), and again shows the sensitivity falling off as the frequency is reduced.

It is sometimes desirable to vary the pattern from the simple cardioid response, and this may be done by applying a time delay which differs from that corresponding to the separation between the hydrophones. Suppose that a time delay equivalent to a separation δ is applied, instead of the true hydrophone separation d. Then equation (10.92b) becomes, for small $\pi d/\lambda$,

$$\hat{M}_{C2} \simeq \frac{2\hat{M}\pi}{\lambda} (d\sin\theta-\delta) \qquad (10.93)$$

By making δ somewhat smaller than d, two nulls may be formed in the pattern, thus providing some control of the response in the rearward direction, as illustrated in *Fig 10.14b* for the case of $\delta = d/3$. This is in fact a particular example of a more general set of curves known as "limaçons", which may be optimised to meet particular requirements by an appropriate selection of δ. The nominal directivity index of a cardioid is 4.8dB, the same as for a dipole. The maximum DI is produced by making $\delta = d/3$, as in *Fig 10.14b*, when the nominal DI = 6dB. The characteristics of these curves again depend on close matching of the individual sensors.

These small directional hydrophones have great attraction for some applications, since they overcome the usual limitations on minimum size to achieve a uni-directional response. They are in fact a simple form of superdirective array, and exhibit the difficulties of such arrays in their sensitivity to deviations from ideal conditions. Their low sensitivity means that typical values of $\hat{M}^2 C_{LF}$ are low, and that problems with electrical self noise may be expected. In addition to electrical self noise, one of the less obvious problems associated with these designs is that they tend to be particularly sensitive to self noise arising from turbulence in the water. This is because the effective length scale of the turbulent pressure variations is much less than for acoustic signals at the corresponding frequency, and the separation between the sensors is then equivalent to a larger value of d/λ for the turbulence than for the acoustic field. Its sensitivity to turbulent pressure fluctuations is therefore much higher than its acoustic sensitivity. Despite their appeal, the practical problems of these small directional hydrophones are thus considerable, and they are not yet in common use.

10-11 SUMMARY AND PRACTICAL ASPECTS

A transducer used as a projector in an active sonar system may also act as a receiver over the relatively narrow band used for the transmissions, but attention must be paid to the effects of any electrical components connected to the transducer (Sect 10.1 and 10.2). The hydrophone sensitivity of a piston–type transducer satisfying the lumped–mass approximation is deduced in Sect 10.3. The practical aspects of such designs are generally as described in Chapter 8.

In many applications, a hydrophone is required to have a uniform response over a broad frequency band, and in these cases it is necessary to avoid any resonances within the band. The limitations arising from thermal and amplifier noise are discussed in Sections 10.5 and 10.6. Design of a hydrophone for measurements down to Sea State 0 levels up to about 30kHz is relatively straightforward, but if a flat response over a broad band up to above 50kHz is wanted, more care is needed to avoid self–noise problems. The resonance frequency of the hydrophone should be made as low as is consistent with the requirement for a flat response curve, and the low frequency roll–off made low enough to avoid thermal noise problems; this may require a high input impedance for the pre–amplifier. The potential power output of the hydrophone is characterised by the value of the parameter $\hat{M}^2 C_{LF}$, which for a piston–type design is determined primarily by the piezoelectric ceramic used and by the resonance frequency (eg equation (10.50)). Trade–offs between the values of $\hat{M}$ and C_{LF} should then be made to ensure that both thermal and amplifier noise are adequately suppressed below the acoustic levels to be measured. A low self–noise pre–amplifier is generally necessary.

Similar considerations should be applied to spherical and cylindrical hydrophones (Sect 10.7 and 10.8), which are especially useful as small probes for measurements in an acoustic field. Spherical hydrophones compare well with the other designs for potential broad band performance, but have rather less flexibility in choice of sensitivity to overcome self noise.

A further source of background noise is the generation of electrical signals because of the response of a hydrophone to accelerations of its housing. The acceleration sensitivity of a piston hydrophone may be reduced by using a balanced design, and by decoupling the element from the source of vibrations (Sect 10.9). It is important to ensure that the resonance frequency of the element on its mounting lies outside the operating band. The reduction of the acceleration sensitivity of spherical or cylindrical hydrophones is more difficult to treat analytically, and is best studied experimentally. In addition to increasing background noise, sensitivity to acceleration may introduce unwanted directionality into the response of small probe hydrophones, and some care is needed in the mounting arrangements to avoid such effects. The control of mechanical resonances, and avoidance of directional response, constitute the commonest practical problems in achieving a broad band pressure sensor with a flat frequency response.

The use of spherical hydrophones to achieve an omni-directional response has an obvious appeal. In practice, it is usual to keep the outer electrodes at earth potential, and it is then necessary to take a lead through the shell to connect to the inner electrode. Considerable care is needed in making sure that this penetrator is strong enough to resist the external pressure but does not introduce too much mass, which would disturb the symmetrical mass distribution. Too rigid a mount for the hydrophone is likely to introduce mounting resonances, and the danger that the sensor will become sensitive to the acoustic particle velocity and hence be directional. These problems can be eased to some degree if a cylindrical element is used, since the connection can be made through an end cap. It is also possible to use a balanced construction, with two ceramic tubes assembled on either side of a nodal plate which can then serve as a convenient location for mounting the element.

The target levels for the acceleration sensitivity of a hydrophone depend on the acceleration levels expected for its particular application. Hydrophones in arrays which are suspended in a shear flow or towed through the water may be particularly subject to vibrations transmitted along the towing cables, and provide an example of how the problems should be tackled by a variety of means. Firstly, the vibration of the cable itself should be minimised by streamlining it to reduce its drag and any strumming effects. Secondly, the vibrations of the cable should be decoupled from the hydrophones by fitting a compliant isolating section between the cable and the array. And thirdly, the acceleration sensitivity of the hydrophones themselves should be minimised by appropriate design of their individual mountings and by using balanced elements. Hydrophone elements in a flow also produce a noise output which is caused by the pressure fluctuations due to turbulent flow over the sensitive surface. This noise contribution can be reduced by increasing the area of the hydrophone to give some averaging of the fluctuations, and by using some form of dome or screen to separate the flow surface from the sensitive face of the hydrophone.

REFS Chapter 10

References

10.1   Mellen,R.H., "The thermal noise limit in the detection of underwater acoustic signals." *J Acoust Soc Am*, <u>24</u>, 478–480, (1952).

10.2   Urick,R.J., *Principles of Underwater Sound*, McGraw–Hill, (3rd Ed), 1983, Ch 7.

10.3   Wenz,G.M., "Acoustic ambient noise in the ocean: spectra and sources." *J Acoust Soc Am*, <u>34</u>, 1936–1956, (1962).

10.4   Woollett,R.S., "Procedures for comparing hydrophone noise with minimum water noise." *J Acoust Soc Am*, <u>54</u>, 1376–1379, (1973).

10.5   Rijnja,H.A.J., "Small sensitive hydrophones." *Acustica*, <u>27</u>, 182–188, (1972).

10.6   Anan'eva,A.A., *Ceramic Acoustic Detectors*, Translated from the Russian, Consultants Bureau, New York, 1965. (Quoted in Wilson,O.B., *An Introduction to the Theory and Design of Sonar Transducers*, US Govt Printing Office, 1985, Ch 8.)

10.7   Lipscombe,L.W., "A design programme to optimise high reliability wideband hydrophones." *Proc IoA*, <u>9</u>, Pt 2, (1987), *Sonar Transducers Past, Present and Future*, 126–133.

10.8   Langevin,R.A., "The electroacoustic sensitivity of cylindrical ceramic tubes." *J Acoust Soc Am*, <u>26</u>, 421–427, (1954).

10.9   Germano,C.P., "Supplement to: The electroacoustic sensitivity of cylindrical ceramic tubes, by R.A.Langevin." *J Acoust Soc Am*, <u>34</u>, 1139–1141, (1962).

10.10 Wilder,W.D., "Electroacoustic sensitivity of ceramic cylinders." *J Acoust Soc Am*, 62, 769-771, (1977).

10.11 Urick,R.J., op. cit., Chapter 11.

10.12 Strasberg,M., "Non-acoustic noise interference in measurements of infrasonic ambient noise." *J Acoust Soc Am*, 66, 1487-1493, (1979).

10.13 Kinsler,L.E., and Frey,A.R., *Fundamentals of Acoustics*, (2nd Ed), Wiley & Sons, 1962, Sect 14.11.

Additional Reading

1   LeBlanc,C., *Handbook of Hydrophone Element Design Technology*, NUSC Technical Document 5813, Naval Underwater Systems Center, New London, Connecticut, (1978).

2   Young,J.W., "Optimisation of acoustic receiver noise performance." *J Acoust Soc Am*, 61, 1471-1476, (1977).

3   Burdic,W.S., *Underwater Acoustic System Analysis*, Prentice-Hall Inc, 1984, Section 3.2.

4   Stansfield,D., "Noise in Broad-Band Hydrophones," AUWE Tech Note 521/75, Admiralty Underwater Weapons Establishment, Portland, 1975.

5   Anan'eva,A.A., "Acoustic non-directional ceramic receiving transducers." *Sov Phys Acoust*, 2, 8-24, (1956); (Reprinted in *Benchmark Papers in Acoustics, Underwater Sound*, Albers,V.M.(Ed), Dowden, Hutchinson and Ross, 1972.)

Chapter 11
HIGHER AND LOWER FREQUENCIES

The previous chapters have dealt mainly with designs operating in the range of about 2 to 20kHz, for which the dimensions of the transducer element are generally small enough compared with a wavelength in the materials for the lumped mass approximation to be assumed with reasonable accuracy. For higher frequencies, the elements become significant fractions of a wavelength, and it is then necessary to take into account more carefully the variations of stress and displacement along the element. The mathematics of the analysis may thus become more complicated, but the design itself has less flexibility than for the lower frequencies, so that there is less opportunity to optimise the parameters. The first part of this chapter considers these designs, for frequencies up to about 500kHz.

The second part of the chapter deals with the design of projectors for operation below about 2kHz. The problems of achieving good efficiency at these low frequencies are associated with the need to generate large displacements, and this probably represents the greatest challenge at present in the field of sonar transducer design. Over the years, many approaches have been proposed, and it is an area in which current developments show promise of significant progress. A comprehensive treatment of the many designs is beyond the scope of this book, but the main features and problems of the various approaches are briefly discussed.

Piezoelectric ceramics may be used as the active material in meeting both of these extensions of the frequency range, and they therefore form the basis for most of the designs discussed in this chapter. In the next chapter, some of the newer materials and designs will be described.

For the transducers described in both this and the succeeding chapter, development has concentrated on achieving a good acoustic output from the transducer itself at the desired frequency. Less attention has generally been paid to optimising the performance of the overall system, partly because the designs have fewer parameters subject to the designer's control than do the Tonpilz designs. The descriptions of the various types of designs in these two chapters reflect this emphasis on the basic problems of satisfying the requirements for acoustic output and operating frequencies rather than those of optimising their design to their function in the complete system. Nevertheless, the methods of analysis through the use of equivalent circuits described in earlier chapters can be applied to these designs, and future developments are likely to improve techniques for optimising the overall performance.

11-1 HIGH FREQUENCY DESIGNS

We start by considering the design of transducers for operation in the range of about 200-500kHz. In this range, the design is normally based on the use of single blocks of piezoelectric ceramic to obtain the required resonance frequency, and the mathematics of such designs have been presented by various authors (eg [11.1]-[11.4]). Only a brief discussion of these treatments will be given here, and for more details the interested reader is recommended to consult one of those texts. The treatments are often applicable to piezoelectric crystals as well as to ceramics, and they have formed the basis for developments of crystal filters and resonators to much higher frequencies than are our concern here. An excellent review of both theoretical and practical aspects of high frequency sonar transducers has been given by Smith and Gazey [11.5].

Equivalent circuits for piezoelectric (and piezomagnetic) materials in various modes of vibration have been derived by Berlincourt et al [11.1]. As an example, consider a piezoelectric plate silvered on the large surfaces, so that the electric field is applied parallel to the thickness. Suppose that this plate is excited in the thickness expander mode, the lateral dimensions being sufficiently large that lateral displacements are constrained. Then the equivalent circuit can be drawn as in *Fig 11.1* [11.2], in which

$$C_0 = \frac{\varepsilon^T A}{t} \qquad\qquad (11.1a)$$

$$C = \frac{C_0}{\alpha^2} = \frac{1}{h^2 C_0} \qquad\qquad (11.1b)$$

$$Z_c = \rho_c c_c A \qquad\qquad (11.1c)$$

$$k_c = 2\pi/\lambda_c \qquad\qquad (11.1d)$$

$$\alpha = Ae/t \quad \text{is the "transformation ratio" [11.6]} \qquad (11.1e)$$

A and t are the cross-sectional area and thickness of the ceramic plate, h is its piezoelectric h-constant, related to the others by $h = g c^D$, and e is its piezoelectric e-constant, related to the others by $e = d c^E$ (See section 3.2). $\rho_c c_c$ denotes the specific acoustic impedance of the ceramic for longitudinal vibrations travelling normal to the thickness. Z_1 and Z_2 represent the loads on the end-faces of the plate.

Fig 11.1 shows the general equivalent circuit for a plate, and analysis may be started by considering a prototype section in which one face has a rigid backing, so that Z_1 becomes infinite. Such a section has resonances corresponding to the section being an odd number of quarter-wavelengths thick. The most practical way of achieving this infinite impedance at the back plane is to couple the back of the plate to a quarter-wave resonator which itself works into a zero impedance, and this is most conveniently effected in general by making the rear element another quarter-wave thickness of the piezoelectric ceramic. The overall element is thus a piezoelectric plate which is a half-wavelength thick, with the rear surface backed by air.

The equivalent circuit for such an element can be derived from *Fig 11.1* by putting $Z_1 = 0$. In the region of the fundamental resonance frequency (ω_S), when the thickness is approximately $\lambda_c/2$, the impedances may be represented by discrete components, and the simplified circuit is then as in *Fig 11.2*.

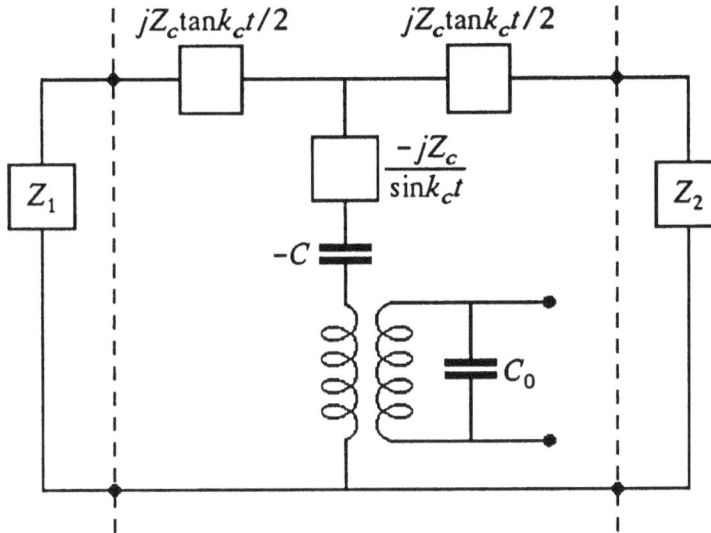

Fig 11.1   Equivalent circuit of piezoelectric plate in thickness mode (electric field parallel to thickness).

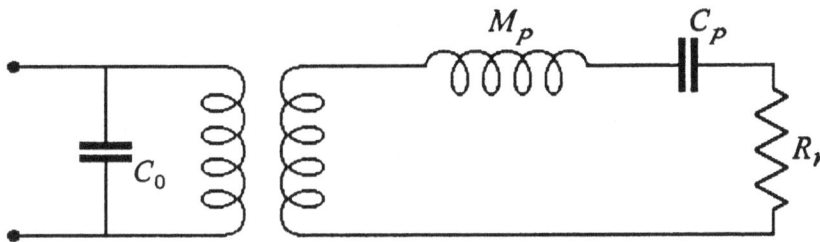

Fig 11.2   Simplified equivalent circuit of air-backed piezoelectric plate for frequencies near resonance.

The components on the right hand side of the transformer in the diagram represent mechanical quantities, the transformer itself representing the conversion from mechanical to electrical terms. Because the displacements are not uniform throughout the plate, the effective dynamic values of vibrating mass and compliance are less than their static values. The calculated values of the components in *Fig 11.2* are

$$M_p = \frac{\varrho_c A t}{2} \tag{11.2a}$$

$$C_p = \frac{2}{\pi^2} \frac{t s_c}{A} \tag{11.2b}$$

where s_c is the compliance coefficient of the ceramic. Thus, the effective dynamic mass is half the static mass, and the effective compliance is $2/\pi^2$ times the static compliance [11.4]. For a radiating surface which is at least $\lambda/2$ across, the radiation resistance is approximately

$$R_r = \varrho c A \tag{11.2c}$$

where ϱc refers to the medium into which the plate radiates (*Fig 6.1*). This simplified equivalent circuit is valid only around resonance; fuller treatments, including consideration of the circuit elements for the higher harmonics of the plate, are given by Berlincourt et al [11.1], and Merhaut [11.3]. They also derive expressions for other modes of vibration.

The main characteristics of importance, so far as the individual elements are concerned, are the resonance frequency and the bandwidth. Values of resonance frequency can be calculated simply from the manufacturers' data sheets, or from the typical values of frequency constants given in Table 3.1 for the appropriate modes. Table 3.1 shows, for example, that an LZT I plate resonant in the thickness mode at 200kHz would have a thickness of $2.08 \times 10^3/200 = 10.4$mm. For a 500kHz resonance, the plate would need to be 4.2mm thick. The motional Q-factor can be derived from the equivalent circuit:-

$$Q_M = \frac{\omega_S M_p}{R_r}$$

$$= \omega_S \frac{\varrho_c A t}{2\varrho c A}$$

$$= \frac{\pi \varrho_c c_c}{2\varrho c} \tag{11.3}$$

(using the relationship for a half-wavelength resonator that $t = \lambda/2 = \pi c_c/\omega_S$.) The value of Q_M given by (11.3) is independent of frequency. For LZT I, which has a typical value of $\rho_c c_c$ = 33.6 × $10^6 kgm^{-2}s^{-1}$ ("Rayls"), the theoretical value of Q_M in water is therefore approximately 34. (For barium titanate the corresponding theoretical value of Q_M is approximately 28.) In practice, other losses associated with the construction, such as energy dissipation in the mounting, reduce the value of Q_M below this calculated value.

The coupling coefficient for a plate vibrating in the thickness mode (k_t) is less than the value of k_{33} for the material itself, because of the lateral clamping effect of the plate and because of the stress distribution through its thickness. For example, for LZT I, k_t is approximately 0.5 (compared with k_{33} = 0.68). These values of k_t and Q_M clearly do not conform to the optimising relationship $kQ_M \simeq 1.2$ of Section 5.3. Several authors have described methods for reducing the mechanical Q-factor to achieve wider bandwidths, and a good review of the topic is given by Silk [11.7]. Two basic methods are applied. In one, the element is backed by a high impedance (and probably lossy) material such as tungsten-loaded epoxy resin, which gives a better match than does air. The second method is to use matching layers between the ceramic and the water, to increase the effective load on the ceramic. It is also possible to apply both of these techniques at the same time. These techniques are more often used with transducers operating at 1MHz or higher, for example for ultrasonic flaw detection, than at the lower frequencies which are common for sonar applications, but application of the technique as low as 5kHz has been described by Van Crombrugge and Thompson [11.8]. Although some of the methods may be advantageous at the high sonar frequencies, there is in general less freedom to optimise the characteristics of these high frequency designs than for the Tonpilz transducers discussed earlier in this book. In most cases, the design problems then lie more in the practical aspects than in any attempts to optimise the basic design.

Practical Aspects.

For ultrasonic applications involving frequencies of 1MHz or higher, it is common to use a single piece of ceramic as the radiator. At rather lower frequencies, of a few hundred kHz, this simple construction becomes less appropriate, and it is more common to use an array of ceramic elements. Suppose, for example, that a rectangular planar transducer with a beam width of 5° × 10° between −3dB points is wanted, operating at 200kHz. This requires an overall array size of approximately

$10\lambda \times 5\lambda$, and since the wavelength in water at 200kHz is 7.5mm, this implies dimensions for the radiating face of 75mm × 37.5mm. We saw above that a half-wave resonator in LZT I at 200kHz would be 10.4mm thick, and a single piece construction would thus need a piezoelectric ceramic plate with a rectangular radiating surface of 75 × 37.5mm and thickness of 10.4mm. Such a construction would however have several objections. Firstly, it is not simple to make and pole a piece of this size consistently. Secondly, and more fundamentally, such a plate would have various transverse and bending resonances with their fundamental frequencies well below the thickness resonance, leading to a strong probability of harmonics at frequencies within the operating range. And thirdly, steering of the beam could be effected only by rotating the transducer itself; electrical beam steering would not be possible.

It is therefore more practical to construct the transducer from a number of individual blocks of ceramic. In order to permit full beam steering, it is desirable to make the separation between the centres of adjacent blocks less than half a wavelength in water (Section 2.2). The block size must then be less than $\lambda/2$ across. Since the speed of sound for longitudinal vibrations in LZT ceramic is roughly twice that in water, the wavelength in LZT is also twice that in water at the same frequency. A block which has a width less than $\lambda/2$ in water must therefore be less than about $\lambda/4$ in the ceramic itself, thus ensuring that any transverse resonances in the block are well above the thickness mode resonance frequency. These blocks could be arranged into an array of the required overall size, and such a construction should avoid the objections mentioned above. However, in the example quoted, each block would now have a front face approximately 3.5mm square, and the array would need about 200 blocks, together with the associated problems of assembling them into an array and making separate connections to each block. This solution, although theoretically ideal, therefore tends to be expensive, and in practice designs between the two extremes are commonly used.

Whether electrical beam steering is required depends on the particular application, and for such relatively small arrays it is often practicable to use mechanical rotation of the array for large steering angles, with electrical steering required only for fast scanning over fairly small angles. If electrical steering over large angles is not needed, the separation between blocks may be allowed to increase towards λ. The beam patterns of arrays of finite size elements were discussed in Section 6.4, where it was noted that the major diffraction lobes which would be expected for large d/λ may be adequately suppressed if the gaps between elemehts are made sufficiently narrow. It is thus reasonable to consider increasing the size of the blocks to just less than λ across. In that case, the width of each block would still be less than $\lambda/2$ for transverse vibrations in the ceramic. Since the lowest transverse resonance occurs when the width is equal to $\lambda/2$ in the ceramic, this resonance should be satisfactorily above the operating frequencies. For the example above, the blocks then become about 7mm square, by 10.4mm thick. It is unlikely that any other resonances in the block will occur below or near the thickness mode resonance frequency, so the presence of other unwanted modes should thus be avoided. About 50 blocks would be used to form the radiating area needed, thus significantly reducing the cost and complexity compared with the "ideal" 200 element array.

It is in principle possible to increase the block size still further, particularly if no electrical steering is to be applied to the array, and such designs have been attempted by several workers, with the aim particularly of reducing the number of elements and the cost of the array. Their success depends on ensuring that all of the unwanted resonances lie outside the operating band of frequencies.

We now need to consider how these piezoelectric ceramic blocks can be assembled into an array. To illustrate this, it is easiest to envisage a transducer with a plane radiating face, using half-wavelength air-backed blocks. The fundamental requirements are that the front faces of the blocks shall lie in a plane, that they shall move uniformly when excited, that the blocks shall be air-backed, and that the whole shall be electrically insulated from the sea water. A satisfactory way of achieving these aims is to cement these blocks to the surface of a sheet of "pressure release" cellular rubber which is itself supported by a rigid backing, and to immerse the whole array in an insulating oil behind an acoustically transparent rubber window, as in *Fig 11.3*. The pressure release rubber is an air-filled cellular rubber in which there is no interconnection between the cells, so that it has a low acoustic impedance and is not flooded by water or the insulating oil. These rubbers can be obtained commercially in various thicknesses, with smooth surfaces. Attaching the ceramic blocks to this rubber thus provides a reasonably flat mounting surface, serves as the air backing to the blocks, and avoids any mechanical coupling through the mounting. Acoustic coupling between the blocks is minimised by inserting thin sheets of pressure release rubber between them, as shown in *Fig 11.3*. This rubber also helps to position the blocks correctly, but must not be too thick, or it will dissipate the acoustic pressure generated at the front face. A similar construction, but

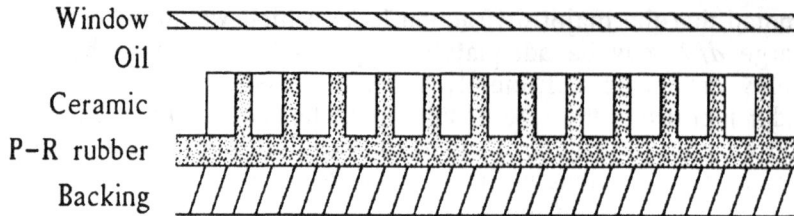

Fig 11.3   Typical piezoelectric transducer assembly for frequencies in the range 200–500kHz.

mounted on a gently curved back plate, can be used to achieve independent control of the radiating area and beamwidth; for example, a convex back plate can be used if a large area is necessary to handle the power, but the required beamwidth is wider than would be produced by a plane array of that area. (as in *Fig 2.1(b)*.)

In the example described above, it is assumed that the ceramic is used in the thickness mode, with the silvered surfaces on the front and rear faces. It is then necessary to make electrical connections to these faces, and this is generally done most reliably by soldering leads to them. Considerable care is necessary for this operation, using special low temperature solder and minimising the heating time, in order to avoid damaging the adhesion of the silvering to the ceramic. The solder "blobs" should be as small as possible, to minimise any loading effect on the ceramic. Nevertheless, it is preferable to make two solder attachments at each surface, to improve reliability. The electrical leads should not be stretched tight between blocks. If the ceramic is used in the (31) mode, the electroded surfaces are on the sides of the blocks, and a technique must be devised to make the connections to these side surfaces.

One problem associated with this method of assembly onto pressure release rubber arises because the rubber tends to compress under hydrostatic pressure. The first effect is to impose stresses on the joints in the assembly, which can cause the joints to fail and the blocks to be less firmly fixed in place. The second effect is to increase the backing impedance presented to the blocks, and the possible mechanical coupling between them, so that the acoustic beam patterns deteriorate. For the usual soft cellular rubber, the effects on the beam pattern arising from this compression can be expected to become significant at pressures in excess of about 1MPa ($\simeq$140psi), which is equivalent to a depth of about 80m. Harder cellular rubbers have been tried, which survive better to higher hydrostatic pressures, but if their collapse pressure is exceeded they are less likely to recover their original characteristics than are the soft rubbers, and their acoustic characteristics are generally less good than for the soft rubbers.

It is in principle possible to use $\lambda/4$ elements mounted on a high impedance metal backing, and this should avoid the difficulties associated with the pressure release rubber. However, the imperfections in the array introduce the probability of mechanical coupling through the mounting, and since the mounting structure is relatively large compared with the elements, many resonances are possible. If these resonances are excited at the operating frequencies, the front face of the array will not be driven as intended, and distortion of the beam pattern may occur. One technique for reducing the coupling between elements is to machine slots in the backing plate between the ceramic blocks. Apart from the added complication in the design of mounting plate, this introduces further problems in supporting what is now a much less rigid structure. Adequate testing by measurements of the beam patterns is essential for such designs.

The fluid most commonly used for filling the transducer is castor oil. This has a density of $0.95 \times 10^3 kg/m^3$, and its speed of sound is 1540m/s at 25°C, so that it is well matched to sea water. It is a good electrical insulator, provided it has not absorbed moisture. The main problems in oil-filling arise from the need to avoid air bubbles within the housing, since even very tiny bubbles may have a large effect because of their resonances. It is therefore essential to de-gas the oil by subjecting it to a vacuum for some time at an elevated temperature, preferably whilst running it over some form of "cascade" to increase its surface area. The increase in temperature is needed to reduce the viscosity of the oil, and the filling of the transducer housing should also be carried out with warm oil under vacuum, taking care to avoid any air pockets within the transducer – except of course for the intentional air cells in the pressure release rubber. The presence of sulphur in the oil should be avoided, since it may attack the silvering on the ceramic. Some of the commercial silicone oils may be used instead of castor oil; they have good insulating properties, and low viscosity, but do not give such a good ρc match to water as castor oil. When oil filling is used for low frequency transducers, the mismatch in ρc is not of any great importance, because the fluid layer is thin compared with a wavelength, but it can be more significant for the higher frequency designs.

An acoustically transparent window needs to be placed in front of the array, to complete the watertight enclosure. This is usually a soft rubber – such as natural rubber – which gives a good ρc matching to sea water. However, some care is needed in selection of the rubber for any particular design, since natural rubber is rather susceptible to the effects of oil and sunlight, particularly when under stress. In some circumstances, it may therefore be better to use a thin window of one of the tougher rubbers. Acoustic characteristics of various possible window materials have been reported by Mikeska and Behrens [11.9]. It is also possible to adopt a quite different approach to the design, in which the blocks are cemented to a thin (solid) front diaphragm. Although this provides the necessary pressure release around the ceramic blocks, there may be strong coupling between elements through the diaphragm unless it is very thin and fragile. Such a design is therefore only suitable for low hydrostatic pressures.

Intermediate frequencies; 20-200kHz.

If the desired operating frequency is reduced below about 200kHz, a $\lambda/2$ resonant ceramic block becomes rather thick and difficult to pole. It is then convenient to adopt a different approach, in which the resonance frequency is reduced by using masses to load the ends of the ceramic blocks. Such a method was first used with crystal transducers, when the end–masses were normally made of glass, but metal end–masses are more commonly used to load piezoelectric ceramic elements. This design allows a fairly simple control of the resonance frequency, using inexpensive metal for the ends and concentrating the ceramic at the centre where it is most effective. The simplest technique is to add end masses which have the same cross section as the ceramic, as in *Fig 11.4*. The equivalent circuit for each end mass has the form of a transmission line as shown in *Fig 11.5*, in which

$Z_{M1} = \rho_1 c_1 A$ for the end–mass under consideration
$k_1 = \omega/c_1$
ℓ_1 is the length of the mass
A is the cross–sectional area of the mass

and Z_r is the radiation loading on the end face of the mass. Equivalent circuits of this form may be substituted for Z_1 and Z_2 in *Fig 11.1*, to derive the complete equivalent circuit for the composite element. Analysis of this complete circuit is best done by means of a digital computer.

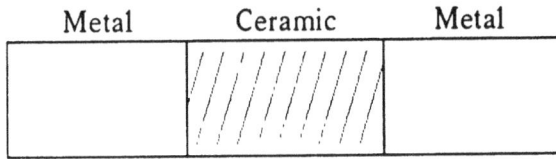

Fig 11.4 Mass loaded piezoelectric element.

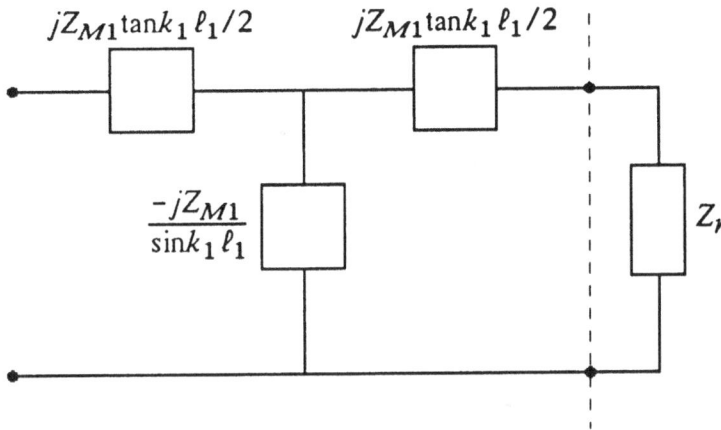

Fig 11.5 Equivalent circuit of end mass.

If the lengths of the end-masses are small compared with a wavelength in the material, the transmission line section can be simplified to give

$$Z_1 \simeq jZ_{M1}\tan k_1 \ell_1 + Z_r$$

$$\simeq j\omega M_1 + Z_r \quad \text{(for small } k_1\ell_1) \qquad (11.4)$$

where M_1 is the actual mass. This corresponds to the lumped mass approximation used previously, and further re-arrangement of the equivalent circuit leads to the derivation of circuits as used in earlier chapters.

Some extra flexibility may be introduced by using a truncated conical head section, as in *Fig 11.6*, to increase the effective water loading. The input impedance of the conical line is given [11.10] by

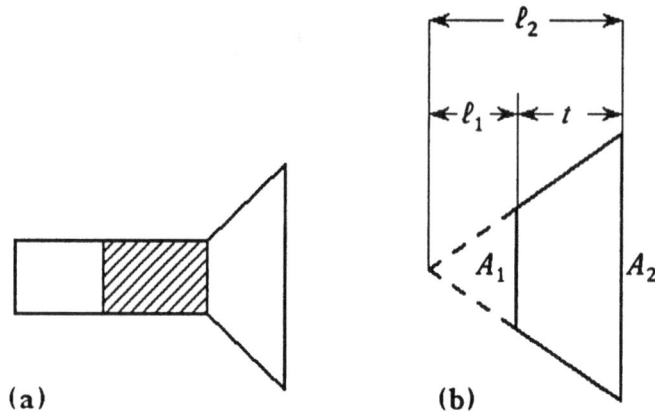

Fig 11.6   End loading using truncated cone: (a) element;
             (b) cone dimensions.

$$Z_1 = R_{11} \left(\frac{R_{12}\sin kt + jR_r \dfrac{\sin k(t-\theta_2)}{\sin k\theta_2}}{R_r \dfrac{\sin k(t+\theta_1-\theta_2)}{\sin k\theta_1 \, \sin k\theta_2} - jR_{12}\dfrac{\sin k(t+\theta_1)}{\sin k\theta_1}} \right) \qquad (11.5)$$

where $R_{11} = A_1\rho_1 c_1$ and $R_{12} = A_2\rho_1 c_1$
          A_1 and A_2 are the surface areas at the rear and front
          of the head, as in in *Fig 11.6b*, and ρ_1, c_1 refer to
          the material of the head itself.
          $R_r = A_2\rho c$, the radiation resistance at the radiating
          surface.
          $k\theta_1 = \tan^{-1}k\ell_1$
          $k\theta_2 = \tan^{-1}k\ell_2$

This expression may be substituted for Z_1 in the equivalent
circuit to represent the load at the head end of the ceramic, whilst
the tail is represented by the circuit for the constant area case.
    Analysis of the equivalent circuits described above may be
used to predict the characteristics of a transducer element. The
calculations are complex, and would become even more complex
if additional factors such as the finite compliance of the joints
were taken into account. Rather than striving for full accuracy in
the analysis, it is generally more practical and efficient to carry

out this analysis only to the accuracy needed to derive the main features of the element, and to complete the design by experimental means. It is worth repeating that the equivalent circuit derived from *measurements* on the transducer, as described in Chapter 4, may be used to evaluate the characteristics of the completed transducer, regardless of any difficulties in *predicting* the characteristics by means of the detailed equivalent circuit.

The **practical aspects** of these designs are similar to those for the higher frequency designs. The elements are usually immersed in an oil-filled casing, with an acoustically transparent window in front of the array. At the lower end of the frequency range, when conical heads become more attractive, it becomes more difficult to rely on attaching the elements to a pressure release rubber backing. For these larger elements, however, it may be more practical to use a nodal mounting within the stack itself. If the node is within the ceramic, it may be desirable to split the ceramic into two pieces and place them either side of a nodal mounting plate, which may then be attached to the main mounting plate. Such an arrangement introduces extra joints in the stack, and it is sometimes preferable to ensure that the piezoelectric ceramic is entirely in front of, or entirely behind, the node. This may be effected, for example, by means of a suitable extension of the tail or head into the stack, although this is likely to be accompanied by a corresponding reduction in coupling coefficient.

The most significant difference from the higher frequency, single block designs is the introduction of joints into the stack. These tend to occur near the centre of the stack, where the dynamic stresses are a maximum. With these relatively small elements, it is usually difficult to fit a centre bolt through the stack and it is therefore important to ensure that the joints are as strong as possible. The power handling capacity of the transducer is difficult to predict reliably, and is best confirmed experimentally.

11-2   LOW FREQUENCY DESIGNS

It is not easy to define the lowest frequency for which a Tonpilz design would be suitable, since there appears no fundamental limit to its range of applicability. However, at 2kHz the wavelength in water is 0.75m, and a single piston would need to have a diameter of 0.3m or more to avoid its radiation resistance falling below about $0.5\rho cA$. If the radiation resistance becomes low, it becomes very difficult to design the low mass piston needed to achieve efficient broad band transmission (Chapter 8). It is therefore easy to see that design for broad band operation becomes increasingly difficult as the frequency is reduced, although the question of how large a piston is practicable is not easy to answer precisely.

Another, and more important, difficulty arises from the need for large amplitudes of vibration at low frequencies. The motional output power from a transducer is given by $u^2 r_m$ (4.29), and the acoustic output power is similarly given by $u^2 R_r$. At the low frequencies where only small values of a/λ are practicable, the radiation resistance (R_r) will be low, and if the power is to remain constant as the frequency is reduced, the piston velocity (u) must increase to compensate for the low R_r. Since the piston amplitude is related to its velocity by $x = u/\omega$, the required piston amplitude must increase even more rapidly than the velocity as the frequency is reduced. In order to achieve reasonable power output from a low frequency source, it is therefore necessary for it to generate high amplitudes of vibration. For example, suppose that we wish to generate 1 watt of acoustic power at 300Hz, using a transducer with a piston radius (a) such that $2\pi a/\lambda = 0.3$. The wavelength at 300Hz is 5m; the piston diameter would thus be just under 0.5m, and its area 0.179m². The normalised radiation resistance for $2\pi a/\lambda = 0.3$ is $R_1 = 0.0443$ (from [6.1] and *Fig 6.1*), and the radiation resistance is thus

$$R_r = \rho c A_p R_1 \qquad\qquad (6.4)$$

$$= 1.5 \times 10^6 \times 0.179 \times 0.0443$$

$$= 11.9 \times 10^3 \text{ kg/s}$$

For an output power of 1 watt, the rms piston displacement is then given by

$$x = u/\omega$$

$$= (1/R_r)^{1/2}/\omega$$

$$= 4.86 \times 10^{-6} \quad (\text{m}) \qquad\qquad (11.6)$$

For a piston area of 1790cm², however, it would be reasonable to expect a power output of at least 1kW, and for this a displacement of $4.86 \times 10^{-6} \times (10^3)^{1/2} = 154 \times 10^{-6}$m would be required – ie 0.154mm. Applying (7.24b) to this example, the approximate stack length for LZT would be

$$\ell_1 = \frac{1.43}{0.3} \left(\frac{1000}{0.179 \times 0.0443}\right)^{1/2}$$

$$= 1693 \text{ mm}$$

This is calculated on the assumption that $kQ_M = 1.2$. Even if Q_M cannot be made low enough to satisfy this relationship, it is clear that the element would have a length of one metre or more, and for frequencies below 300Hz the element would become even larger.

This example illustrates how a Tonpilz transducer would have to become very large to generate substantial power at low frequencies. This arises fundamentally from the high stiffness of the piezoelectric ceramics, which produce only small strains despite the large piezoelectric stresses. They are therefore not well suited to generating large displacements, and the only way to produce large amplitude vibrations with such designs is to use a long stack. One of the problems which then arises is the possibility of flexural resonances in the element itself, which cause a rocking motion of the piston. Also, the low radiation resistance emphasises the need for a lightweight piston, if the mechanical Q_M is to be made reasonably low. These problems may be tackled by using several stacks to drive the same piston, but such a design poses considerable difficulties in the mechanical arrangement, and does little to reduce the size and weight of the element. Thus, although use of the Tonpilz design is reasonably straightforward for frequencies down to about 2kHz, the question of how much lower in frequency it can usefully be applied depends on the details of any particular requirement.

The basic requirement for generating power at low frequencies is thus a mechanism which can produce large displacements efficiently. To do this in a moderate size, it should be less stiff than a piezoelectric ceramic used in the thickness mode. The low internal losses of some of the piezoelectric ceramics have led numerous workers to investigate their use in other modes which are more compliant than the simple thickness mode. These depend for example on using some form of lever (or "impedance transformer") to amplify the mechanical displacement, or on using the ceramic in a flexural mode. Amongst the former type are several forms of "flextensional" transducer, whilst the second type is typified by the flexural disc designs. Other approaches have used driving methods which are essentially less stiff than the ceramics, such as electromagnetic, or hydraulic, systems. The general characteristics of these designs will be discussed in the remainder of this chapter. Some of the modern developments in materials are also applicable to these low frequencies, and are the subject of Chapter 12.

11-3   FLEXURAL DISC TRANSDUCERS

The use of metal diaphragm springs has already been mentioned in Chapter 8. They constitute one example of how metals may be used in a bending mode in order to achieve a relatively low stiffness, and a similar principle is used in flexural disc transducers. If a load is applied to one face of a thin circular disc supported at its edges, one side of the disc is subjected to tensile stresses whilst the other is in compression. Thus, if the disc was composed of two piezoelectric discs cemented together in such a way that a compressive stress in the top half generated a positive voltage, and a tensile stress in the lower half also generated a positive voltage, as in *Fig 11.7*, the two voltages resulting from an applied load would reinforce each other. This is the basic action of the flexural disc transducer acting as a hydrophone. The action as a projector is also easily understood. If a field is applied to the upper ceramic, its dominant effect is to generate a lateral stress in the plane of the disc. If the connections are such as to produce an expansion of the lower half and a compression of the upper, then the result is to cause a bending of the pair of discs. This may be achieved by the arrangement shown in *Fig 11.7*, with the applied field in the same direction for each disc, but the polarisation in opposite directions. Alternatively, it may be achieved by having the polarisation of the discs in the same

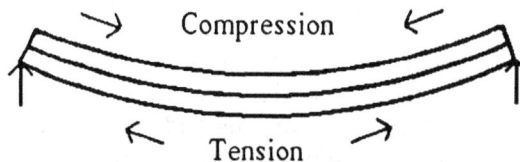

Fig 11.7   Flexural bilaminar disc element.

direction and the applied field in opposite directions. In this second case, the piezoelectric discs are electrically in parallel, and their capacitances add, whilst for the first case the capacitances are in series and the resultant capacitance is much less.

This basic design represents one technique for converting a small displacement in the ceramic into a larger displacement of the face radiating into the water. There are two important requirements in the design. Firstly, the shear stress on the joint between the two ceramic discs is high, and a good bond of high shear strength and stiffness is necessary. This is usually effected by means of a thin epoxy resin joint, but some care is needed to avoid problems which may be caused by differential thermal expansion during the curing process. Secondly, the disc must be mechanically supported and air-backed. If the bilaminar disc were driven without any support, it would vibrate with a nodal ring having a radius about 2/3 of the radius of the disc [11.11]. The pressure generated by the central part would then be counteracted by the pressure generated in the opposite phase by the outer part, and the resultant output would be small. This effect can be prevented by supporting the disc either at the edge or at the centre, the edge support being generally more convenient for underwater transducers. Even when the disc is supported at the edge to avoid this phase reversal over the front face, the pressure generated at the rear face is in opposition to that at the front face. Since these flexural disc transducers are usually small compared with a wavelength in water, there could be some cancellation between radiation from the front and rear of the disc, and it is therefore necessary to ensure a mis-match between the impedances at the front and rear faces. This is usually effected by arranging for the rear face to be air-backed.

These design requirements can be met by a double bilaminar disc transducer, as shown diagramatically in *Fig 11.8*. The two

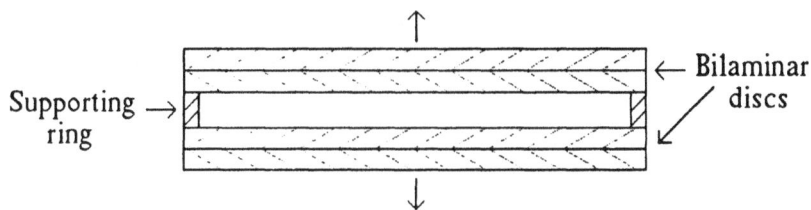

Fig 11.8 Double bilaminar disc.

(a)   Supported edge

(b)   Clamped edge

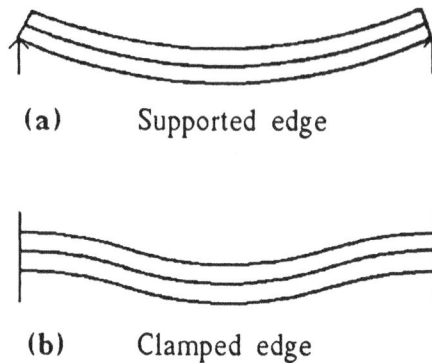

Fig 11.9   Effects of edge conditions.

flexural discs are arranged to radiate in anti-phase, so that the whole transducer acts as a volume source. The centre is normally air-filled to provide the necessary pressure release, and the spacer ring supports the edges of the discs. Although this basic design is simple, care is needed with some of the details. Electrical connections need to be made to the discs, and the whole assembly made water-tight. As the discs bend, their edges move inwards slightly, and the supporting ring should have some radial compliance to accomodate this movement. Also, the spacer ring should support the discs without clamping them. The reason for this may be understood by considering the difference between the simply supported case shown in *Fig 11.9a* and the clamped edge case in *Fig 11.9b*. If the edge of the disc is clamped into its mounting, an applied load will produce a curvature near the edge which is opposite to that at the centre. The resultant piezoelectric output from a bilaminar disc is then greatly reduced (theoretically to zero), and the resonance frequency increased, compared with the supported edge disc. It is therefore important that the supporting ring should restrict axial vibration without entirely preventing rotation of the disc edge about its support or small radial movement of the edge. Design of this spacer ring is critical in controlling the performance and consistency of such transducers, but accurate theoretical analysis is difficult, and experimental confirmation of the transducer's characteristics is usually necessary.

The basic design parameters of these flexural disc transducers have been derived and tabulated by Woollett

[11.12]. For bilaminar discs made of PZT-4 (a LZT I material), with overall thickness t and radius a, he gave the following theoretical values for a supported edge disc in a plane baffle:-

Resonance frequency in air, f_{Sa} (Hz)	$705\ t/a^2$	(11.7a)
Coupling coefficient	0.406	(11.7b)
Clamped capacitance (F)	$97.4 \times 10^{-9} a^2/t$	(11.7c)
f_{Sa}/f_{Sw} (f_{Sw} = resonance frequency in water)	$\{1+0.103a/t\}^{1/2}$	(11.7d)
Radius/wavelength in water	$0.470(f_{Sw}/f_{Sa})t/a$	(11.7e)
Motional Q_M	$7.02(f_{Sa}/f_{Sw})^3 \eta_{ma}$	(11.7f)
Field-limited acoustic intensity (W/m²)	$760 \times 10^4 \left(1+0.103\dfrac{a}{t}\right)\left(\dfrac{t}{a}\right)^2 \eta_{ma}^2$	(11.7g)
Max. static pressure differential (Pa)	$2.8 \times 10^7 (t/a)^2$	(11.7h)
LF hydrophone sensitivity (V/Pa)	$0.00974 a^2/t$	(11.7j)

The maximum static pressure differential is based on an assumed tensile strength of the material of 5000psi (34.5MPa). These relationships are based on the assumption that the discs are thin, and their validity becomes doubtful for $t/a > 0.2$.

Several modifications of this basic design are possible. A metal plate can be inserted between the piezoelectric ceramic discs, thus forming a trilaminar flexural element. This plate can be extended beyond the edges of the ceramic, to give a convenient method of mounting the element. Since the stress in the flexed plate is greatest at the outer plane surfaces, and lower near the neutral axis through the central plane, the ceramic in a trilaminar disc is concentrated where it is most effective. Woollett [11.12] has shown that the effective coupling coefficient of a trilaminar element may thus be slightly higher than for the bilaminar disc, if a brass centre plate about equal in thickness to each of the ceramic discs is used.

Another simple configuration is a bilaminar element in which one of the ceramic discs is replaced by a metal plate. If the metal is of the same thickness as the ceramic, and has the same acoustic properties as the ceramic, the value of k_e^2 will be halved. It would be most convenient to use the ceramic on the inside surface of the metal, since a pair of these discs would then form a self-contained watertight element. However, the effect of external pressure on such a design is to produce a tensile stress in the ceramic, and the weakness of the ceramic in tension would thus lead to severe limitations in depth capability. It is therefore more usual to use the ceramic on the outer surface of the metal, in which case the application of external pressure causes compression of the ceramic. Since the ceramic is much stronger in compression than in tension, the transducer should thus be able to withstand much higher hydrostatic pressures. It is also possible to use one bilaminar disc instead of the two shown in *Fig 11.8*. In that case, the second disc is replaced by an inactive metal housing, to contain the air which is needed for pressure release purposes. The assembly now radiates from only one side, but if the element is small compared with a wavelength, as is often the case, the radiation pattern is effectively omni-directional. For these single-sided elements, some modification is in principle needed for the design relationships (11.7), which were based on the assumption of an element in an infinite rigid baffle. When the double-sided element is used, the symmetry of the radiation from each disc means that there is a virtual baffle extending outwards in the plane mid-way between the discs, so that the assumption is justifiable. When only the single sided element is used, the assumption is not valid, but in practice the discrepancies are unlikely to be important.

The limitations on hydrostatic pressure can be tackled by filling the element housing with gas at a pressure which balances the external pressure. Such an arrangement may be achieved by means of a large air-filled bag, which collapses as the pressure is raised, retaining an adequate compliance within the element. However, the need for the large gas-filled bag, and the associated acoustic effects which it may cause, detract from the basic simplicity of the element. An alternative is to use an automatic sytem for balancing the gas pressure (eg a "Scuba" valve), but this involves considerable extra complication, and the release of gas as the ambient pressure is reduced. Instead of using gas, it is possible to fill the housing with an oil of low acoustic impedance. This reduces the static pressure differential across the disc, but introduces significant problems in ensuring satisfactory acoustic performance, because of the impedance at the rear surfaces of the discs. However, careful choices of the oil and dimensions of the cavity have allowed some workers to achieve satisfactory performance at high pressures.

Waterproofing of the element can present considerable practical problems. Insulation of the piezoelectric ceramic discs from the sea water can be effected by coating the element with a material such as rubber or polyurethane plastic. However, the stiffness and mass of such a layer may have significant effects on the resonance frequency and the coupling coefficient of the disc, especially when the disc is thin. In general, the thickness of the insulation should be thin compared with that of the ceramic, if the effects are to be kept small. There is then a conflict between this desire to keep the insulation thin and the wish to have an adequate thickness to ensure reliable watertightness.

Experimental results for flexural disc transducers were quoted by Woollett [11.12], who drew attention to this influence of the waterproofing layer on the disc. In his example, a polyurethane layer reduced the resonance frequency by over 20%, and the coupling coefficient from 0.203 to 0.164. The other major practical consideration in realising the theoretical values is the degree to which the supported-edge condition can be achieved. Expressions given by Woollett showed that a clamped-edge disc would have a resonance frequency more than twice that of a supported-edge disc of the same dimensions, and a coupling coefficient of zero. Great care is therefore needed in the design of the mounting arrangement to ensure that it approximates as well as possible to the assumed supported-edge conditions. Woollett used an empirical correction in which he took the effective radius as 92% of the actual radius to achieve better agreement with theory.

These flexural disc transducers are very convenient as small size sources of modest power at low frequencies. Values of Q_M are typically in the range 10-15. Because of the inter-relationships between their parameters, it is not easy to control their characteristics to optimise the performance as described previously for the Tonpilz designs. Nevertheless, their simple construction leads to relatively low cost, and for many applications which are not too demanding they therefore constitute an economical and compact solution to the problem of generating sound efficiently at low frequencies. They are also quite extensively used as low frequency hydrophones, for which their low cost and high sensitivity make them well suited.

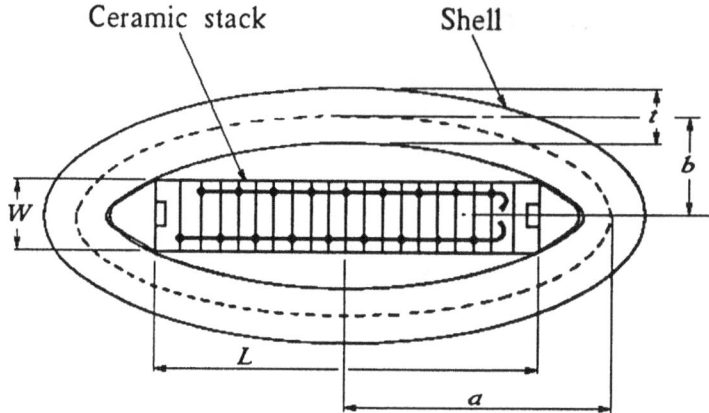

Fig 11.10   Flextensional transducer: basic mechanism; part of Fig 6.5 in paper by Oswin and Dunn, in Hamonic and Decarpigny (Ed), *Power Sonic and Ultrasonic Transducers Design*, Springer–Verlag, 1988.

11-4   FLEXTENSIONAL TRANSDUCERS

Instead of using the piezoelectric ceramic itself in a flexural mode, it is possible to devise structures in which the high stress but low strain generation of the ceramic in the thickness mode is transformed into larger displacements by means of some type of lever action. Several different types of designs make use of this basic mechanism, and are known collectively as flextensional transducers. They are used primarily for applications requiring resonances within the range of about 300Hz to 3kHz, and generally offer potentially higher power outputs than the flexural disc transducers described in the previous section, although at higher cost. A review of their status in 1980 was given by Brigham and Glass [11.13].

The basic mechanism is illustrated in *Fig 11.10*. A stack of piezoelectric ceramic operating in the thickness mode is connected to a surrounding elliptical shell. When the stack extends, along the major axis of the ellipse, the shell moves inwards along the minor axis, thus producing a large volume displacement overall. *Fig 11.10* shows only a two-dimensional section of the transducer, and it may be converted into a three-dimensional structure in various ways. It may be rotated about an axis along the stack, to form what is known as a Class I flextensional transducer - although a practical realisation of such a device described by Royster [11.14] was in fact polygonal rather than ellipsoidal in shape. Alternatively, it may be rotated about an axis normal to the axis of the stack, as is essentially the arrangement for the ring-shell projector developed by McMahon et al [11.15]. In this case, the ceramic is also extended into the third dimension, to form either a ceramic disc which drives the shell through the radial stress in the disc, or a ring around the circumference joining the two half-shells [11.16].

Yet another version is one in which the elliptical shell is extended out of the plane of the paper to form an elliptical cylinder, which is capped on the top and bottom. (In practice, the shell may be only an approximation to a true ellipse.) This is known as a Class IV flextensional design, and has received generally more attention than the other types (see eg [11.13],[11.17]-[11.19]). Calculation of the design characteristics has been tackled by analytical methods and also by finite element techniques [11.13, 11.20], but the results involve a number of parameters and hence do not lead to any simple design equations. In general terms, the resonance frequency of such a transducer depends principally on the major and minor axes, wall thickness, and material properties of the shell, with the stack itself having a lesser influence. *Fig 11.11* shows results obtained by Oswin et al [11.17] for transducers with shells made of aluminium (N8(5083)). *Fig 11.11a* shows the relationship between resonance frequency, shell thickness, and semi-major axis, for a given ratio of major/minor axis (a/b = 2.5). The variation of resonance frequency with semi-major axis and a/b is shown in *Fig 11.11b*. The curves show that the smallest, lightest design for a given resonance frequency is achieved by using a thin shell with the maximum allowable eccentricity (a/b). The bandwidth is also dependent primarily on the parameters of the shell. Maximum eccentricity leads to the maximum bandwidth, but has the lowest power output, whilst least bandwidth and highest power occurs for the least eccentric shape. The maximum pressure which an elliptical shell can withstand is also dependent on its shape and thickness, and is therefore related to its resonance frequency. *Fig 11.12* illustrates the variation of maximum operating depth with semi-major axis, wall thickness, and resonance frequency, again for an aluminium shell of eccentricity a/b = 2.5 [11.17]. Design for any particular application involves a careful balancing of the interactions illustrated by these diagrams. Brigham and Glass [11.13] proposed as a rule of thumb that the eccentricity (a/b) should not have a value greater than about three. Oswin

(a)

FREQUENCY   WALL
Hz   THICKNESS
mm

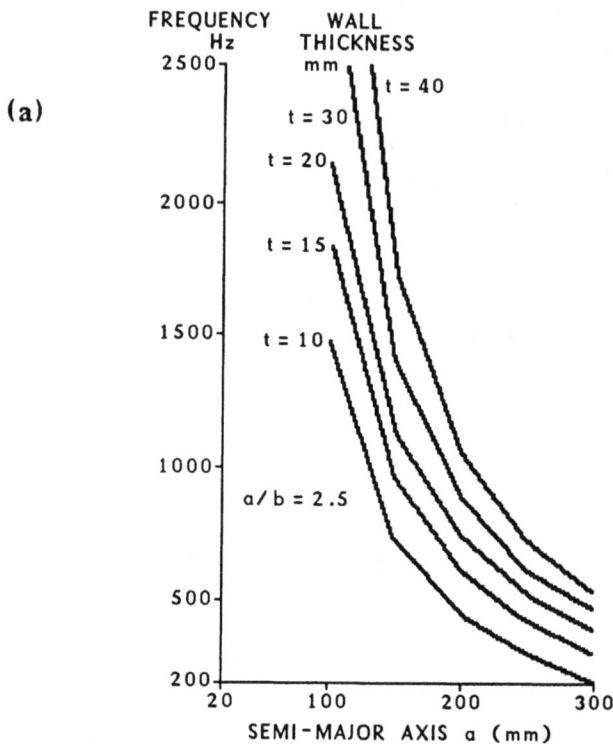

Fig 11.11   Flextensional transducers: relationships between parameters, for aluminium shell: (a) Variation of resonance frequency in water with semi-major axis and shell thickness (a/b = 2.5);

suggested that 500Hz represented an approximate lower limit for a flextensional transducer using an aluminium (N8) shell, if an operating depth of 100m was needed. The use of glass reinforced plastic for the shell helps to achieve a low resonance frequency and wide bandwidth, but needs a suitable and consistent manufacturing technique.

The size of these flextensional transducers is generally much less than a wavelength in water at their resonance frequency. They therefore radiate approximately omni-directionally in the plane perpendicular to their axis. As with other omni-directional sources, although they may be assembled into arrays, it is not easy to produce a source which radiates only in one direction. The field needed to generate a given power output can be

FREQUENCY
Hz

(b)

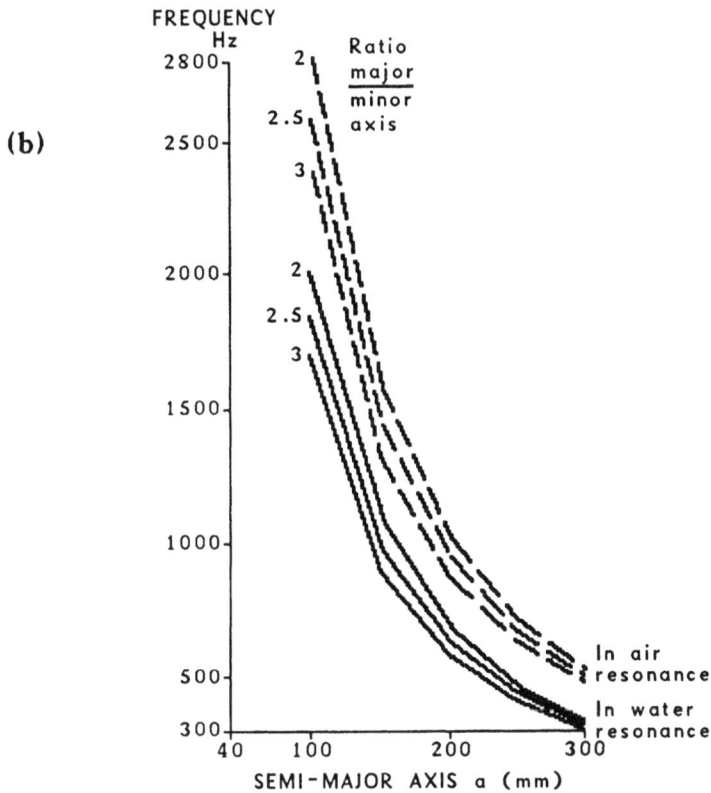

(b) **Variation of resonance frequency with semi-major axis and eccentricity ratio a/b (t = 15mm); from Inst Acoust in paper by Oswin and Turner, in Proc Inst Acoust 6(3), 95-100.**

calculated from the measured equivalent circuit by the methods of Chapter 7 (eg equation (7.5)), and the field-limited power output can thus be estimated. In general, power handling is good, and Brigham et al. [11.13] quoted as an example a power output of 330W/kg (150W/lb) for a transducer resonant at 1kHz. The maximum theoretical value of the effective coupling coefficient was given as 0.64 for an element using a PZT 4 stack, but practical values are reduced by factors such as the stiffnesses of the shell and joints, and values of k_e down to 0.3 or even lower are common. Values of Q_M down to two appear possible. Marshall et al [11.19] compared the theoretical characteristics of Class IV flextensional transducers with those of flexural disc designs, and concluded that the Q_M for a

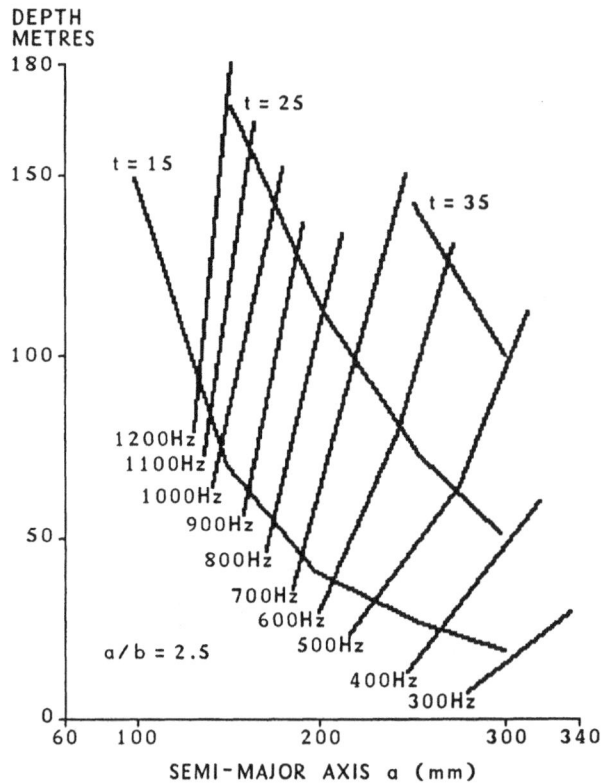

Fig 11.12   Variation of maximum depth with semi-major
axis, shell thickness, and resonance frequency, for
aluminium shell with a/b = 2.5; from Inst Acoust
in paper by Oswin and Turner, in Proc Inst
Acoust 6(3), 95-100.

flextensional transducer would be lower by a factor of 2–4 than
for a flexural disc of similar size. They concluded also that a
flextensional could be considerably smaller than a flexural disc
transducer when hydrostatic pressure was the limiting
requirement.

In order to achieve substantial power output from such a
transducer, a compressive pre-stress needs to be applied to the
stack, just as for the Tonpilz type of design. This is usually
done by applying pressure to the minor axis of the shell, thus
extending the major axis, and inserting the stack into the
extended shell. On release of the pressure, the relaxation of the

shell applies the necessary force to the stack. If a GRP shell is used, some of the pre-stress may be applied by wrapping the fibre around the shell under tension. Unfortunately, hydrostatic pressure applied to the shell also compresses the minor axis and extends the major axis, thus reducing the pre-stress in the stack and possibly even opening up a gap between stack and shell. Although convenient, this technique of applying pre-stress by means of the shell therefore requires careful assessment of the design stresses and accurate control of the tolerances in the assembly. The junction between the stack and the shell can give rise to considerable stress concentrations, particularly when the shell is deformed by the external pressure. Detailed calculation of the static stresses in the shell under hydrostatic pressure is therefore needed, usually involving finite element techniques in order to deal with realistic shapes. The possible effects of fatigue during high power operation also need to be taken into account.

If a higher operating pressure is required, for a constant resonance frequency, *Fig 11.12* illustrates how a thicker and larger shell must be used, to keep the stresses within safe levels. At very high hydrostatic pressures the transducer size may then become large enough to cause some directionality in

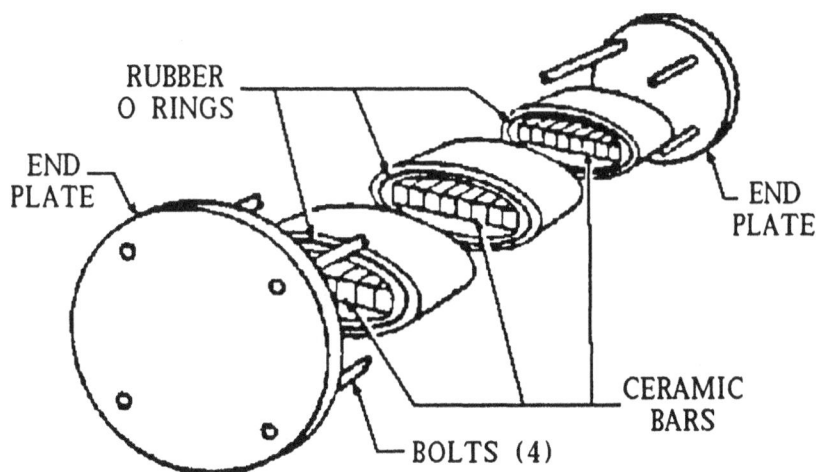

Fig 11.13 Exploded view of three-module, Class IV, flextensional transducer; from *J Acoust Soc Am*, in Brigham and Glass, *JASA*, 64, 1046-1052, (1980).

the radiation pattern. The possibility of using an oil-filled design to reduce the pressure differential was considered by Brigham and Glass [11.13], who concluded that the air-filled design was generally preferable. The size is most effectively reduced by using GRP for the shell instead of aluminium. One advantage of the ring-shell design described by Armstrong et al. [11.16] is that it can be fitted with a gas-filled bladder inside the shell to provide some pressure compensation.

A single element may incorporate several stacks in parallel on the major axis, to increase the power handling capacity, and several elements may be assembled together to form a longer transducer, as illustrated in *Fig 11.13*. Care is needed in design of the end plates, to ensure watertightness without introducing excessive clamping or mechanically induced stresses. If the material of the shell is subject to corrosion in sea water, it is normally protected by encapsulating it in rubber, which may also be used to help in achieving watertightness of the assembly.

Examples of the characteristics of practical designs are given by Oswin et al. [11.21], who quoted the following results for transducers having a major axis of 150mm, minor axis 56mm, and wall thickness 16mm. The depth of the shell was 100mm. One set of values is for a design using an aluminium shell, and the other for a glass reinforced plastic shell.

	Aluminium shell	GRP shell
Mass	3 kg	2.2 kg
Resonance frequency	3 kHz	1.5 kHz
Projector sensitivity S_V	134 dB	125 dB
Voltage for 200W output	1000 V	2818 V
Efficiency	60%	50%

Transducers with resonance frequencies down to several hundred Hertz have also been designed. An example quoted in [11.20] was of a transducer using a GRP shell with dimensions 190 × 480 × 530mm and weighing approximately 45kg. Its resonance frequency was 390 Hz, bandwidth 65 Hz (to the −3dB points), and maximum source level about 210dB re 1 μPa at 1m. Oswin [11.21] discussed the potential advantages of using the newer magnetostrictive materials as the driving mechanism in these flextensional transducers, in order to reduce the overall size of the transducer plus matching network.

In summary, flextensional transducers constitute useful sources within the band of about 300-3000Hz. They can provide appreciable power from a compact size, potentially up to high operating pressures, and have wide bandwidth, compared with other designs such as the air-backed flexural discs, but at rather higher cost. The design process needs considerable care, to allow for the complex interactions between parameters, and studies to develop improved design methods are therefore receiving much attention.

11-5   OTHER LOW FREQUENCY DESIGNS

Various other designs have been used to generate power at low frequencies, these designs having their own characteristics which may make them especially suitable for particular applications. They will be described only briefly here.

Free-flooded rings.

If a piezo-ceramic ring is driven in its circumferential mode whilst completely immersed in water, the radiation from the outer surface is opposed by that from the inner and the end surfaces. For low pressure applications, the unwanted radiation may be prevented by fitting end-caps to exclude water from the inside of the rings. The seal between the end-caps and the ring must allow the ring to vibrate freely, and for low frequency sources the overall size is large, so that the end-plates become rather heavy. A more convenient arrangement is to coat the inner and end surfaces with cellular rubber, to provide pressure release, and radiation then takes place only from the outer surface. This provides a very simple source, but is restricted to fairly shallow depths, where the cellular rubber acts as an effective pressure release. For greater depths, it would appear that the cancellation effects which would exist if the pressure release rubber were removed would render the design of little use. However, it is possible to make the height of the ring such that the cancellation effects are not too great, and the free-flooded ring then forms a very simple radiator. It has the great advantage that its output is virtually independent of hydrostatic pressure, since no pressure-release mechanism is involved. Its radiation pattern is omni-directional in the plane perpendicular to the axis of the ring, whilst its directivity in the other planes depends on the length of the cylinder, approximately as for a linear source.

The resonance frequency of the ring occurs when its circumference is equal to a wavelength for longitudinal vibrations in the material of the ring. If the ring is silvered on the inner and outer surfaces, and poled in the radial direction, its resonance frequency is thus given by

$$\omega_S = c_1/a \qquad (11.8)$$

where a is the radius of the ring and c_1 is given by

$$c_1^2 = 1/\rho s_{11}^E \qquad (11.9)$$

For a thin-walled ring of height h and thickness t, the low frequency capacitance is calculated as for a parallel plate capacitor; ie,

$$C_{LF} = 2\pi a h \varepsilon/t \qquad (11.10)$$

where $\varepsilon = \varepsilon_{33}^T \varepsilon_0$ is the free dielectric constant of the ceramic. The mass of the ring is readily calculated, as

$$M = 2\pi a t h \rho$$

but it is not so simple to predict the radiation impedance of a ring radiating from all surfaces. It is therefore usually necessary to adopt an experimental approach to establish the remaining acoustic parameters of the ring.

Since the speed of sound in LZT ceramic is just over twice that in water, the diameter of a ring transducer is about two-thirds of a wavelength of sound in water at its resonance frequency. (For barium titanate, the ring diameter is roughly one wavelength in water.) At 1kHz, the wavelength in water is 1.5m, and a LZT ring resonant at 1kHz would thus be about 1m in diameter, rather larger than is practical by present manufacturing methods. In practice, readily producible rings are currently restricted to about 150mm diameter, corresponding to a resonance frequency of about 6.7kHz. This is the resonance frequency in air, and the effective dynamic mass of the water can cause a significant reduction in the resonance frequency when the ring is operated in water. Nevertheless, it is clear that a single free flooded ring of this type is not suitable for frequencies of 1kHz and below, despite its attractions for use above about 4kHz.

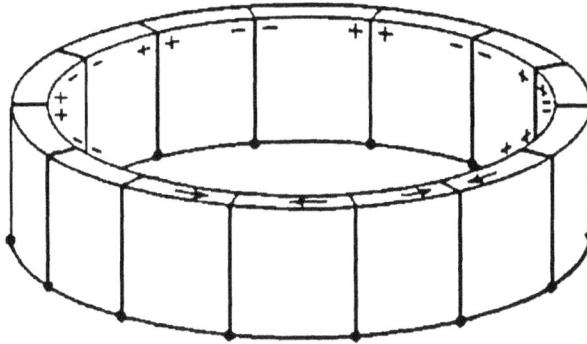

Fig 11.14 Segmented piezoelectric ceramic ring transducer.

This size restriction may be overcome by assembling the ring from a number of segments, as illustrated in *Fig 11.14*. In this case, the electrodes are positioned between each ceramic segment, and the segments are poled around the circumference. Poling is arranged in alternating directions around the ring, with an even number of segments, and the segments are then generally connected in parallel. This arrangement (often described, for obvious reasons, as a **"barrel-stave"** construction) allows very large rings to be constructed, although careful control of the tolerances is necessary to obtain a well-consolidated assembly with uniform joints. (The term "barrel-stave" is also sometimes used to describe a version of the Class I flextensional transducer.) The piezoelectric ceramic is used in the thickness mode, so that the coupling coefficient of the transducer should be substantially higher than for the single piece ring with radial poling. The calculated resonance frequency is given by equations similar to (11.8) and (11.9), except that s_{11}^E should be replaced by s_{33}^E. For best accuracy, a correction term should be added to allow for the effects of the joints, but an experimental approach is again most commonly adopted. When driven to high power, the joints and ceramic are subjected to considerable tensile stresses, and some means of pre-stress is necessary. This is usually effected by wrapping glass fibre under tension around the circumference of the ring, and consolidating the whole in epoxy resin. This method of glass-fibre wrapping to apply pre-stress is also applied to rings made as a single piece, if high power is required from them. As with the centre bolt through a ceramic stack, the stiffness of the fibre wrapping applying the stress causes some degradation of the coupling coefficient.

A further development of the concept is to replace alternate ceramic pieces by metal. This reduces the cost to some extent, but its main advantage is that it introduces the possibility of adjusting the characteristics of the ring by a suitable selection of material and dimensions. In particular, the resonance frequency may be reduced by allowing the metal segments to protrude into the inner space, thus increasing the effective mass of the ring. As would be expected, the substitution of metal for ceramic also carries with it the penalty of reducing the coupling coefficient of the transducer. Hamonic [11.22] has described a transducer in which a ring is driven by a number of Tonpilz stacks mounted on the inside of the ring.

The radiation loading in water depends on the dimensions of the ring, and its material properties – specifically its Poissons ratio. If the height of the ring becomes a significant fraction of a wavelength in water, the water within the ring may act as a cavity resonator. This effect may be used to give some control of the resonance frequency, and to improve the acoustic properties. It is also possible to arrange several rings co-axially, with gaps between the rings, and make use of the acoustic interaction between the rings to control the characteristics. Although considerable effort has been devoted to analysing such arrays mathematically, it is probably simpler and more straightforward to base the design on experimental measurements such as those by McMahon [11.23].

Both of these designs require electrical insulation from the sea water, and this is usually effected by encapsulating the ring in an insulating material. This must not be too stiff or lossy, to avoid degrading the acoustic properties, and rubber or a flexible epoxy resin are often used. A simple arrangement which may be suitable for experimental purposes is to immerse the ring in oil in an acoustically transparent enclosure.

These insulation problems are largely avoided by making the rings of magnetostrictive material instead of piezoelectric ceramic [11.24]. This can be done for low frequencies by winding a scroll of nickel alloy strip into a ring, and winding a coil around the ring to generate a magnetic driving field around the circumference. This is probably the commonest remaining use of magnetostrictive alloys for transducers, because of the simplicity of its basic construction. However, it does require the specialised techniques involved in magnetostrictive transducers, and is subject to their limitations of power and coupling coefficient. Although the need for electrical insulation of the ring itself is eliminated by adopting these designs, it is important to remember that the insulation of the wire carrying the current to generate the magnetic field must remain good, and adequate protection for this wire must therefore be provided. (See Section 12.2.)

Non-resonant transducers.

Any transducer can of course be driven at low frequencies, even well below resonance, but in general its output will be severely limited by the small amplitudes and low efficiencies which can be achieved. It is however possible to tackle the problem of generating large amplitude vibrations by direct mechanical means. This is particularly appropriate at the lowest frequencies, in the range of 1–50Hz; at these frequencies, amplitudes become sufficiently large for it to be practical to use an electric motor to drive pistons in the wall of a watertight housing by using cams and levers (eg [11.25]). Although the frequency may be changed by varying the speed of the motor, it is not easy to change the amplitude of the oscillation, and these designs are thus limited in their versatility. They are nevertheless particularly useful for applications requiring appreciable power at frequencies below about 20Hz.

At audio frequencies, it is natural to consider electromagnetic designs such as moving coil transducers as possible sources, since they are widely used as loudspeakers to generate the large displacements needed in air. For low power wide bandwidth applications such as standard sources for calibration purposes, they are indeed very useful. Bobber [11.26] described two designs for such applications, working over the range 40Hz–20kHz. The moving coil of the J9 transducer is mounted on a very compliant suspension, which allows large amplitudes of vibration of the 57mm diameter piston. In order to maintain the coil accurately between the pole pieces, it is necessary to balance the internal air pressure against the external water pressure by means of an air-filled collapsible rubber bag. This compensation system works satisfactorily down to a depth of about 25m, and beyond that a more active pressure compensation method is needed. The maximum power input is 20W at frequencies above 200Hz, but the efficiency is less than 1%, a figure which is quite typical for a moving coil transducer working well away from resonance.

The J11 design has a larger piston (100mm diameter) and can handle electrical inputs up to 200W, but with a more restricted frequency band. These designs need careful handling to avoid damage to the suspension, and more rugged versions have been produced to make them more suitable for sea trials. They represent good examples of transducers which are capable of low power outputs over a wide frequency band at shallow depths; although various trade-offs are possible between power output, frequency band, and size, they typify what can be achieved with non-resonant moving coil designs.

Resonant bubble transducer.

The broad-band but low efficiency characteristics of the moving coil design can be exchanged for higher efficiency but lower bandwidth characteristics by transforming the radiation loading. This was effected by Sims [11.27], who used a gas-filled bag at the face of a moving coil transducer to provide a compliance which resonated with the water loading. The increase in effective radiation loading then resulted in a much higher efficiency around this resonance. Such a design provides a compact low frequency source of modest power, but has a narrow bandwidth, and a centre frequency which varies significantly with operating depth as the pressure in the bag changes. It is therefore most suitable for operation at a specified constant depth.

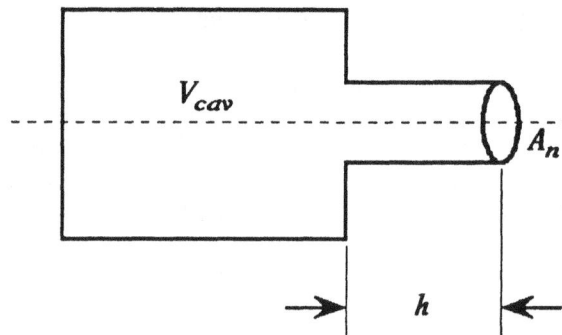

Fig 11.15 Basic Helmholtz resonator.

Helmholtz resonator.

Another method for obtaining a resonant radiation load is to use a Helmholtz resonator. The classical Helmholtz resonator consists of a rigid enclosure of volume V_{cav}, with a short neck of length h and cross-sectional area A_n (*Fig 11.15*), the outer end of the neck being open to the surrounding medium. The compliance of the fluid in the cavity can resonate with the mass of the fluid in the neck, to form a simple mass-spring system. This is well known in air from the resonance which can be set up by blowing over the neck of a bottle. A similar resonance may be observed in water by using a transducer in the wall of the cavity to excite the resonance, and this can form the basis of a compact low frequency source of sound.

The stiffness of the fluid in the cavity (K_{cav}) can be derived by calculating the increase in pressure caused by a displacement of the fluid in the neck, giving the result

$$K_{cav} = \rho_0 c^2 A_n^2 / V_{cav} \qquad (11.11)$$

in which ρ_0 and c are the density and sound speed of the fluid in the cavity and neck [11.28]. The mass of fluid in the neck is $\rho_0 A_n h$, but the effective vibrating mass is larger than this, to allow for contributions from the fluid adjacent to the ends of the neck. We have seen in Chapter 6 (equation (6.8)) that the associated mass of water for a small piston (of area A_n) at low frequencies is equal to that contained in a cylinder of area A_n and length $8a/3\pi$, where a is the radius of the neck. This is equivalent to adding an end correction of $8a/3\pi = 0.85a$. Adding this contribution for each end of the neck, the total effective mass of the fluid in the neck is given by

$$M_n = \rho_0 A_n \{h + 1.7a\}$$

This end correction was derived from the expressions in Chapter 6 for a piston in a baffle. If the piston is in the end of an unbaffled tube, the end correction has the value $0.61a$, and observed values in practice are approximately $0.75a$, between the two cases [11.28]. A more probable value for the effective mass is therefore

$$M_n = \rho_0 A_n \{h + 1.5a\} \qquad (11.12)$$

The resonance frequency of the resonator is then given by

$$\omega_S^2 = K_{cav}/M_n$$

$$= \frac{c^2 A_n}{V_{cav}(h + 1.5a)} \qquad (11.13)$$

The theoretical value of the motional Q-factor [11.28] is calculated by using the value of M_n (11.12) and the radiation resistance for a small piston (6.6), assuming the radiation to account for all the energy dissipation. Thus,

$$Q_M = \omega_S M_n/R_r$$

$$= 2\pi(\rho_0/\rho)\{(h+1.5a)^3 V_{cav}/A_n^3\}^{1/2} \qquad (11.14)$$

where ρ is the density of the surrounding fluid. In the simplest case, for which the fluid in the cavity is the same as the surrounding fluid, $\rho_0/\rho = 1$, and it is of interest to note that the density does not then enter explicitly into either of these equations. The resonance frequency for a cavity filled with water is therefore greater than that for the cavity filled with air only by the ratio of their speeds of sound – ie about three, – and their theoretical values of Q_M are equal. Calculated values of Q_M are generally high, values well above 20 being common. A low resonance frequency is obtained by using a long neck with a small area, combined with a large cavity, but these also result in a high Q_M. Because the dependence of Q_M on V_{cav} is not so strong as for the other factors, the best way of achieving a low resonance frequency whilst minimising the increase in Q_M is to make the volume as large as practicable. However, the assumption that the cavity acts purely as a compliance is equivalent to assuming that its dimensions are small compared with a wavelength at its resonance. This is a reasonable approximation if the dimensions are less than $\lambda/4$, and this therefore represents an upper limit to the size of the cavity. Within this limit, the shape of the cavity is not important.

The method of exciting the resonance is to insert a transducer into the wall of the cavity. For example, if the cavity is cylindrical in shape, it is simple to drive it with a piston or flexural disc in the base of the cylinder. Greater power input can be achieved by using an arrangement described by Hanish [11.29], in which piezoelectric ceramic rings form the cylindrical walls of the cavity. Another arrangement described by Hanish used a spherical shell transducer as the cavity itself. In the absence of the cavity loading, the transducer for all such cases would be operating in its stiffness-controlled region below resonance, and the cavity resonance is used to optimise the radiation loading on the transducer in order to obtain good efficiency at the required operating frequency.

Woollett [11.30] considered an example of a Helmholtz resonator transducer to operate at 100Hz. The water-filled cavity was cylindrical, with diameter 1m and length 1m, and with a neck of length 0.42m and ID 0.3m. The driver was a flexural disc transducer covering the base of the cylinder. For this hypothetical arrangement, the calculated Q_M was 220, neglecting internal dissipation. In practice, of course, the fluid is subject to viscous losses which become more important as the area of the neck becomes smaller and the velocity of the fluid flow through the neck becomes larger. The losses are not easy to calculate accurately, but in addition to reducing the Q_M they also reduce the efficiency. Effective coupling coefficients for these designs are also low. However, despite these disadvantages, the Helmholtz resonator constitutes one of the few feasible solutions to the problem of designing a compact low frequency sound source which can operate down to great depths.

The power may be limited by the stresses within the driving transducer, or by the pressure within the cavity. Because of the large volume flows through the neck necessary to generate high power, quite high alternating pressures may exist within the cavity, and these bring with them the possibility of cavitation inside the cavity, with the consequent limitation on the output power. As usual with cavitation limitations, this is likely to be more significant near the surface than at deeper depths.

Another of the assumptions in the simple treatment above is that the walls of the cavity are acoustically rigid. This is readily achieved for a Helmholtz resonator in air, since even quite thin metal walls have a much higher acoustic impedance than air. In water, however, the walls must be much thicker to present a high impedance to the water. The combined requirements for large cavity volume and thick walls conflict with the potential advantages of the Helmholtz resonator as a small and lightweight low frequency source. The volume of the cavity may be reduced by using a more compliant fluid within it, and various approaches were described by Woollett [11.30]. One of the more interesting, which is suitable for operating depths of about 200-1000m, is the use of **compliant tubes** [11.31] in the cavity. These are hollow metal or plastic tubes, of elliptical cross-section, filled with air. As for the casing of a flextensional transducer, they can be designed to have a relatively high compliance and high collapse pressure, and can therefore be used to increase the effective compliance of the cavity. They have been used for similar purposes in other oil-filled transducers, and for making low impedance (pressure release) baffles.

Helmholtz resonators thus appear to offer some promise for compact low frequency sources. However, in practice, the weight needed to achieve rigidity in the walls, together with the acoustic limitations of bandwidth and efficiency, have limited their usefulness.

Hydroacoustic Projectors.

Another type of low frequency transducer which has received much attention in recent years is the hydroacoustic projector, developed by Boyoucos [11.32]. In this design, high pressure fluid is used to drive the radiating piston, thus providing a mechanism which is capable of large amplitudes of vibration. A diagram of the basic system is shown in *Fig 11.16* (from

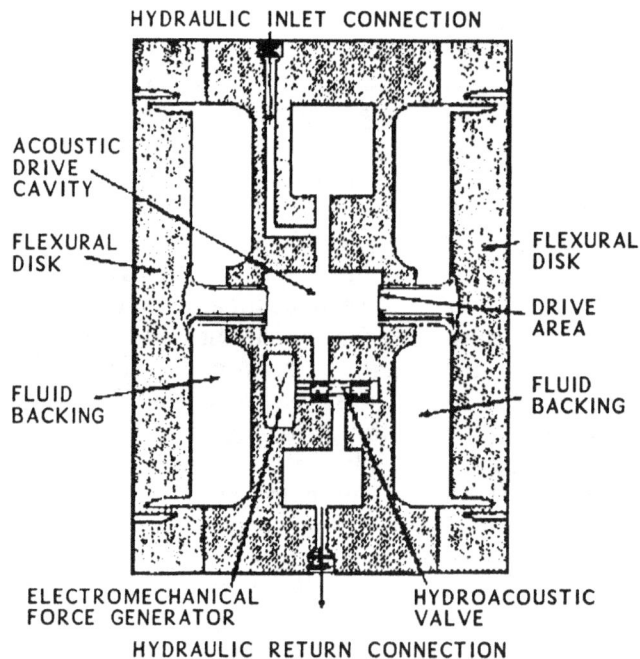

Fig 11.16   Hydroacoustic transducer; from IEEE, in Woolett, R.S., *IEEE Trans on Sonics and Ultrasonics*, SU-15, 218-229, (1968). © 1968 IEEE.

[11.33]). A flow of high pressure fluid is modulated by means of an electromechanically controlled valve, which thus causes pressure oscillations in the acoustic drive cavity. These exert alternating forces on the flexural discs, which radiate into the water. The electromechanical valve is often a piezoelectric stack driving a spool valve, and the function of the piezoelectric element is thus to provide the control for the transducer, whilst the power is derived from the high pressure driving fluid. The various components of the system are designed to resonate at the desired frequency, and adequate energy transfer over a fractional bandwidth of about 20% is achievable.

The maximum force on the pistons is determined by the maximum safe working pressures in the rest of the system, which can reach 35–70MPa (5–10kpsi). Considerable power outputs can be achieved, though the transducers are quite large and heavy. For example, a power output of 35kW at 750Hz from a transducer with a piston diameter of 0.64m has been reported [11.33]. These hydroacoustic transducers can thus serve as effective high power sources at low frequencies, and many versions have been built over a number of years, where size and weight have not been critical. They are best suited for frequencies below a few kiloherz. They are not suitable for use as hydrophones.

Other methods.

Many other methods have been used to generate low frequency signals. For applications where impulsive sound is acceptable, perhaps the most common method is to use explosive sources. Their characteristics are well reviewed by Urick [11.34] and Hanish [11.35]. Analysis of the waveform emitted by small (eg 450g) explosive charges shows that most of the energy is concentrated below 1kHz, and such charges can therefore certainly be regarded as compact low frequency sources. They are non-directional, and emit a short high-power pulse which can give good range resolution. However, great care is obviously needed in their deployment, they are single pulse devices, and their pulse characteristics are not accurately controllable, so that their use is restricted to particular applications.

Other devices have been developed to overcome the single pulse limitation of explosive charges, though retaining the transient nature of the waveform. An electrical discharge under water acts as a small impulsive source, but is capable of being repeated for a number of pulses [11.36]. The sound generated by the impact of one metal on another has also been used, some form of air-gun often being employed to drive the components together. Water gun impulsive sources are more compact than air guns, and have been extensively used up to 200Hz [11.37]. Barger and Hamblen have also described an air gun impulsive source in which high pressure air is momentarily vented to the surrounding water to produce an air-filled cavity which radiates a transient waveform [11.38]. In the "Sonar Thumper", an impulsive waveform is generated by the repulsion of an aluminium plate by the eddy-currents which are induced in the plate by a large coil embedded in a plastic matrix [11.39]. These types of impulsive sources may be controlled more accurately than the explosive sources, and several sources can thus be used in an array, with the input timings adjusted to provide improved source level and wave shape.

Although these impulsive sources have their own particular advantages for some applications, especially where a wide range of frequencies is wanted, the main line of development has been aimed at sources of controlled sinusoidal signals. Amongst the approaches which have been used are variable reluctance transducers of both monopole and dipole types, flexural (bender) bar transducers, and spherical shell radiators assembled from many pieces of piezoelectric ceramic. In general, these have been superseded by the designs described above, but the interested reader could find brief descriptions of the earlier methods included in several review papers [11.33], [11.40-11.42]. Some of the new materials, to be discussed in the next chapter, may also be applicable to these low frequency sources.

REFS Chapter 11

References

11.1   Berlincourt,D.A., Curran,D.R., and Jaffe,H., "Piezoelectric and Piezomagnetic Materials and their Function in Transducers," in Mason,W.P.(Ed), *Physical Acoustics*, Vol 1 - Part A, Academic Press, 1964, p170.

11.2   Gooberman,G.L., *Ultrasonics*, English Universities Press, 1968, Ch 3.

11.3   Merhaut,J., *Theory of Electroacoustics*, McGraw-Hill, 1981, Ch 2.

11.4   Wilson,O.B., *An Introduction to the Theory and Design of Underwater Transducers*, US Govt Printing Office, 1985, Ch V.

11.5 Smith,B.V., and Gazey,B.K., "High–frequency sonar transducers: a review of current practice," *IEE Proc*, 131, (Part F), 285–297, (1984).

11.6 Tucker,D.G., and Gazey,B.K., *Applied Underwater Acoustics*, Pergamon Press, 1966, Ch 5.

11.7 Silk,M.G., *Ultrasonic Transducers for Nondestructive Testing*, Adam Hilger, 1984, Ch 3, Sect 3.6.

11.8 Van Crombrugge,M., and Thompson,W.Jr., "Optimisation of the transmitting characteristics of a Tonpilz–type transducer by proper choice of impedance matching layers," *J Acoust Soc Am*, 77, 747–752, (1985).

11.9 Mikeska,E.E., and Behrens,J.A., "Evaluation of transducer window materials," *J Acoust Soc Am*, 59, 1294–1298, (1976).

11.10 Olsen,H.F., *Elements of Acoustical Engineering*, Van Nostrand, 1947.

11.11 House,R.N., and Kritz,J., "An analytical study of the vibrating free disk," *Ultrasonic Engineering*, UE–7, 76–84, (1960).

11.12 Woollett,R.S., "Theory of the Piezoelectric Flexural Disk Transducer with Applications to Underwater Sound," USN Underwater Sound Laboratory, USL Research Report No 490, (Dec 1960).

11.13 Brigham,G., and Glass,B., "Present status in flextensional transducer technology," *J Acoust Soc Am*, 64, 1046–1052, (1980).

11.14 Royster,L.H., "The flextensional concept: a new approach to the design of underwater acoustic transducers," *Applied Acoustics*, 3, 117–126, (1970).

11.15 McMahon,G.W., and Armstrong,B.A., "A pressure–compensated ring–shell projector," *Conf Proc, Transducers for Sonar Applications*, Inst of Acoust, Birmingham, 1980.

11.16 Armstrong,B.A., and McMahon,G.W., "Discussion of the finite element modelling and performance of ring shell projectors," *IEE Proc*, 131 (Part F), 275–279, (1984).

11.17 Oswin,J.R., and Turner,A., "Design limitations of aluminium shell Class IV flextensional transducers," *Proc IoA*, 6, Pt 3, 95–100, (1984).

11.18 Butler,J.L., and Peirce,T.J., "A desktop computer program for a flextensional transducer," *Proc IoA*, 9, Pt 2, 31–41, (1987).

11.19 Marshall,W.J., Pagliarini,J.A., and White,R.P., "Advances in flextensional transducer design," *IEEE Oceans*, 124–129, (1979).

11.20 Hanish,S., *A Treatise on Acoustic Radiation*, Vol III, Naval Research Laboratory, 1985, Chapter 5.

11.21 Oswin,J.R., and Maskery,A., "Magnetostriction in flextensional transducers," *Proc IoA*, 9, Part 2, 23–30, (1987).

11.22 Hamonic,B., "Application of the finite element method to the design of power piezoelectric sonar transducers," in Hamonic,B., and Decarpigny,J.N.,(Eds), *Power Sonic and Ultrasonic Transducer Design*, Springer–Verlag, 1988.

11.23 McMahon,G.W., "Performance of open ferroelectric ceramic cylinders in underwater transducers," *J Acoust Soc Am*, 36, 528–533, (1964).

11.24 Camp,L., *Underwater Acoustics*, Wiley, 1970, Chapter 6.

11.25 Wilcox,H.A., "Non–resonant acoustic projector project," Naval Ocean Systems Center, NOSC TR 579, (1980).

11.26 Bobber,R.J., *Underwater Electroacoustic Measurements*, US Government Printing Office, 1970, Chapter 5, Section 5.10.

11.27 Sims C.S., "Bubble transducer for radiating high–power low–frequency sound in water," *J Acoust Soc Am*, 32, 1305–1308, (1960).

11.28 Kinsler,L.E., and Frey,A.R., *Fundamentals of Acoustics*, (2nd Ed), Wiley & Sons, 1962, Chapter 8.

11.29 Hanish,S., op. cit., Vol II, Chapter 3.

11.30 Woollett,R.S., "Basic problems caused by depth and size constraints in low frequency underwater transducers," *J Acoust Soc Am*, 68, 1031–1037, (1980).

11.31 Toulis,W.J., "Acoustic refraction and scattering with compliant elements," *J Acoust Soc Am*, 29, 1021–1033, (1956).

11.32 Boyoucos,J.V., "Hydroacoustic transduction," *J Acoust Soc Am*, 57, 1341–1351, (1975).

11.33 Woollett,R.S., "Power limitations of sonic transducers," *IEEE Trans on Sonics and Ultrasonics*, SU–15, 218–229, (1968).

11.34 Urick,R.J., *Principles of Underwater Sound*, McGraw–Hill, (3rd Ed), 1983, Chapter 4.

11.35 Hanish,S., op. cit., Vol I, Chapter XI.

11.36 Bjørnø,L., "A comparison between measured pressure waves in water arising from electrical discharges and detonation of small amounts of chemical explosives," *Journal of Engineering for Industry, Trans of the ASME*, 92, 29–36, (1970).

11.37 Safar,M.H., "On the calibration of the water gun pressure signature," *Geophysical Prospecting*, 33, 97–109, (1985).

11.38 Barger,J.E., and Hamblen,W.R., "The air gun impulsive underwater transducer," *J Acoust Soc Am*, 68, 1038–1045, (1980).

11.39 Van Reenan,E.D., "A Complete Sonar Thumper Seismic System," in Gaul,R.D., Ketchum,D.D., Shaw,J.T., and Snodgrass,J.M.(Eds), *Marine Sciences Instrumentation*, Vol I, Plenum Press, 1962 (pp 283–288).

11.40 Woollett,R.S., "Ultrasonic transducers: 2.Underwater sound transducers," *Ultrasonics*, 3, 243–253, (1970).

11.41 Hueter,T.F., "Twenty years in underwater acoustics: Generation and reception," *J Acoust Soc Am*, 51, 1025-1040, (1972).

11.42 Massa,F., "Sonar transducer developments during the period of World War II and beyond," *Proc IoA*, 9, Part 2, *Sonar Transducers Past, Present and Future*, 1-22, (1987).

Additional Reading.

High Frequency Designs.

1 Meeker,T.R., "Thickness mode piezoelectric transducers," *Ultrasonics*, 10, 26-36, Jan 1972.
2 Bradfield,G., "Ultrasonic transducers 1. Introduction to ultrasonic transducers," Part A, *Ultrasonics*, 8, 112-123, (1970)., Part B, *Ultrasonics*, 8, 177-189, (1970).
3 Katz,H.W., *Solid State Magnetic and Dielectric Devices*, Wiley & Sons, 1959, Ch 3.
4 Hanish,S., op. cit., Vol III,
5 Kikuchi,Y.,(Ed), *Ultrasonic Transducers*, Corona Publishing Co. Ltd., 1969, Ch 7.

Helmholtz Resonators.

6 Woollett,R.S., "Underwater Helmholtz Resonator Transducers: General Design Principles," Naval Underseas System Center, NUSC Tech Rep 5633, (1977).
7 Henriquez,T.A., and Young,A.M., "The Helmholtz resonator as a high-power deep-submergence source for frequencies below 500Hz," *J Acoust Soc Am*, 65, 1556-1558, (1980).
8 Young,A.M., and Henriquez,T.A., "Underwater Helmholtz resonator transducers for low-frequency, high power applications," *Proc IoA*, 9, Part 2, 52-57, (1987).
9 Dowling,A.P., and Ffowcs Williams,J.E., *Sound and Sources of Sound*, Ellis Horwood, 1983. (Chapter 6)

Flextensional Transducers.

10 Oswin,J., and Dunn,J., "Frequency, Power, and Depth Performance of Class IV Flextensional Transducers," in Hamonic and Decarpigny (Ref [11.22])

General.

11 Sherman,C.H., "Underwater sound - A review. 1.Underwater sound transducers," *IEEE Trans on Sonics and Ultrasonics*, SU-22, 281-290, (1975).

Several of the references above also provide good reviews of the approaches to the design of high or low frequency transducers. These are generally evident from their titles.

Chapter 12
NEW MATERIALS
AND CONCEPTS

12-1   INTRODUCTION

Piezoelectric ceramics have been much the most popular active material for underwater transducers for over 30 years, because of their advantages over the preceding crystal and magnetostrictive nickel devices. They do however have some disadvantages, which limit their usefulness in some areas. Their ceramic nature means that they are both brittle and stiff. Because of their weakness under tension, it is necessary to apply a compressive stress, for example by means of a centre bolt as described in Chapter 7, and although this technique gives good results, it does introduce extra design complications and cost. The ceramic elements may also be vulnerable to damage from mechanical shock. The high stiffness of the ceramic limits the strain which can be produced, unless very high electric fields can be applied, and breakdown due to arcing across the ceramic is one of the main problems. This limitation becomes particularly serious for low frequency sources, where the required amplitudes become large, and has led to the need for the alternative designs discussed in Chapter 11.

Although the dielectric constants of the piezoelectric ceramics are relatively high, their electrical impedances at low frequencies are high, and they are therefore not readily matched to typical amplifier outputs. Transducers based on piezoelectric ceramics also tend to be relatively heavy, partly because of the density of the ceramic itself and partly because of the associated metal components. Ceramic processing techniques allow piezoceramic pieces to be made in a variety of shapes and sizes, but their dimensions are generally limited to less than about 100mm unless exceptional care is taken in their manufacture and handling. The thickness in the poling direction is usually limited to about 15mm. These limitations can in some circumstances introduce extra costs in the transducer design, or even performance penalties.

Finally, variations arise between different samples or batches of piezoelectric ceramics, partly because of the aging inherent in their internal structure, and partly because of variations in the production processes involved. The consequences of these variations are generally not too serious, provided that due allowance for them is made in selecting the material and in the system design, but adequate quality control and testing is necessary.

Thus, although piezoelectric ceramics have considerable advantages over crystals, they do possess their own particular limitations. Various new materials or concepts have therefore been investigated in recent years, and these are the subject for this chapter. They include new magnetostrictive materials, composite materials, piezoelectric polymers, and acousto-optic sensors.

12-2 MAGNETOSTRICTIVE TRANSDUCERS

We start this section by describing the main features of magnetostrictive transducers of the traditional type, since this forms the foundation for any discussion of subsequent developments. Magnetostrictive alloys of **nickel** were commonly used during the 1940s, and their use in sonar transducers is well described in [12.1, 12.2]. (The principal characteristics of several magnetostrictive materials are listed in Table 3.2.) The basic mechanism depends on the generation of a mechanical strain by the application of a magnetic field. As for the electrostrictive effect, this magnetostrictive effect obeys approximately a square law, and some biasing field is needed to produce a linear relationship between applied field and strain.

Unfortunately, there is no simple equivalent in the magnetostrictive alloys to the poling process in piezoelectric ceramics, and they generally have relatively low remanent field values. It is thus necessary, for most high power applications, to apply a steady magnetic bias field to the alloy in order to achieve the optimum operating point. By analogy with the piezoelectric ceramics, these magnetostrictive materials are sometimes described as "piezomagnetic" when they are suitably biassed to produce a linear magneto-mechanical relationship.

There are two main methods for applying this magnetic bias field.

a) A permanent magnet may be used. This apparently simple method is somewhat more complicated than it appears, since the magnet forms a part of the magnetic circuit. In order to avoid eddy current losses in the magnet, due to the applied alternating field, it should thus be laminated or made of a high resistivity material such as a ferrite. The magnetic circuit for the alternating field may then be separated from that for the d.c. biassing circuit. It is not easy to apply a very large bias field by this method, so it is generally restricted to applications requiring only modest powers. For these, it constitutes an economical and relatively efficient design.

b) A bias field may be applied by passing a steady electrical current through a coil around the element. This may be either the coil carrying the alternating current or a separate winding. In the usual case, where limited space makes it desirable to use the same winding for both bias and excitation fields, it is then necessary to separate the d.c. and h.f. electrical circuits by using a capacitance and blocking inductor in the network. The generation of high powers requires quite high fields, and the power needed to produce the bias field causes a significant reduction in the overall efficiency.

For low power transmission, or merely for receiving, it is possible to operate at remanence, but only the high-cobalt alloys, such as permendur, have a high enough remanent field to be safely used for transmission at any significant power.

Since the resistivity of these alloys is quite low, an alternating field applied to a solid sample of the alloy would generate eddy currents within the metal, which would thus dissipate energy by Joule heating. This eddy current generation is reduced, as in transformer cores, by dividing the alloy into thin laminations. The thickness of the laminations is determined

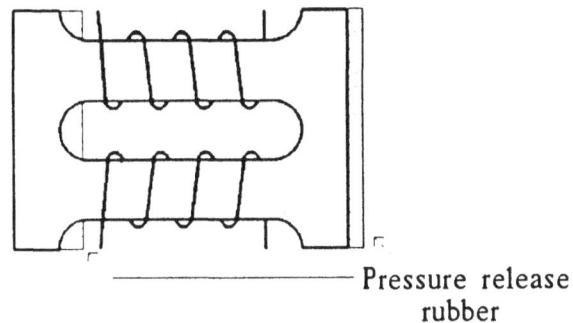

————————————————————— Pressure release
                                        rubber

**Fig 12.1 Magnetostrictive   element   using   "window"
laminations.**

by the frequency of operation, and the resistivity and
permeability of the metal, and can be calculated from the value
of the parameter $F_c t^2$ given by (3.11) and listed in Table 3.2.
For example, nickel laminations for a transducer operating at
50kHz should be not more than 0.18mm thick, and for a 20kHz
transducer 0.29mm. Because such thin laminations are very
flexible, they need to be consolidated into a stack which is thick
enough to avoid flexing resonances. This process needs to be
carried out whilst retaining the insulation between laminations on
which the eddy current prevention depends, and this is usually
achieved by covering each lamination with an insulating coating.
This process of building up the element from such thin
laminations introduces considerable extra complexity and cost
into the production methods.

   Two types of magnetostrictive elements have been most
commonly used – longitudinal and ring. The **longitudinal
element** often takes the form of an element built up from
"window" laminations as illustrated in *Fig 12.1*. The magnetic
field is applied to the narrow sections by coils wound around
them as indicated, and this then resembles the Tonpilz design in
effectively having end masses driven by a thinner central
section. The elements are often completely immersed in water,
and radiation from the tail mass and the rear of the front mass
is prevented by applying pressure-release (cellular) rubber to
those surfaces. In this type of design, the bias field is usually
applied electrically, but it is possible instead to split the rear
mass and insert into it a ceramic permanent magnet. Once the

elements have been built up from the individual laminations, the assembly is therefore relatively simple. However, the efficiency is typically about 50% or lower, the characteristics vary with bias field, and the coupling coefficient cannot exceed about 0.3 for nickel stacks. The use of pressure release rubber limits the operating pressure to about 1MPa, equivalent to about 80m depth.

Several metallic alloys were developed to improve on the magnetostrictive properties of nickel. One which was fairly commonly used was **Permendur**, composed of 49%Co:49%Fe:2%V. This has a higher remanent field than nickel, and can thus be used for modest power transmission without any applied bias field, although its best performance is still achieved when it is suitably biassed. It also allows rather thicker laminations than for nickel stacks (See Table 3.2). However, its coupling coefficient is little higher than for nickel, and its efficiency and pressure limitations are also similar. The difficulties caused by eddy currents become more severe as the frequency is raised, and metallic magnetostrictors are therefore rarely used above 100kHz. At the low frequency end, the practical restrictions on lamination size limit the Tonpilz designs generally to frequencies above a few kHz. Within this range, the magnetostrictive alloy transducers were widely used for industrial applications such as ultrasonic cleaning, and for a time also in sonar systems.

The next step was the development of the **magnetostrictive ferrites**. Many of the problems with magnetostrictive elements arose from the effects of eddy currents, and the consequent need for laminating the stack. In transformers, this had been tackled by the introduction of magnetic ferrites, which were ceramic materials with good magnetic properties and much higher resistivity than the metallic alloys. As a result, the need for laminations was virtually eliminated. Several laboratories developed ferrite compositions which were particularly aimed at giving good magnetostrictive properties, whilst retaining their high resistivity [12.3]. The characteristics of a typical composition, Ferroxcube 7A1, are listed in Table 3.2. Coupling coefficients were generally about the same as for the metallic alloys, and the main advantage of the ferrites was the great reduction in eddy currents, giving higher efficiencies and avoiding the need for laminations in the element. However, the ceramic nature of the ferrites brought with it the problem of brittleness under tension, as for the piezoelectric ceramics. The reduction in eddy current problems was therefore accompanied by increased difficulties in avoiding fracture of the ferrite material itself.

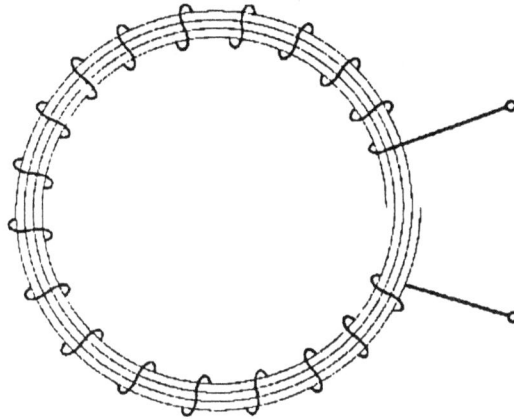

Fig 12.2   Magnetostrictive scroll transducer.

The use of metallic magnetostrictive alloys for Tonpilz designs had largely declined as a result of the introduction of barium titanate ceramic, with a coupling coefficient as high as for the alloys, higher efficiency, and more flexible and pressure tolerant constructions. The investigations into magnetostrictive ferrites roughly coincided with the development of the piezoelectric lead zirconate titanate compositions, with their greatly enhanced coupling coefficients, and the ferrites have therefore never been widely used for underwater transducer applications.

The second type of element is the **free-flooded ring**, the acoustic features of which are similar to those of the piezoelectric ceramic rings described in Section 11.5. This type of construction is relatively simple, as it is usually made by winding the material around a mandrel, to form a "scroll", as in *Fig 12.2*. This avoids the difficulties of assembling stacks of laminations, although the practical problems of maintaining insulation between the layers of the scroll remain. The magnetic field is applied by means of a coil wound toroidally around the scroll; the bias and driving fields may be applied by using two separate coils, or by using the same coil for both. Pressure release rubber is often used to prevent radiation from the inner surface of the scroll, and sometimes also from the top and bottom. It is nevertheless possible to obtain good performance from rings with no pressure release rubber, if the dimensions are correctly chosen, as with the piezoelectric rings described in Section 11.5. An example of the design of a scroll transducer, together with an analysis of its equivalent circuit, is given by Camp [12.4].

The diameter of a magnetostrictive scroll is a little larger than that of a barium titanate ring of the same resonance frequency, and is roughly equal to a wavelength of sound in water at its resonance. The advantage of the scroll transducer is the ease of insulating the electric circuit from the sea water, although its electroacoustic characteristics are basically inferior to those of a piezoelectric ring. For frequencies above about 4kHz, piezoelectric ceramic rings can be made as a single piece, and it is relatively easy to insulate them from the water by means of a resin or polyurethane coating. Piezoelectric rings are therefore usually preferred for these frequencies. For lower frequencies, it is not practicable to make single piezoelectric rings of a large enough diameter, and it is necessary to assemble ceramic rings from a number of pieces, as described in Section 11.5, which involves a much more expensive construction. It is however possible to wind large scrolls, which are suitable for these low frequencies, and these provide an economical alternative. Woollett [12.5] quoted an example of a very large scroll resonant at 350Hz and capable of radiating 100kW, and although applications demanding such large rings are rare, scrolls of more modest size constitute probably the most common current use of magnetostrictive sonar transducers.

Rare earth-iron magnetostriction.

Research into magnetostrictive materials led to the discovery, in 1963, of large magnetostriction effects in some rare earths. These effects were confined to low temperatures, but it was found that the addition of elements such as Fe, Co, or Ni, could raise the Curie point appreciably, and in the early 1970s giant magnetostriction effects were obtained in rare earth-iron alloys at room temperature [12.6]. A wide range of alloys has now been investigated, including binary, tertiary, and quaternary compositions of rare earths such as terbium, dysprosium, and holmium, generally with iron. The interest in these materials is due to the very large mechanical strain which can be produced when a magnetic field is applied. One alloy which has been optimised for its magnetostrictive properties is known as Terfenol D, with a composition $Tb_{0.27}Dy_{0.73}Fe_2$. The saturation strain for this material can be as high as 1500×10^{-6}, some 50 times larger than that of nickel and its alloys. This value can be compared also with a strain of 110×10^{-6} produced in LZT I by a field of 4×10^{5}V/m (4kV/cm), which is at the high end of the fields normally used in practice. The optimum coupling coefficient of Terfenol is 0.6, rather less than the value of 0.68 for LZT I (See Tables 3.1 and 3.2).

Properties of these polycrystalline samples are very sensitive to the magnitude and direction of the applied field, and to the orientation of the grains in the material. Improved properties can be obtained if the grains can be preferentially oriented, but this requires expensive processing methods. The low incremental permeability and relatively high resistivity of the rare earth–iron alloys give a value of $F_c t^2$ which is appreciably higher than for nickel, and the thickness of the laminations can therefore be increased. However, current materials are brittle and difficult to machine, so laminated elements are not easy to construct.

Several designs of experimental transducers have been made using thin rods of these alloys [12.7], and Oswin and Maskery [12.8] have drawn attention to their possible use in flextensional transducers. Results for these experimental transducers reflected the limitations in the available sizes and shapes of material, and values of efficiency were relatively low, although power outputs higher than for a similar sized piezoelectric design appeared feasible. Merely replacing piezoelectric ceramic in a transducer by a rare earth–iron element appears to have no overall advantage, and the magnetostrictive material shows up to best advantage for low frequency applications, where its large potential strain is of most use.

The large magnetostrictive strains available from these materials have obvious appeal for application to transducers. However, these strains can only be obtained by applying very large magnetic fields, of about 16kA/m, compared with fields of less than 1kA/m for nickel. The volume of conductor needed to apply this field, and the associated cooling for the winding and core, constitute serious problems in the design of transducers using these rare earth–iron alloys, and current materials are unlikely to displace the piezoelectric ceramics for general sonar transducer use. However, continuing investigations of these alloys may well reduce their limitations, and allow them to fulfil their promise, particularly for low frequency transducers where the need for large amplitudes is most acute.

Metallic Glasses.

Research into improved magnetostrictive materials has led also to the development of metallic glasses. These are amorphous alloys of a transition metal such as iron, nickel, or cobalt, with a metalloid such as boron, silicon, or carbon. They are produced as very thin ribbons, some 20-50µm thick by 25mm wide, by very rapid quenching of the molten alloy onto a cooled cylinder. The rapidity of the quenching preserves the amorphous structure of the molten alloy, as in a glass, but the material also retains the strength and ductility of the metal. This exceptional combination of properties gives rise to their name of metallic glasses.

When annealed in a transverse magnetic field, some of these metallic glasses exhibit strong magnetostrictive effects along the length of the ribbon. Values of coupling coefficient are commonly above 0.7, and a value of 0.95 has been measured for a commercially available material called METGLAS 2605SC (Allied Chemical Corporation), with a composition $Fe_{81}B_{13.5}Si_{3.5}C_2$ [12.9]. In contrast to the rare earth-iron compounds, these alloys can only be manufactured as thin samples, and the bias field required to optimise the properties is very low; at about 60A/m, it is less than 1/10th of that for nickel. The saturation strain is only about 50×10^{-6}, similar to that for nickel, and this constitutes a significant limitation to their use for high power applications. However, the value of the hydrophone figure of merit ($g_{33}d_{33}$) is quoted as more than four times greater than that for lead zirconate titanate ceramic, and it seems likely that present metallic glasses will be better suited to hydrophone applications than to projectors. The history of these materials for magnetostrictive transducers is still quite short, and if the other characteristics can be improved to take better advantage of the exceptionally high coupling coefficient, they may become more widely adopted.

12-3   LEAD NIOBATE AND LEAD TITANATE

For most of the piezoelectric materials, an expansion generated along one axis is accompanied by contractions along the other axes, so that the overall volume change is small. Hence, if a simple block of the ceramic is driven electrically whilst immersed in a fluid, it radiates only poorly. In the receiving mode, similar cancellation occurs between the voltages generated along the three axes, with the result that the **hydrostatic coefficient** (g_h) is usually much smaller than for the 33 or 31 modes. Values given in Table 3.1, using equation (3.5), show that g_h for the lead zirconate titanate ceramics is typically 1/8th of g_{33}. The pressure sensitivity of a simple block of piezoelectric ceramic immersed in water (with suitable electrical insulation!) is therefore relatively low, and it is for this reason that some form of pressure release needs to be incorporated, to isolate the sides from the acoustic pressure.

A few materials show anomalous behaviour, in having much less cross-coupling to the transverse mode, and these can then possess much higher hydrostatic sensitivities. This is true, for example, of lithium sulphate and tourmaline crystals, and these have been used for many years for standard hydrophones. They do however have the usual disadvantages of the crystalline materials, and the use of more easily fashioned ceramics is preferable. Lead metaniobate ($PbNb_2O_6$) and lead titanate are unusual ceramics which provide such an opportunity, because of their high hydrostatic coefficients. Modifications of the ceramics are often made, to optimise the characteristics, and Table 12.1 shows the main parameters of basic lead metaniobate [12.10] and a commercial version (EDO Corporation's EC-82), compared with a lead zirconate titanate hydrophone composition, LZT II. The main properties of commercially available versions of lead titanate, such as EDO EC-97, or Toshiba C-24, are typified by the figures listed for C-24.

TABLE 12-1

	$PbNb_2O_6$	EC-82	LZT II	C-24	
ε_{33}^T	225	300	1700	209	
k_{33}	0.38	0.39	0.70	0.54	
d_{33}	85	85	420	68	10^{-12}m/V
d_{31}	-9	-10	-160	-3.2	10^{-12}m/V
g_{33}	42.5	30	28	37	10^{-3}Vm/N
g_{31}	-4.5	-3.5	-12.0	-1.7	10^{-3}Vm/N
g_h	33.5	23	4.0	33.6	10^{-3}Vm/N
$g_h d_h$	2.24	1.50	0.24	2.07	10^{-12}m²/N
Density	6.0	5.8	7.7	6.9	10^3Kg/m³
Curie Temp	570	400	370	255	°C

The hydrostatic g-coefficients of both lead niobate and lead titanate are more than six times higher than for the LZT. The hydrophone figure of merit ($g_h d_h$) determines the potential performance of a hydrophone (Section 3.2), and despite the relatively low dielectric constants of both materials, their figures of merit show a clear advantage over that for LZT in the hydrostatic mode. Although lower than for LZT in the thickness mode, they are comparable with that for LZT used in the 31 mode, as in a cylindrical hydrophone. Exposure of the ceramic to hydrostatic pressure is less likely to cause depoling than the stresses in a more conventional LZT design, and a hydrophone using lead metaniobate or lead titanate in the hydrostatic mode therefore has considerable potential for deep submergence operation. For some applications, the high Curie temperature of the metaniobate has advantages, but for underwater use this is unlikely to be significant. Lead metaniobate ceramic has been available for some time, but has not been widely used in practice for sonar applications. Lead titanate suffered from some difficulties in manufacture, but is now more readily available and receiving increasing attention.

The main advantage of these ceramics for low frequency hydrophones lies in the simplicity of the construction, since the ceramic needs no air backing, and it is relatively easy to mount a simple free-flooded cylinder or disc of the ceramic. The usual precautions against acceleration response and water leakage are of course still necessary.

12-4   CERAMIC-ELASTOMER COMPOSITES

Much research has been conducted in recent years into composite materials in which piezoelectric ceramic is mixed with various elastomers. This research is based on two areas of interest. Firstly, such composites can have high hydrostatic coefficients, and represent an alternative to the use of lead metaniobate or titanate for this purpose. Secondly, the combination of elastomer with ceramic can markedly reduce the stiffness compared with the ceramic itself, and even in some cases produce a piezoelectric material which is flexible and suitable for covering a large area. Both features could be combined to simplify the design of large hydrophone arrays.

The original approach was simply to mix small particles of piezoelectric ceramic into an elastomer such as silicone rubber or polyurethane. Unfortunately, such a material is difficult to pole effectively, because most of the poling voltage is lost across the elastomer rather than the ceramic. The alternative, of dispersing into the elastomer small pieces of ceramic which are already poled, is an intricate and expensive process, and not readily adapted to production. These investigations led to renewed interest in how the properties of composite materials depend on their internal structure. Two phases may be mixed in a number of ways, which are characterised by their "connectivities". These are represented by a pair of numbers, the first indicating the number of axes for which the ceramic is continuous, and the second the corresponding figure for the elastomer. Ten possible arrangements exist, ranging from 0-0 to 3-3, and several of them have now been studied [12.10]. The dispersion of particles in a matrix is described as having 0-3 connectivity, since the matrix is connected along all three axes, whilst the ceramic particles have no connectivity. (This system for describing the connectivities now appears to have become the standard, although some authors have used the reverse notation.)

One way of reducing the poling difficulties which arise for the dispersed particles is to ensure connectivity of the ceramic between the electrodes. This has been effected by aligning an array of thin ceramic rods perpendicular to the electrodes, the interstices being filled with a flexible polymer. This structure has 1-3 connectivity, and is probably the most extensively studied. In addition to being relatively easy to pole, it would be expected that the stress would be taken by the ceramic rods, whilst lateral coupling would be reduced by the soft elastomer, so that a high g_h should result. In practice, the response is significantly affected by the Poisson's ratio effect of the elastomer, which transfers the axial stress to the sides of the rods, thus enhancing the g_{31} effect and reducing the value of g_h. This problem can be tackled by foaming the elastomer, but this introduces an undesirable depth dependence. An alternative is to incorporate some controlled stiffness in the transverse plane, for example by means of a mat of glass fibres. In such cases, the fibre connectivity is described by adding a third suffix, to give eg a 1-3-1 material. Although good properties can be obtained for some of these 1-3 composites, fabrication of other than small experimental samples poses considerable difficulties.

Samples of 3-1 and 3-2 composites have been made for research purposes by drilling holes in LZT blocks, perpendicular to the poling direction. Samples of 3-3 composite have been prepared by using a lost wax process which replicates the structure of natural coral, and by a less time-consuming method which involves burning out small plastic spheres from a mixture, and back-filling the pores with silicone rubber or epoxy resin ("Burps"). The difficulties of poling the 0-3 composites appear to have been overcome in the production of Piezo-Rubber (NTK Technical Ceramics Division, NGK Spark Plugs Corporation), which is manufactured as a thin (0.5mm) flexible sheet [12.11]. Theoretical treatments of the properties of these composite materials have been derived by various workers, and experimental data for various samples are given by Cannell [12.10] and Ting [12.11]. A few examples of the results, taken from [12.10], are shown in Table 12.2. These show that the characteristics of these materials are very dependent on the elastomer used, and the details of the processing technique, but suggest that hydrophone figures of merit in the hydrostatic mode may range up to $20\times10^{-12}\text{m}^2/\text{N}$ or more, some 5-10 times higher than for lead metaniobate.

TABLE 12-2

Property (Units)	ε_{33}^{T}	d_{33} pm/V	g_{33} mVm/N	d_h pm/V	g_h mVm/N	$g_h d_h$ m^2/N $\times 10^{-12}$
3-3 "Burps" Silicone rubber	300	250		260	100	26
3-3 "Burps" Epoxy resin	500	175		120	27	3.2
2-3 Epoxy resin	290	290	114	329	128	42
1-3 Epoxy resin	450	290	73	222	56	12.3
0-3 Silicone rubber	120	50		30	30	0.9

These composites have lower densities and speeds of sound than do the piezoelectric ceramics, and thus offer extra possibilities for the design of small, lightweight hydrophones. Those with greater proportions of polymer can be mechanically flexible, and some of the compositions could in principle be made in large samples, so that large arrays of simple construction may be achievable. However, much further work needs to be done to achieve the optimum balance of properties and manufacturing costs.

12-5 PIEZOELECTRIC POLYMERS

Many materials exhibit piezoelectric effects, but generally these effects are not large enough to be useful in transducers. Thus, although numerous organic polymers had been known for many years to be piezoelectric, they aroused no great interest until the discovery by Kawai in 1969 of strong piezoelectric activity in **polyvinylidenefluoride (PVDF)** (sometimes known as PVF_2). This discovery generated much research into the understanding and improvement of the properties, and into their potential applications. As a result, PVDF is now a well-developed material, with widespread use in the commercial field and for ultrasonic transducers. Its development for underwater transducers has been rather slower, perhaps because it has been difficult to demonstrate clear advantages for the material over a significant range of sonar applications.

A good review of the physical and piezoelectric properties of PVDF has been given by Sessler [12.12]. The material itself is a semicrystalline polymer which can readily be formed into thin sheets. Four crystalline phases are known to exist, of which only two, the α and β phases, are important for this discussion. Cooling the material from its molten form produces α-phase crystals. This phase is non-polar and non-piezoelectric. It may however be transformed by suitable treatment into the β-phase, which is polar and hence intrinsically piezoelectric. This transformation is commonly effected by stretching a sheet of the α-phase material obtained by cooling from the melt. As with the piezoelectric ceramics, this material then needs to be converted to a linear piezoelectric by a poling process. Manufacture of piezoelectric PVDF from the non-piezoelectric α-phase film thus involves the following series of processes:-

a) uniaxial stretching at 60-65°C to 3-5 times its original length, converting the polymer to its β-form.

b) clamping and annealing of the material at 120°C, to stabilise the film.

c) poling the film by applying a strong electric field across its thickness, to cause partial alignment of the dipoles. This may be done by electroding the film and then applying a field of 50-80kV/mm at about 100°C for an hour, followed by cooling to room temperature with the field maintained. An alternative process involves electroding only one surface, and using a corona discharge onto the film to achieve polarisation.

Once poled, the material shows good stability with time at room temperatures, but decays rapidly above 80°C. Typical values of some of its principal characteristics at room temperature are listed in Table 12.3 (from [12.12, 12.14]), compared with LZT II hydrophone ceramic. The subscript 1 refers to the direction of stretching.

TABLE 12-3

	PVDF	LZT II	
d_{33}	−25	420	10^{-12}m/V
d_{31}	20	−180	10^{-12}m/V
g_{33}	−250	28	10^{-3}Vm/N
g_{31}	200	−12	10^{-3}Vm/N
g_h	80	4.0	10^{-3}Vm/N
$g_{31}d_{31}$	4	2.2	10^{-12}m²/N
k_{31}	0.16	0.32	
ε_r	11	1700	
Density	1.76	7.7	10^3Kg/m³

The d-coefficient is much lower than for LZT ceramic, but the low dielectric constant leads to a high value for the g-coefficient, and the hydrophone figure of merit (gd) is higher for PVDF than for LZT II in the 31-mode. PVDF is thus primarily suitable for hydrophone applications. It is nevertheless worth noting that very high electric fields can be applied to PVDF without causing depoling, and the maximum achievable strain for PVDF is some ten times larger than for ceramics, so that it may have some potential for use as a projector. The main advantages of PVDF lie in its mechanical properties. It is flexible, rugged, has low density, and can be made in large sheets relatively easily. These properties are in marked contrast to those of the piezoelectric ceramics and thus offer the possibility of quite different design approaches.

PVDF was first available as Kureha KF Piezofilm from the Kureha Chemical Industry Co. of Japan, with thicknesses ranging from about 6 to 30μm. A number of applications are described by Sessler [12.12], many of them being devices used in air for the consumer market. An example of the application of PVDF to an underwater hydrophone design is given by Sullivan and

Powers [12.13]. Their technique was to construct a flexural disc design in which the PVDF was bonded to a plastic disc. Since the film is so thin, and its elastic modulus is relatively low, the mechanical characteristics of the flexural disc are determined primarily by the plastic disc, and the PVDF acts largely as a strain gauge. The toughness of the PVDF allows it to withstand large strains, and it is therefore not so susceptible to fracture under hydrostatic pressure as the ceramic in the bilaminar discs described in Section 11.3.

The hydrophone described by Sullivan and Powers illustrates one of the factors which results from the small thickness of the available Kureha PVDF film. Despite the low dielectric constant of the material, the capacitance of the film is quite high; for example, a $1cm^2$ sample of $30\mu m$ film would have a capacitance of about 0.32nF. But the associated penalty is that the pressure sensitivity is low, since that is proportional to the thickness of the film. In order to increase the sensitivity, it is usual to use several layers of film bonded together and connected electrically in series. This trade-off between sensitivity and capacitance corresponds to that described in Chapter 10 for ceramic hydrophones, but its importance is greater for PVDF than for the ceramics because of the low sensitivity of the thin film. This process allows the full potential of the high figure of merit to be realised, but at the expense of considerably increased assembly time. Several other developments making use of PVDF for underwater transducers have been described by Ricketts [12.14], including designs using PVDF film in the thickness mode, wrapped around cylinders, or on compliant tubes. PVDF has a high hydrostatic coefficient, and designs using the hydrostatic mode have therefore been investigated. PVDF has also been used in the thickness mode as a projector for some high frequency medical ultrasonic applications.

Thicker PVDF sheet has been produced by Thorn-EMI, thus reducing the assembly problems associated with the thin film. This material is available in thicknesses up to about 0.5mm, and in two forms. One of these forms contains many needle-shaped cavities ("voids") within the material, the proportion of voids being controllable during the manufacture. The effect of these voids is to enhance the hydrostatic coefficient, values of $g_h \simeq 250mVm/N$ and $g_h d_h \simeq 5 \times 10^{-12}m^2/N$ being quoted by McGrath [12.15]. The non-voided form has greatly increased stiffness in the direction of stretching, and a $g_h d_h$ product of $0.7 \times 10^{-12}m^2/N$. The potential of improving the properties of PVDF by forming copolymers with monomers such as trifluoroethylene has been investigated, and an improvement of some 50% in the $g_h d_h$ product compared with thin film PVDF has been reported [12.16]. A possible advantage of these copolymers is that they do not need to be stretched, and hence can be moulded directly into the required shapes.

PVDF has received considerable attention for ultrasonic applications and for various audio-frequency transducers in air. For underwater systems, its use is primarily in hydrophones, where its advantages lie mainly in its flexibility, low density, and availability in pieces having large area at potentially low cost. These could allow it to replace conventional ceramic designs for applications where these advantages are important. Disadvantages of the material include a significant pyroelectric coefficient, some variability of parameters arising from manufacturing techniques and aging, and a temperature range restricted to below 80°C. Experience with this material is steadily increasing, and its position appears well established for suitable applications.

12-6   FIBRE-OPTIC HYDROPHONES

The interaction between acoustics and optics forms the basis for a range of hydrophones, of which those using optical fibres are the most common. These differ from the previous techniques by representing an entirely new approach to sensing acoustic pressure, not just a variation on existing methods. Only the main features of these hydrophones will be described in this section, and for a fuller treatment the reader is referred to reviews such as those by Bjørnø [12.17], or Bucaro et al [12.18].

Optical fibres have been developed primarily for communications applications, and are now widely available at moderate cost. The fibres are sensitive to a number of externally applied fields, including pressure, and much research has been conducted since the early 1970s to investigate their characteristics as sensing devices. This interest stemmed from their potential advantages in providing sensors which could be geometrically very flexible, light weight, resistant to high temperatures and electrical interference, and compatible with optical fibre communication systems. As a result of this research, several basic mechanisms have been developed for use as hydrophones. Optical fibres can also be used to measure various other quantities, and the techniques involved have many common features.

A typical fibre-optic cable has a glass core, surrounded by a cladding, and protected by a plastic outer coating. The cladding has a lower refractive index than the core, so that light is kept within the core by total internal reflection at the boundary. Light travels along the core in the form of modes, analogous to electromagnetic propagation in wave-guides, and the cables are classified as "single-mode" or "multiple-mode" fibres. Single-mode fibres are typically about 3μm in diameter, and multi-mode 40μm or more. Such thin fibres are exceptionally strong and flexible, and retain their strength if they are adequately protected from the environment by the outer sheath. Both types are used for sensors, although the single-mode fibres offer the best prospects for hydrophones.

When pressure is applied to a fibre, it may affect the optical path through changes produced in properties such as the path length or refractive index. Although these changes may appear very small, their effects in a single-mode fibre can become large enough, when integrated over an adequate length of cable, to cause a measurable change of phase. Measurement of this phase change forms the basis for probably the most common type of fibre-optic hydrophone. The measurement is usually carried out by means of a Mach-Zehnder interferometer, in which the optical beam is split into two fibres, one a reference fibre and the other a sensing fibre. The sensing fibre is exposed to the acoustic field, which modulates the phase of the optical signal, whilst the reference fibre is isolated from the field. The two beams are then recombined, detected by a photo-detector, and demodulated to give an output at the acoustic frequency. At low frequencies, where the sensor dimensions are small compared with the acoustic wavelength, the calculated phase shift in the sensing arm has a simple expression, with one contribution from the axial strain of the fibre and another from the fractional change in refractive index. Most importantly, the resultant phase shift is proportional to the length of fibre exposed to the field. As with other interferometers, the output becomes non-linear if the phase shift exceeds π, and a linearising element (such as a phase or frequency modulator) needs to be included in the reference arm.

Detection of the phase shift can be carried out with great sensitivity, a precision of 10^{-5}-10^{-6} radians being achievable. Since the optical wavelength is of order 1μm, this phase shift corresponds to a change of 10^{-11}-10^{-12}m in the optical path length. In order to obtain the required acoustic pressure sensitivity, a fibre length of 10m or more (wrapped into a small coil) is usually exposed to the acoustic field, giving the potential for detecting acoustic signals down to about 10^{-4}Pa (40dB re 1µPa), even at frequencies down to a few hundred Hz. Realising this sensitivity demands nearly ideal performance of all the components of the system, and this is not always easy to achieve. For example, a temperature difference of only 10^{-3}°C between the sensing and reference fibre will change the path length enough to spoil the performance, unless a method of compensation is included. Problems can arise from the effects of the acoustic (or other) field on the optical cables connecting the sensor to the rest of the system, particularly if the reference arm is not exposed to identical effects. One possible solution is to put the interferometer near to the sensor, preferably within the sensor assembly [12.19]. Because of the sensitivity of optical fibres to other influences, much care is necessary to ensure that the sensor responds only to acoustic pressures at the sensing position.

The sensitivity of the fibre to acoustic pressure may be increased by suitable choice of the outer coating. If the coating has a low Young's modulus and low Poisson's ratio, the stress in the coating can be transferred to the core, thus enhancing the stress in the core and producing an order of magnitude increase in acoustic sensitivity. This is achieved by using a plastic coating, which also has the advantage that it provides some thermal isolation of the fibre. Such a cable is flexible and allows a variety of shapes of array to be employed. One arrangement is to position the reference fibre close to the sensing fibre, which makes it possible to measure acoustic pressure gradients, and reduces the problems of balancing the sensing and reference paths.

These phase sensitive fibre-optic hydrophones have been developed from experimental laboratory equipments to reasonably rugged operational devices, and continuing work will doubtless introduce further improvements.

Single-mode fibres permit the propagation of two orthogonally polarised modes, which normally have identical propagation constants. If an anisotropic stress is applied to the fibre, birefringence may be caused, thus producing a difference in propagation between the two modes. Measurement of the resulting phase difference between the modes is used in **polarisation sensors** to give a measure of acoustic pressure. One method for applying an anisotropic stress is to wrap the fibre around a compliant cylinder. Vibration of the cylinder under the influence of the acoustic field imposes an alternating modulation of the stress on the fibre, and thus provides the basic mechanism for this type of hydrophone. One advantage of the technique is that only one fibre is needed, which reduces the susceptibility to environmental conditions and leads to a simple construction.

In addition to these two techniques based on measurement of phase differences, methods have been devised which depend on acoustic modulation of the intensity of light transmitted along the fibre. These can use multi-mode fibres, which are generally less expensive than single-mode. **Intensity sensors** are often based on the effect of "microbending" on the fibre. If a fibre is bent, light is lost into the cladding, and the associated transmission loss for light propagating along the fibre becomes large if the bending radius of curvature is small. In one version of this "fibre microbend sensor", a fibre was subjected to a number of opposing microbends by distorting it between two corrugated plates. The plates were exposed to the acoustic pressure, and transferred the acoustic pressure to the fibre by squeezing it between the corrugations, the distance between the corrugations in the plates being selected to optimise the effect. The acoustic signal thus produced a modulation of the intensity of the light transmitted along the fibre. A reference channel is needed to compensate for any drift with time of the optical source. The sensitivity of this type of sensor is less than for the phase sensitive design, and a minimum detectable signal of about 60dB re 1μPa for frequencies above 500Hz is reported [12.17].

Numerous other techniques for optical hydrophones have been investigated, and it is difficult at present to foresee which of the possible methods will be most successful. The optical path in the fibre is sensitive to many other influences, such as temperature, magnetic field, and acceleration, and one of the fundamental difficulties with all the designs of fibre-optic sensors is to enhance the sensitivity to acoustic fields whilst suppressing the response to the other factors.

In addition to the fibres themselves, various other devices are needed to produce a complete hydrophone system. The light source needs to be compact and low noise, with a good coherence length. Semiconductor laser diodes are small and have high efficiency, but their noise characteristics are poor compared with gas lasers, and efforts to reduce line broadening are often necessary. Stable and low loss couplers are needed for some of the systems, and some involve very careful assembly of components to within tolerances of a few μm. These elements add considerably to the complexity and cost of the system, despite the relatively low cost of the optical fibres themselves.

Fibre-optic hydrophones are potentially capable of high sensitivity and have considerable flexibility in possible geometries. The main problems lie in the technical engineering aspects, to convert laboratory devices into equipment which can be manufactured at low enough cost and is rugged enough for general use at sea. These problems present serious challenges, and it remains to be seen how well they can be overcome. Another feature which is receiving study, and needs to be solved if fibre-optic hydrophone are to reach their potential, is that of multiplexing a number of hydrophones onto a common fibre channel. If all of these aims can be achieved, it should be possible to link the optical hydrophones into an optical processing element, and thus obtain a wholly optical system.

12-7 SUMMARY

In this chapter, a wide variety of transducer materials and techniques has been described, illustrating the breadth of activity in this field. Most of the developments are directed at hydrophone applications, and each has its own special features which may be of benefit for particular uses. In general, all are capable of measuring acoustic levels down to the ambient noise levels in the sea, so sensitivity in itself is not a difficulty. Although the values of hydrophone figure of merit differ, the differences are not generally so large as to lead to any overall conclusion on the suitability of the materials. The choice between materials therefore becomes dependent on other factors, such as mechanical restrictions, or cost, or degree of acceptable technical risk. A summary of the main characteristics and probable applications is indicated in Table 12.4.

Magnetostrictive transducers continue to be heavy, although the rare earth-iron alloys show some promise, especially for low frequency sources. Piezoelectric polymers have considerable attraction for large hydrophone arrays, despite their limited piezoelectric properties, and technical risks are relatively low, although their potentially low cost remains to be demonstrated in practice. The high hydrostatic coefficients of the lead titanate and lead metaniobate compositions make them useful as simple low frequency hydrophones, now that the material is more readily available. The combination of good hydrostatic coefficient and flexibility make the ceramic-elastomer composites promising for large area arrays, but much work remains to be done to achieve consistent and economical manufacture. The advantages of the novel fibre-optic hydrophones for underwater use are still

TABLE 12-4

Material	Main feature	Tech. risk	Application
Rare earth-iron	Large strain	Medium	LF projectors
Metallic glass	High coupling	High	
Niobates/ titanates	High hydrostatic coefficient	Low	Deep hydrophones
Composites	High hydrostatic coeff.,flexible	Medium	Large area arrays
Piezo-polymers	Flexible, low density	Low	Large area arrays. Towed arrays
Fibre-optic	Flexible, novel	High	Large arrays

not clear, although they show promise for use in arrays with large numbers of elements if a satisfactory multiplexing technique can be implemented.

This range of new materials provides many opportunities for selecting materials to optimise the properties for particular applications. However, none of the current materials has a combination of properties which would make it likely to displace the present piezoelectric ceramics for general purpose use in a wide range of underwater transducers.

REFS Chapter 12

References

12.1   Nat. Def. Res. Comm., Div. 6, Summary Technical Report, Vol 13, *Design and Construction of Magnetostriction Transducers*, 1946.
12.2   Kikuchi,Y.(Ed), *Ultrasonic Transducers*, Corona Publishing Co. Ltd., 1969, Chapter 5.
12.3   van der Burgt,C.M., and Stuijts,A.L., "Developments in ferrite ceramics with strong piezomagnetic coupling," *Ultrasonics*, $\underline{1}$, 199-210, (1963).
12.4   Camp,L., *Underwater Transducers*, Wiley Interscience, 1970, Chapter 6.

12.5   Woollett,R.S., "Power limitations of sonic transducers," *IEEE Trans. on Sonics and Ultrasonics*, SU-15, 218-229, (1968).

12.6   Clark,A.E., "Magnetostrictive Rare Earth-Fe_2 Compounds." Chapter 7 in *Ferromagnetic Materials*, Vol 1, Wohlfarth,E.P.,(Ed), North Holland Pub. Co., 1980. (Reproduced also in Hamonic et al, [11.22])

12.7   Butler,J.L., and Ciosek,S.J., "Development of two rare-earth transducers," in *Proceedings of 25-26 Feb 1976 Workshop on Magnetostrictive Materials*, (NRL Report 8137, June 1977), p165. Also, Smith R.R., and Logan,J.C., "Design of a transducer using rare-earth magnetostrictive materials," op. cit., p175. Also, Akervold,O.L., Hutchins,D.L., Johnson,R.G., and Koepke,B.G., "Rare-earth magnetostrictive transducer and material development studies," op. cit., p183.

12.8   Oswin,J.R., and Maskery,A., "Magnetostriction in flextensional transducers," *Proc IoA*, 9, Part 2, Sonar Transducers Past, Present, and Future, 23-30, (1987).

12.9   Modzelewski,C., Savage,H.T., Kabacoff,L.T., and Clark,A.E., "Magnetomechanical coupling and permeability in transversely annealed Metglass 2605 alloys," *IEEE Trans. on Magnetics*, MAG-17, 2837-2839, (1981).

12.10   Cannell,D.S., "The development of piezoelectric ceramics for transducers," *Proc IoA*, 9, Part 2, 159-170, (1987).

12.11   Ting,R.Y., "Evaluation of new piezoelectric composite materials for hydrophone applications," *Ferroelectrics*, 67, 143-157, (1986).

12.12   Sessler,G.M., "Piezoelectricity in polyvinylidenefluoride," *J Acoust Soc Am*, 70, 1596-1608, (1981).

12.13   Sullivan,T.D., and Powers,J.M., "Piezoelectric polymer flexural disk hydrophone," *J Acoust Soc Am*, 63, 1396-1401, (1978).

12.14   Ricketts,D., "Recent developments in the USA in the application of PVF_2 polymer in underwater transducers," *Proc IoA*, 6, Part 3, 46-54, (1984).

12.15   McGrath,J.C., "Research at Thorn-EMI Central Research Laboratories on piezoelectric polymer materials," IEE and IoA Colloquium on *New Materials for Sonar Transducers*, Digest No 1986/60, (1986).

12.16   Suttle,N.A., "New piezoelectric polymers," *GEC Journal of Research*, 5, 141-147, (1987).

12.17   Bjørnø,L., "Adaptation of Fiber Optics to Hydrophone Applications," in *Adaptive Methods in Underwater Acoustics*, Urban,H.G.,(Ed), D.Reidel Pub. Co, 1985, pp 629-641.

12.18 Bucaro,J.A., Lagakos,N., Cole,J.H., and Giallorenzi,T.G., "Fibre-Optic Acoustic Transduction," in Mason,W.P., and Thurston,R.N.,(Eds), *Physical Acoustics*, Vol 16, Academic Press, 1982.

12.19 Jones,R.E., Neat,R.C., and Hale,P.G., "Optical fibre sensors for passive hydrophones" IEE and IoA Colloquium on *New Materials for Sonar Transducers*, Digest No 1986/60, (1986).

Additional Reading

1   Berlincourt,D.A., Curran,D.R., and Jaffe,H., "Piezoelectric and Piezomagnetic Materials and their Function in Transducers," in Mason W.P.,(Ed), *Physical Acoustics*, Vol 1A, Academic Press, 1964.

2   *Proceedings of 25-26 Feb 1976 Workshop on Magnetostrictive Materials*, Timme,R.W.,(Ed), NRL Report 8137, 1977.

3   *Proceedings of Colloquium on New Materials for Sonar Transducers*, 24 April, 1986, IEE and IoA, Digest No. 1986/60.

4   Savage,H.T., and Spano,M.L., "Theory and applications of highly magnetoelastic Metglas 2605SC," *J Appl Phys*, 53, 8092-8103, (1982).

5   Neppiras,E.A., "New magnetostrictive materials and transducers - I," *J Sound Vib*, 8, 408-430, (1968). Also Pt II, *J Sound Vib*, 8, 431-456, (1968).

6   Walmsley, P., "Metallic glass as a hydrophone material," *Proc IoA*, 6, Part 3, 38-45, (1984).

7   Sessler,G.M., "What's new in electroacoustic transducers," *IEEE ASSP Magazine*, 1, 3-13, (Oct,1984).

8   Bobber,R.J., "New Types of Transducer," in Bjørnø,L.(Ed), *Underwater Acoustics and Signal Processing*, Reidel, 1985, p243.

9   Taylor,G.W., Gagnepain,J.J., Meeker,T.R., Nakamura,T., and Shuvalov,L.A.(Eds), *Piezoelectricity, (Ferroelectric and Related Phenomena, Vol 4.)*, Gordon and Breach Science Publishers, 1985.

Appendix A
SYMBOLS

Most of the symbols used in this book are listed here. Reference is generally made to the section or equation (indicated within brackets) where the symbol is first used. Some symbols which occur only in a limited section of text, and are defined where they occur, are omitted from this listing. In attempting to use symbols which are familiar to workers in the appropriate fields involved, it has proved impossible to avoid some clashes or duplications without introducing undue complexity in the nomenclature. It is hoped that the meaning of those symbols where ambiguity is possible will be readily evident to the reader from their context.

A	acceleration response factor	(10.73)
A	area of piston or array	(2.4)
A_b	effective cross−sectional area of bolt	(7.18)
A_c	cross−sectional area of ceramic	(7.2)
A_n	cross−sectional area of neck	(11.11)
A_p	piston area	Sect 1.3
A_1, A_3,etc	network parameters	(5.18),(5.20)
a	network coefficient	(5.18)

a	piston radius	(6.4)
a	radius of neck	(11.12)
a	semi–major axis of shell	Sect 11.4
B	susceptance	(1.6)
B	acceleration response factor	(10.73)
B_{in}	input susceptance	(5.1)
B_S	susceptance at f_S	(4.4)
b	network coefficient	(5.18)
b	ratio of stack stiffness/bolt stiffness	(7.19)
b	external radius of sphere	(10.51)
b	semi–minor axis of shell	Sect 11.4
C	capacitance	Sect 1.1
C_A, C_B	capacitances at f_A, f_B	(9.5)
C_B	compliance of centre bolt	(8.2)
C_c	compliance of ceramic rings	(8.2)
C_{HF}	capacitance at high frequency	Sect 9.2
C_J	compliance of joints, etc	(8.2)
C_{LF}	capacitance at low frequency	Sect 3.2, 4.3
C_p	effective compliance of plate	(11.2)
C_S'	$C_c + C_J$	(8.7)
C_0	clamped capacitance	Sect 4.3
C_1	series capacitance	Sect 4.3
C_2, C_3, C_4	filter capacitances	(5.22–5.32)
C_{12}, C_{23}	capacitances in coupled resonator network	(5.49)
c	speed of sound	Sect 1.2
c	elastic stiffness coefficient	Sect 3.2
D	electric displacement	Sect 3.2
DI	directivity index	(2.4)
D_1, D_2	array interaction factors	(6.17)
d	inter–element spacing	(2.1)
d	separation between source and receiver	(9.11)
d	piezoelectric strain constant	Sect 3.2
d_1	network parameter	(5.24)
E	electric field strength	Sect 3.2, (7.8)

E_b	Young's modulus for centre bolt	(7.19)
E_e	effective Young's modulus	(7.2)
E_p	Young's modulus for piston	(8.10)
E'_{11}	elastic modulus normal to poling direction	(10.53)
e	piezoelectric stress constant	Sect 3.2
e_E	amplifier noise spectral level	Sect 10.6
e_p	piezoelectrically generated voltage	(10.1)
e_T	thermal noise spectral level	(10.26)
F	force	Sect 1.1
F	frequency in kHz	(9.12)
F_b	static force in bolt	(7.17)
F_C	roll-off frequency (kHz)	(10.33)
F_c	characteristic frequency for M/S materials	(3.11)
$\hat{F}_c$	peak dynamic force in stack	(7.16)
F_i	force on i^{th} piston	Sect 6.2
F_p	total force in stack due to acoustic pressure	(10.83)
F_S	resonance frequency in kHz	(7.23)
F_0	amplitude of alternating force	Sect 1.3
F_0	acoustic force on piston	(10.81)
F_1,F_2,F_3	forces in springs	(10.67)
$F(\theta)$	directivity function	(6.20)
f	frequency	Sect 1.1
f_A,f_B	measurement frequencies	Sect 9.2
f_f	flexural resonance frequency	(8.11)
f_m	frequency of maximum admittance	Sect 4.3
f_n	frequency of minimum admittance	Sect 4.3
f_p	parallel resonance frequency	Sect 4.3
f_S	series (motional) resonance frequency	Sect 4.3
f_0	frequency of maximum conductance	Sect 4.2
f_1,f_2	half-conductance frequencies	Sect 4.3
G	conductance	(1.6)
G	flexural resonance frequency parameter	(8.11)
G_{air}	peak G_m in air	(4.21)
G_{in}	input conductance	(5.9)
G_m	motional conductance	Sect 4.2

G_{max}	maximum value of G	Sect 4.2
G_{mS}	motional conductance at f_S	Sect 4.3
G_r	motional equivalent of conductance	(6.9)
G_S	$1/R_S$	(5.14)
G_{wat}	peak G_m in water	(4.21)
G_1	normalised value of G_r	(6.10)
g	piezoelectric stress constant	Sect 3.2
g_h	piezoelectric hydrostatic coefficient	Sect 3.2
H	flexural resonance frequency function	(8.10)
H_0, H_1	Hankel functions	(6.11)
h	piezoelectric strain constant	Sect 3.2
h	effective thickness of piston	(8.11)
h	thickness of diaphragm	(8.14)
h	length of neck	(11.12)
I	acoustic intensity	Sect 1.2
I_r	acoustic intensity at range r	(2.6)
I_c	cavitation threshold	(2.9)
i	current	Sect 1.2
i_m	current in motional arm	(4.29)
J	reciprocity parameter	(9.10)
J_1	Bessel function	(6.20)
j	$\sqrt{-1}$	
K	spring stiffness constant	(1.9)
K	stiffness of stack	(7.1)
K	$(2\pi/\lambda)\sin\theta$	(6.25b)
K'	$(2\pi/\lambda)(\sin\theta - \sin\theta_0)$	(6.30)
K_{cav}	stiffness of fluid in cavity	(11.11)
K_0	Boltzmann's constant	(10.23)
	$(1.38 \times 10^{-23} \text{J/deg K})$	
K_1, K_2, etc	spring stiffnesses	Sect 8.6
K_{13}	K_1/K_3	(10.73)
K_{23}	K_2/K_3	(10.73)
k	coupling coefficient	(3.5a),(4.5),(4.12)
k	$2\pi/\lambda$	(6.3)
k_e	effective coupling coefficient	(8.6)
k_p	radial mode coupling coefficient	(3.6)

k_1, k_2	functions determining diaphragm characteristics	(8.14)
k_{12}, k_{13}, etc	filter parameters	(5.25),(5.26)
L	inductance	Sect 1.1
L	length of line array	(2.5)
L	length of element	(6.33)
L	length of cylinder	(10.61)
L_0	parallel inductance	(5.1)
L_1	effective radiating mass	Sect 8.2
L_2	rear mass	Sect 8.2
L_2, L_3, L_4	filter inductances	(5.22–5.32)
l_c	length of ceramic stack	(7.2)
M	hydrophone sensitivity (dB)	Sect 1.2
$\hat{M}$	hydrophone sensitivity (linear)	Sect 1.2
$\hat{M}_a$	acceleration sensitivity in acoustic terms	(10.86)
$\hat{M}_C$	sensitivity of cardioid hydrophone	(10.92)
$\hat{M}_D$	hydrophone sensitivity of doublet	(10.89)
M_e	effective vibrating mass	(4.30),(7.3)
M_n	effective mass for Helmholtz resonator	(11.12)
M_p	effective mass of ceramic plate	(11.2)
M_r	associated water mass	(6.8)
M_S	mass of stack	(8.1)
$\hat{M}_0$	average $\hat{M}$ for dipole pair	(10.90)
M_1, M_2	vibrating masses	(7.3)
$\Delta \hat{M}$	difference in $\hat{M}$ for dipole pair	(10.90)
m	mass	(1.9)
N	number of elements	Sect 2.2
N_P	radial mode resonance frequency constant	Sect 3.2
N_1, N_2	mutual impedance parameters	(6.14),(6.15)
N_{31}	ω_1 / ω_3	(10.79)
n	number of piezoelectric rings	(7.6)
n_1, n_2, n_3	$\omega/\omega_1,\ \omega/\omega_2,\ \omega/\omega_3$	(10.70)
P_E	equivalent amplifier noise pressure	Sect 10.6
P_N	ambient noise pressure	(10.35)
P_{TR}	equivalent thermal noise pressure	(10.33b)
p	acoustic pressure	Sect 1.2

p_T	thermal noise pressure for ideal hydrophone	(10.23)
p_{TR}	equivalent thermal noise pressure (linear)	(10.28)
p_0	amplitude of alternating pressure	Sect 1.2
Q_c	charge on ceramic	Sect 1.3, (3.8)
Q_E	electrical Q-factor	(4.4)
Q_M	mechanical Q-factor	(4.3),(4.11)
Q_0	$R_0/\omega_0 L_0$	(10.1)
Q_1	$\omega_1 L_1/R_1$	(10.1)
R	resistance	(1.6)
$\tilde{R}$	normalised radiation resistance $(R_r/\rho c A_p)$	(7.15)
R_e	dielectric loss resistor	Sect 4.3
R_i	internal mechanical losses	(4.21)
R_N	nominal resistance	(5.5)
R_r	radiation resistance	(4.21),(6.2)
$\langle R_r \rangle$	average radiation resistance for array	(6.17)
R_S	source resistance	Sect 5.2
R_0	parallel resistance	Sect 10.1
R_1	series resistance	Sect 4.3
R_1	radiation resistance function	(6.4)
R_{11}	self radiation resistance	Sect 6.2
r	resistive damping coefficient	(1.9)
r	range	(2.6)
r_m	motional resistance in mechanical terms	(4.29),(7.3)
r_0	Fresnel range	Sect 2.3,6.4
S	strain	Sect 3.2
SL	source level	(2.7)
S_I	currrent projector sensitivity (dB)	Sect 1.2
$\hat{S}_I$	current projector sensitivity (linear)	Sect 1.2
S_V	voltage projector sensitivity (dB)	Sect 1.2
$\hat{S}_V$	voltage projector sensitivity (linear)	Sect 1.2
s	compliance coefficient of ceramic	Sect 3.2
T	stress	Sect 3.2
T	temperature	(10.23)

T_f	limiting fatigue stress	(8.9)
T_m	mean applied stress	(8.9)
T_{max}	maximum stress in plate	(8.14)
T_t	tensile strength	(8.9)
t	time	(1.1)
t	ceramic thickness	(7.6),(10.52),(11.1)
U_j	normal velocity of jth piston	(6.13)
u	velocity	(1.2)
V	voltage	Sect 1.2,(10.1)
V_a	output voltage due to acceleration	(10.75)
V_c	volume of ceramic	(10.42)
V_t	volume of transducer	(7.25)
W	bandwidth factor	(5.3)
W_a	acoustic power	(2.6)
W_{aS}	acoustic power at resonance	(7.5)
W_e	electrical input power	(2.8)
W_m	motional power	(4.29)
W_{max}	maximum power from source	(5.14)
W_0	power/unit area of piston	(7.11)
W_0'	power/unit area of array	(7.13)
X	reactance	(1.6)
X_r	radiation reactance	(6.2)
$\langle X_r \rangle$	average radiation reactance for array	(6.17)
X_1	radiation reactance function	(6.5)
X_{12}	mutual radiation reactance	(6.15)
x	displacement	(1.1)
x	f/f_S	(9.4)
x_r	motional reactance in mechanical terms	(4.36)
x_0	amplitude of alternating displacement	(1.1)
x_0,x_1	ω/ω_0 $(=f/f_0)$, ω/ω_1	(10.1)
x_0	axial displacement of casing	(10.69)
Y	admittance	(1.6)
Y_e	blocked electrical input admittance	Sect 4.8
Y_{in}	input admittance	(4.19)
y	frequency variable	(5.4)

Z	impedance	(1.7)
Z_c	$\rho_c c_c A$	(11.1)
Z_D	impedance of drive circuit from radiation load	(6.18)
Z_{IN}	mechanical input impedance	(8.3)
Z_i	impedance of ith piston	(6.13)
Z_{ij}	mutual radiation impedance	Sect 6.2
Z_{in}	input impedance	(4.19)
Z_M	mechanical impedance	Sect 1.1
Z_M	combination of L_1, L_2, and R_1	Fig 8.3
Z_M'		Fig 8.4
Z_m	mechanical impedance at output terminals	(4.32)
Z_r	load impedance	(4.33)
Z_r	radiation impedance	(6.1)
$\langle Z_r \rangle$	average radiation impedance	(6.16)
Z_1	$\omega Z/K_1$	(10.70)
Z_1, Z_2	loads on end-faces	Fig 11.1
Z_{11}	self radiation impedance	Sect 6.2
Z'_{11}	self radiation impedance of piston without baffle	(6.16)
z	displacement	(8.14)
α	measurement parameter	(9.5)
α	transformation ratio	(11.1)
α_1, α_3,etc	network parameters	(5.21),(5.27)
β	matching factor	(5.6)
β	ratio of ext/int radii	(10.51),(10.62)
β'	$\omega C_p R_p'$	(10.27)
β_S	R_N/R_S	(6.18)
γ	cone angle of piston	(8.11)
δ	time delay parameter	(10.93)
$\tan \delta$	dielectric loss factor	Sect 3.3
ε	dielectric constant (absolute) (superscripts indicate conditions)	Sect 3.2
ε_{33},etc	relative dielectric constant (subscripts indicate axes)	Sect 3.2

ε_r	relative dielectric constant (general)	Sect 3.2
ε_0	permittivity of free space $(8.85 \times 10^{-12}\,\text{F/m})$	Sect 3.2
η_{ea}	electrical–acoustic efficiency	(2.8)
η_{em}	electrical–mechanical efficiency	(4.22)
η_{ma}	mechanical–acoustic efficiency	(4.21)
θ	phase angle of impedance	Sect 1.1
θ	phase angle of admittance	(5.10)
θ	phase angle of output voltage	(10.3)
θ	acoustic incidence angle (rel to broadside)	Sect 2.2
θ_0	steering angle	(2.1)
θ_1	angle of first diffraction lobe	(2.1)
$\theta(-3\text{dB})$	beam width between -3dB points	(2.2),(6.22)
λ	acoustic wavelength	Sect 2.2
λ_c	wavelength in ceramic stack	Sect 7.3
λ_s	acoustic wavelength at resonance (f_s)	(10.47)
μ_r	reversible permeability (for M/S material)	(3.11)
ρ	density	Sect 1.1
ρ	reflection coefficient	(5.15)
ρ_c	density of ceramic	(7.22)
ρ_e	resistivity (of M/S material)	(3.11)
$|\rho|_{max}$	maximum permitted value of $|\rho|$ within pass band	(5.17)
σ	Poisson's ratio	(3.6),(8.10)
φ	transducer transformation ratio	(4.31)
φ	applied phase shift	(6.28)
φ	electro–mechanical transformer ratio	Fig 8.3
χ	packing factor of array	(6.16)
Ω	frequency variable $(\omega/\omega_s - \omega_s/\omega)$	(4.18)
ω	angular frequency $(2\pi f)$	(1.1)

ω_B	(angular) width of pass band	(5.17)
ω_S	angular resonance frequency	(4.10)
ω_{S0}	resonance frequency of ideal system	(8.5a)
ω_0	frequency defined by $\omega_0{}^2 = 1/L_0 C_0$	(10.1)
ω_1	frequency defined by $\omega_1{}^2 = 1/L_1 C_1$	(10.1)
ω_2	frequency defined by $\omega_2{}^2 = K_2/M_2$	(8.13)

INDEX

www.ingramcontent.com/pod-product-compliance
Lightning Source LLC
Chambersburg PA
CBHW080139220326
41598CB00032B/5117